WATER

A Source of Conflict or Cooperation?

Editor
Velma I. Grover
Natural Resource Consultant
Hamilton, Ontario
Canada

Science Publishers

Enfield (NH)　　　Jersey　　　Plymouth

CIP data will be provided on request.

SCIENCE PUBLISHERS
An Imprint of Edenbridge Ltd., British Isles
Post Office Box 699
Enfield, New Hampshire 03748
United States of America

Website: *http://www.scipub.net*

sales@scipub.net (marketing department)
editor@scipub.net (editorial department)
info@scipub.net (for all other enquiries)

© 2007, Copyright Reserved

ISBN 978-1-57808-511-8

Published by Science Publishers, NH, USA
An Imprint of Edenbridge Ltd.
Printed in India

International Joint Commission
Canada and United States

The Rt. Hon. Herb Gray, P.C., C.C., Q.C.
Chair, Canadian Section

Commission mixte internationale
Canada et États-Unis

Le très hon. Herb Gray, C.P., C.C., C.R.
Président, Section canadienne

Foreword

I am pleased to provide a foreword for "Water, a Source of Conflict or Cooperation". This is an important subject as we are in a period of time when there is growing concern about the availability of a sufficient supply of clean water for use by humans for drinking and sanitation and other uses needed for modern civilization like industry and agriculture.

There are differences of view about the fundamental question in the book's title. Some aruge we'll soon reach a tipping point when the supply of water is less than that required by an ever-expanding world population. Others argue there is sufficient water for this purpose if the existing supply is properly managed, if that supply can be expanded through use of technology which, more efficiently and cheaply than at present, can convert the salt water of the oceans to fresh water, or by conservation. There are those also who argue with regard to conservation that, particularly in developed countries, water for all purposes could be used more efficiently and less used for purposes like golf courses and lawns particularly in the southwest United States.

This subject can also be looked at in the context of two or more countries having access to the same body of fresh water. This gives rise to issues of management to maintain its drinkability, swimability and fishability, as well as its use for navigation and industry. These issues could be the basis for conflict between the countries involved. However, what is meant by "conflict" in the book's title? Is this conflict in the sense of negotiation resulting in resolution by agreement, or is it conflict in the sense of resort to weapons and war?

www.ijc.org

234 Laurier Avenue W., 22nd Floor,
Ottawa, ON K1P 6K6
Phone: (613) 992-2417 Fax: (613) 947-9386
grayh@ottawa.ijc.org

234, Avenue Laurier Quest, 22e étage
Ottawa (Ont.) K1P 6K6
Téléphone: (613) 992-2417 Télécopieur: (613) 947-9386
grayh@ottawa.ijc.org

As I said, in a speech given recently to the Royal Society of Canada at a symposium on "Water in Canada and the World: Rising tensions in the 21st century: issues and solutions", Professors Yoffe and Wolf of the University of Oregon concluded, that in the past, tensions between countries over water most often led to cooperation and solutions, not conflict, at least not in the sense of military conflict. For example, in their paper "Water, Conflict and Cooperation: Geographical Perspectives" they said that *"Accounts of conflict related to water indicate that only seven minor skirmishes have occurred in this century, [the 21st] and that no war has yet been fought over water. In contrast, 145 water-related treaties were signed in the same period"*.

One can ask, however, are Yoffe and Wolf's findings and conclusions valid predictors for a future marked by growing world populations and the supply of fresh water likely to be influenced by the factors I've indicated as well as climate change? It could still be argued, however, that the experiences of the past can offer lessons and guidance for the future.

The Boundary Waters Treaty between Canada and the United States signed in 1909 and the International Joint Commission it created have been an effective means of managing the shared fresh waters forming much of the Canada-US boundary. Also, the Great Lakes Water Quality Agreement signed by Canada and the United States has been an effective framework for progress in cleaning up the Great Lakes although there is still more to be done. This agreement has facilitated these results without conflict between the two countries.

This book can make a valuable contribution to a very important debate about whether water, especially fresh water, is a source of conflict or cooperation and therefore should be of help both to scholars and policy makers.

The Rt. Hon. Herb Gray, P.C., C.C., Q.C.

Preface

The end of the colonial era coincided with the emergence of international agencies such as the World Bank, Asian Development Bank and the various arms of the United Nations Organization. Most governments in the newly liberated third world countries set in place poverty reductions programs. International agencies contributed and aided such programs. Poverty reduction is becoming all the more important since poverty has been linked to poor health, food insecurity, environmental degradation and environmental scarcity (even leading to violent conflicts) in different parts of the world. All of these problems are interlinked and can be attributed, at least in part, to population growth.

Environmental scarcity is caused by degradation and depletion of renewable resources because of increased demand caused not only by population growth but also by increase in per capita consumption. This coupled with unequal distribution of wealth and power results in some groups of the society getting a disproportionately large slice of resources, while others get slices that are too small to sustain livelihood. Increased demand of resources because of industrial revolution and population explosion has inverse effect upon supply because of reduction in the total supply due to degradation of the environment and the limited environmental resources - reduction in the size of the resource "pie". Most of the literature suggests that conflicts due to environmental scarcity will happen at a small scale (confined to a region), if they happen at all.

Lately, international security and environmental conflicts have been researched in terms of development and its linkages with poverty and global economy. It is believed that the localized war(s) in the South can be caused because of scarcity of natural resources.

In the past imbalances of natural resources have led to violent conflicts in some natural resource sectors; however, not many wars have been fought for water (see water chronology at www.worldwater.org/conflictchronology.pdf). What makes the water sector so different from other natural resources and what lessons can we learn from it? The challenge is to develop institutional capacity, a culture of cooperation and avoid costly, time consuming crises, that threaten lives, regional stability, and ecosystem health. It was mainly with these goals in mind that I decided to edit this book on "Water: source for conflict or cooperation".

This book contains chapters which discuss water conflict in different scenarios such as trans-boundary issues, multiple stakeholders conflict, conflicts due to water scarcity caused by either unequal distribution of water, or, change in water quality due to climate change or change in precipitation pattern. The book will interest environmental scientists, policy makers, aid agencies, NGOs, social scientists amongst others.

Hamilton   **Velma Grover**
Canada
February 2007

Contents

List of Contributors

Anthony Turton
Gibb SERA Chair in IWRM
Environmentek
Council for Scientific and Industrial Research (CSIR)
African Water Issues Research Unit (AWIRU)

Desheng Hu
Economist, Professor of Law, Zhengzhou University, P. R. China

Dr. Atiq Rahman
Bangladesh Centre for Advanced Studies (BCAS)

Dr. Irna van der Molen
University of Twente, Netherlands

Eberhard Weber
The University of the South Pacific, School of Social and Economic
Development, Department of Geography, Suva, Fiji Islands

Haseen Khan, P. Eng.
Manager, Water Resources Management Division, Department of
Environment and Conservation, P.O. Box - 8700, Confederation Building,
West Block, 4th Floor, St. John's NL A1B 4J6, Canada

Håvardå Hegre & Hans Petter Wollebækæ Toset
International Peace Research Institute Oslo (PRIO),
Norwegian University of Science and Technology,
and University of Oslo

Imoh J. Ekpoh
University of Calabar, Department of Geography, Calabar, Nigeria

Michael Scozzafava
U.S. Environmental Protection Agency

Naho Mirumachi
University of Tokyo, Tokyo, Japan

Nils Petter Gleditsch
PRIO, Hausmanns gate 7, 016818 Oslo, Norway.

Piet K. Kenabatho
University of Botswana, Department of Environmental Science,
Private Bag UB00704, Gaborone, Botswana.

Santosh Kumar
University of Melbourne, Department of Mathematics and Statistics,
Parkville, Victoria 3010, Australia.

Shlomi Dinar
Department of International Relations and Geography
Florida International University, 3000 N.E. 151st Street, AC I 323A
North Miami, FL 33181

Syed U. Hussainy
Faculty of Engineering, Science and Technology, Victoria University
PO Box 14428, Melbourne City, MC8001, Australia

Umoh T. Umoh
University of Botswana, Dept. of Environmental Science,
Private Bag UB00704, Gaborone, Botswana

Velma I Grover
Natural Resources Consultant
Adjunct Professor, York University, Toronto, Canada

William James Smith, Jr.
Department of Environmental Studies, University of Nevada, Las Vegas

Young-Doo Wang, John Byrne and Joon-Hee Lee
Center for Energy and Environmental Policy, University of Delaware

Introduction

1

Introduction

Velma I. Grover

Natural Resources Consultant
Adjunct Professor, York University, Toronto, Canada
vgrover@can.rogers.com

"Rivers have a perverse habit of wandering across borders . . . and nation states have a perverse habit of treating whatever portion of them flows within their borders as a national resource at their sovereign disposal."

John Waterbury, *Hydropolitics of the Nile Valley*, 1979

The first thing which comes to mind on hearing the word conflict is violence. The word 'conflict' generally reflects negative dynamics – even in the discourse of the UN, the term 'conflict' refers to armed hostilities between or within states. But this is an inaccurate and misleading perspective mainly because conflicts range from political, social, economics, institutional, and military combats – but not all of them are violent. Usually, conflict is inevitable and omnipresent in all societies that have diverse groups based on ethnicity, religion, ideology, class or interests. The main reason for conflict is not just the mere presence of diversity but the unequal access to power and resources giving rise to competition and conflicts (though it is not always violent in nature). The response to a conflict depends on the understanding of the situation and how it is interpreted. If conflict is inherently seen as destructive then the efforts are directed towards suppressing or eliminating it rather than understanding the basic cause of the conflict and solving it. But if conflict is viewed as normal and unavoidable, then one looks at the ways to solve it or explore how to manage it constructively and cooperatively (Nathan, 2001). As a matter of fact, conflict can be healthy also, if it keeps the society or its members on their toes and the adrenals active.

Conflict results from competition–competition between different desires within a human being or between at least two parties. A party may be a person, a family, a lineage, or a whole community; or it may be a class of ideas, a political organization, a tribe or a religion. Conflict is occasioned by incompatible desires or aims and by its duration which may be distinguished from strife or angry disputes arising from momentary aggravation.

Most environmental disputes arise because people have different views over what constitutes good policy for the environment. A utility may propose to build a power-generating dam, but the farmers and conservationists fight it because of its effect on irrigation and wildlife downstream. People take opposing views because they have different stakes in the outcome. Another example of conflict could be from the perspective of farmers who support dams, because the dam provides more water for irrigation, but the fishermen oppose the dams because dams are harmful for fishing.

As summarized by Buckles (1999), the use of natural resources is susceptible to conflict for a number of reasons. First, natural resources are embedded in an environment or interconnected space where actions by one individual or group may generate effects far off-site. For example, the use of water for irrigation in the upper reaches of the Calico River, Nicaragua, pitted upstream landowners and communities against downstream communities in need of water for domestic use and consumption.

Second, natural resources are also embedded in a shared social space where complex and unequal relations are established among a wide range of decision-makers in society – agroexport producers, small-scale farmers, ethnic minorities, government agencies, etc. As in other fields with political dimensions, those keypersons with the greatest access to power are also best able to control and influence natural resource decisions in their favor (Peet and Watts, 1996).

Third, natural resources are subject to increasing scarcity due to rapid environmental change, increasing demand, and their unequal distribution (Homer-Dixon and Blitt, 1998). Environmental change may involve land and water degradation, overexploitation of wildlife and aquatic resources, extensive land clearing or drainage, or climate change. Increasing demands have multiple social and economic dimensions, including population growth, changing consumption patterns, trade liberalization, rural enterprise development, and changes in technology and land use. Natural resource scarcity may also result from the unequal distribution of resources among individuals and social groups or ambiguities in the definition of rights to common property resources. As noted by Homer-Dixon and Blitt

(1998), the effects of environmental scarcity such as "constrained agricultural output, constrained economic production, migration, social segmentation, and disrupted institutions ... can, either singly or in combination, produce or exacerbate conflict among groups."

Fourth, natural resources are used by stakeholders in ways that are defined symbolically. Land, forests, and waterways are not just material resources people compete over, but are also part of a particular way of life (farmer, rancher, fisher, and logger), an ethnic identity, and a set of gender and age roles. These symbolic dimensions of natural resources lend themselves to ideological, social, and political struggles that have enormous practical significance for the management of natural resources and the process of conflict management (Chevalier and Buckles, 1995). Multi-stakeholder analysis of problem areas and conflicts is a key step in catalyzing recognition of the need for change. Multi-stakeholder analysis is a general analytical framework for examining the differences in interests and power relations among stakeholders, with a view of identifying who is affected by and who can influence current patterns of natural resource management. Problem analysis from the points of view of all stakeholders can help separate the multiple causes of conflict and bring a wealth of knowledge to bear on the identification and development of solutions. Particular attention is paid to gender-based and class-based differences in problem identification and priority setting because in many societies these differences are systematically suppressed or ignored.

Having discussed different causes of conflict, let us concentrate on water and causes of water conflict – the subject matter of this book. Water is one of the natural resources which flow freely from one country to the other without acknowledging man-made physical boundaries and without requirements of visas to cross the borders to enter neighboring countries. Since water is one of the basic requirements for life – no form of life can exist without water – all the countries want sovereign rights over the water flowing through their land. This is bound to cause some friction between the multiple users in same country as well as users in the different countries along the river or watershed. In other words, conflict is inherent in river-basin management because of the presence of multiple-stakeholders. There is not only a conflict between different users – there are more than a billion people currently without any assured and regular supply of drinking water; close to two billion people lack proper water-based sanitary facilities[1]. Almost 80 % of all diseases and 30 % of all unnatural deaths in the Third

---

[1] The UN Millennium Development Goals (2000) and the World Summit on Sustainable Development (Johannesburg, 2002) pledged to halve the proportion of people without access to safe drinking water and basic sanitation by 2015. To achieve these targets, an additional 1.5 billion people will require improved access to water supply. This means providing services for another 100 million people each year (274,000/day) between 2000 and 2015.

World are due to waterborne diseases and pollution, and are attributed to lack of proper sanitary facilities. Water scarcities are also leading to a serious decline of agriculture and industries in many yet-to-be-developed nations. It is becoming increasingly clear that water will soon become the most crucial resource constraint for economic development and social welfare in the arid and semi-arid regions of the Third World where it is not already so. Global environmental changes are likely to further accentuate the potential for acute conflict over shared waters. This book will discuss how water can act as a source for conflict or cooperation between countries sharing the same river or watershed, with literature reviews and case studies from different countries and continents.

Most of the literature related to water security, insecurity, transboundary waters deals with 'water wars' a term that emerged after Dr. Ismail Serageldin, Vice-President of Environment Branch of the World Bank remarked in 1995 that the wars of 21st century would be over water[2]. This quote led to the publication of numerous scholarly articles, books, newspaper articles and got media attention on the subject for a while. Probably influenced by such statements and literature prompted the former UN Secretary General, Kofi Annan, to state, "Fierce competition for fresh water may well become a source of conflict and wars in the future", in March 2001. However, a year later–Kofi Annan (February 2002) changed his stand and said "But the water problems of our world need not be only a cause of tension; they can also be a catalyst for cooperation. If we work together, a secure and sustainable water future can be ours." This is the message of this book. An effort is made in different chapters in the book to show that water conflicts do lead to tensions and a way has to be found to come up with creative and durable solutions.

Some scholars such as Wolf, Dinar and others emphasize that water can be used as a tool for cooperation among neighboring countries and is not necessarily a cause of war and violence. It is clear that incidents over water rights and water shortages do happen but none of the incidents has led to major wars among nations – for example there is always tension between Bangladesh and India over the quantity of water flow in the river Ganges during different seasons, but there has never been a war between the two nations over water. It is attributed to policies such as development aid to have joint projects across borders from a donor agency can help prevent war over water. For example, India and Pakistan have signed treaties – tied with aid from the World Bank and these treaties (over water) have survived three wars over territory between the two nations, however no war has ever been fought over water.

---

[2]Quoted in *New York Times*, 10 August 1995

Basically, research points out that the only recorded incident of an outright war over water was 4,500 years ago between two Mesopotamian city-states, Lagash and Umma, in the region of present southern Iraq. Conversely, between the years 805 A.D. and 1984 A.D., countries signed more than 3600 water-related treaties. An analysis of 1831 international water-related events over the last 50 years has revealed that two-third of these encounters were of cooperative nature (Postel and Wolf, 2001). Figure 1 shows an index5 of conflictive and cooperative events from 231 river basins.

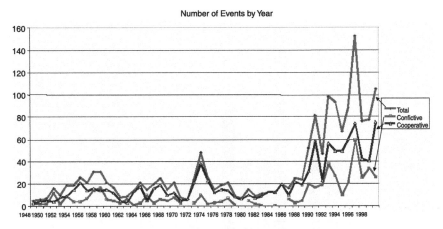

Fig. 1. Conflict-related and cooperative-related reported events in international basins
Source: Dinar, 2001

It can be clearly seen in Figure 1 that cooperative efforts between countries far exceed the number of conflicts. Two observations in Figure 1 are worth mentioning. First, the number of cooperative events consistently exceeds the number of conflictive events for all except four years. Second, there is a trend of increasing conflictive and cooperative events in the last decade. Are these responses due to increased scarcity? Are both cooperative and conflictive events at the international arena a reflection also of domestic reforms in the water and other sectors? (Dinar, 2001). Another question, which arises is that in case of war does the country which loses the war, also lose the right over water use?

Figure 1 depicts 1831 events selected from 231 river basins. There were 214 international river basins in the 1978 Register of International River Basins, covering 47% of the world's land surface (excluding Antarctica). The latest update by Oregon State University database lists 261 international basins, covering 45.3% of the world's land surface. The difference of 47 international basins in the 1998 update as compared to 1978 Register can be attributed to some geo-political changes (for example the break up of

USSR resulted in some national basins getting international basins status); due to better access to maps a few more international basins have been identified; and lastly the 1978 registry did not include island states[3]. Even though the 1996 Register shows 261 river basins, only 231 river basins were selected for the study. In spite of the fact, there are so many shared river basins, hardly any war has been fought over the shared resource.

This is further reflected more clearly in Figure 2 (Gupta, 2005).

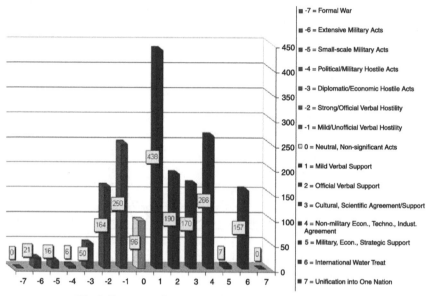

Fig. 2. Spectrum of experience in international water basins

Historically, cooperation over water has been the norm. One-fourth of water-related interactions during the last century has been with some form of agitation–only verbal antagonism. In some cases even some form of military action. But the general conclusion is if information such as water flow, water quality, and quantity, new projects such as dam building, information during natural calamities such as floods etc., is shared between different countries and some technical projects carried out to build trust among nations to cooperate further, chances of war over water are slim.

Generally, it has been observed that water-related violence is more limited at the subnational level, generally between tribe, water-use sector, or state rather than between countries (though examples of war between countries, even in history, exist). Some of the places where war has been fought over water in recent times (not going too back into history and not

---

[3]http://www.transboundarywaters.orst.edu/publications/register/ter_paper.html#table1

including the Middle East) are internal water conflicts such as: interstate violence and death along the Cauvery river in India. There were also fights between indigenous people and European settlers in North America over water rights; California farmers blowing up a pipeline meant for Los Angeles (rural vs urban user conflict) (Wolf, 1998).

Close examination of the water war claims in the media and the actual observation of the number of times there is real violence over water, reveals that the occurrences of violence are quite few as compared with the number of treaties signed for cooperation among countries sharing water. This is reinforced by the more scholarly endeavors of the Water Conflict Chronology[4] and the Transboundary Freshwater Dispute Database[5]. Violent actions over water do exist and occur on all scales in all corners of the planet, even in places where water is abundant, but the actual number of violent events is low.

The origin of the problem of water in general and transboundary water in particular can be traced back to two major legacies of 20th century population explosion and technological revolution that have taken a toll on our water supply. More and more freshwater sources are being used up and contaminated, mainly because modern technologies have allowed us to harness much of the world's water for energy, industry and irrigation– but often at a terrible social and environmental price - and many traditional water conservation practices have been discarded along the way[6]. As water flows across borders in case of shared river/watershed, it flows across countries with different political agendas and different socio-economic and cultural backgrounds, which can lead to conflicts. Water is distributed unevenly across countries and continents (Elhance, 1992–93). With the result some regions in the world have an abundance of water while the others have scarcity of water – some regions suffer from floods while others are drying up under the spell of drought. Besides, increase in water consumption (which can be attributed to increase in human population, growth in agriculture, industries, etc.) and decrease in water quantity (due to climate change) has added stress to the water supply (although, earth is composed of approximately 70% of water, only 3% is freshwater and of this only 1% is available for human consumption). When water availability is getting limited, allocation of water among different users and uses can be highly contested. Also, decreasing water quality poses a health risk which aggravates water scarcity, and can again be a source for potential conflict. In addition to these factors, climate

---

[4] Water Conflict Chronology: www.worldwater.org/conflict/, The Pacific Institute

[5] Transboundary Freshwater Dispute Database: www.transboundarywaters.orst.edu, Oregon State University

[6] http://www.greencrossinternational.net/en/programs/confprevention/wfp/press/intro.htm

change is also adding stress to watershed by changing the precipitation pattern across countries which is not only leading to severe droughts and floods but also creates 'environmental refugee' problems. These kind of environmental problems spur migrations to neighboring cities of countries and can lead to political destabilization and violence at the places receiving these migrants. This can be attributed to higher pressure on already scarce resources in addition to increased poverty, unemployment rates, etc.

The increasing stress in one country also becomes a cause of national concern and security issues for the neighboring countries. Unfortunately, water is one of the most politicized natural resources, which is needed for the very survival of human beings and also used for development. Some of the present water treaties between two states or countries, especially dealing with maintaining certain levels of water quality and quantity, do not have provisions of changes caused by climate change and can be a point of conflict. Water is also a subsidized commodity leading to questions such as charging for water or giving it free since it is a basic human right. Water is one of the unique elements which exists in all three different states – solid, liquid and gas–requiring unique solutions. Also, just as water flows across nations ignoring all the boundaries, and exists in more than one form the evaluation of water resources transcend the analysis of any single discipline, it requires interdisciplinary analysis and solutions. Science is needed to measure the quantity and quality of water resources; while engineering is required for infrastructure needed for water storage, supply and water treatment; economics is necessary when the demand and supply of water aspects is dealt with regard to costing and trading issues; moreover legal, economic, institutional and political aspects are needed for good governance of water. With all these issues surrounding water, it has become one of the debatable topics of whether it can be a source of war or cooperation between the two (or more) countries. Both schools of thought have their theories and have traced historically how many conflicts have taken place over water. This book looks at water from different angles – as a source for cooperation and/or conflict and has various case studies to support the theories. I have not traced the history exhaustively of either of the two schools of thought because various authors have looked at it while describing their points of view. The book gives various basins case studies by authors supporting both sides.

If we look at the major indicators for conflict, as cited in literature, they are: water shortage because of climate change, water stress because of increased population, higher level of development (increasing demand and increased competitive users), greater dependence on hydropower, dams or development per se and some of the "Creeping" changes such as general degradation of quality and climate change induced hydrologic variability. None of these indicators suggests that it will lead to escalated tension between nations, but may be small conflicts among different users. Only cases such as unilateral actions, miscommunication or total lack of communication, blocking information flow (or water flow), lack of training for conflict resolution will lead to conflict among nations over water (Gupta, 2005), as discussed above.

As mentioned above, one of the important points for cooperation is sharing data on water quality, quantity, and water flow etc., environmental disasters such as floods. Flood water travels fast and causes serious damage to the population and economy downstream, if information is not shared between neighboring riparian countries. At the moment, flood planning is very piecemeal and country-specific, but may be better flood control measures can be achieved if planning is done across the countries sharing the water resources. As sharing of information is important, another important factor is building trust to share information. One of the ways to build trust between the riparian neighbors is to have common technical projects.

Another reason for conflict among countries sharing the water source is being upstream or downstream country. Traditionally, all the countries have sovereign rights over the land and resources within the country but transboundary raises the question mark on the sovereign rights over the flowing water when it is in a particular country. Historically, it can be observed that all the major civilizations were established around water and even today most of the big cities (especially big capital cities and commercial centers) are situated along water ways – the question can be raised whether the earlier wars to occupy territories were just limited to war over territories or the main cause was to get better land along the river! But we do not have concrete evidence or an answer for this question. Since water is the life line of any kind of development in the country, all the countries want to maximize their share of water. Water is not only crucial for consumption, sanitation, hydropower production, transport, economic activities (industries) and livelihoods such as fishing but in developing countries it is needed very much for food security 'agriculture'. Almost 75% of water in developing countries is used in the agricultural sector, which is the backbone of economy (especially rural economy) in a developing country to reduce poverty and promote socio-economic development.

The main issue in international water quantity disputes (especially among upstream, downstream countries) is that there are no internationally acceptable criteria for allocating shared water. Another factor is the question of equity. In case of upstream and downstream countries, upstream countries have more control over the water source and in a negotiation over the use of shared water prefer to emphasize on the principles of sovereign rights of the country over water and also rights based on hydrography[7], while the downstream countries try to stress on 'equitable use' principles of international laws. The downstream country claims often depend on the climate of the region. In a humid region, the countries are interested in the "doctrine of absolute riparian integrity", which means that every riparian is entitled to the natural flow of a river system crossing its borders. In an arid climate area, the downstream countries often have older infrastructure and they are interested in defending it and so they are more interested in the principle that rights are acquired through older use, referred to as "historic rights", that is "first in time, first in right" (also called chronology rights). Both these conflicting doctrines of hydrography and chronology clash along some international river basins. For example, downstream countries such as Iraq and Egypt often receive less rainfall but developed earlier than upstream countries such as Turkey and Ethiopia, therefore, claim historic rights over water use while upstream countries argue in favor of sovereignty rights (Wolf, 1999).

According to Homer-Dixon (1999), the war between upstream and downstream countries is possible only when the downstream country is highly dependent on the water for its national well-being and the upstream country threatens to restrict substantially the river's flow; and, most importantly, if the downstream country believes that it is militarily stronger than the upstream country.

"Transboundary water resources tie up all the states sharing a river basin into a tightly-knit and highly complex web of environmental, economic, political, and security interdependencies. This is because any manipulation of a shared river and its water flow by any riparian state inevitably has economic, environmental and security impacts on other riparian states. The uncertainties inherent in estimating future water supplies and demands – within each riparian state and across the basin – combined with the unpredictability of interstate relations and domestic politics, confront states with very complex issues and dilemmas in conducting hydropolitics with their neighbors. The choices that can be and are made by states in conducting hydropolitics with their neighbors depend on the *unique* combinations of the geographic features of the specific basins with a multiplicity of historical, political, economic, social, strategic, and cultural factors and circumstances specific to each basin" (Elhance, 2000).

Since, water can be a potential reason of conflict, countries have tried to establish rules to share the transboundary water resource – the Helsinki rules and 1997 UN Water courses Convention (Sherk et al.) where rules

---

[7] Hydrography principle means that the country from where a river or aquifer originates and how much of that territory falls within a certain state determines how much water is available to the country.

have been established to share the water (equitable sharing can be questionable as to who determines the reasonable and equitable share of water), rules for pollution prevention, etc. While the International Law Association in its Helsinki rules has adopted the principle of equitable and reasonable use as the governing principle of water law and also allows use of water which can cause significant harm, the UN Watercourses Convention differs in its approach which does not allow for a use which can cause significant harm (although States could argue that read together, Articles 5-7 mean this) (Sherk et al.). In fact, the relationship between "reasonable and equitable use", and the obligation not to cause "significant harm", is a more subtle manifestation of the argument between hydrograhpy and chronology (Wolf, 1999). *"Not surprisingly, up-stream riparians have advocated that the emphasis between the two principles be on "equitable utilization", since that principle gives the needs of the present as the same weight as those of the past. Likewise, down-stream riparians (along with the environmental and development communities) have pushed for emphasis on "no-significant harm", effectively the equivalent of the doctrine of historic rights in protecting pre-existing use"* (Wolf, 1999).

One of the chapters in the book, by **Dinar**, "Water Wars? Conflict, Cooperation, and Negotiation over Transboundary Water", emphasizes the claim that the next war will be about water has been echoed in the popular press and in some policy-orientated writings, yet, the track record of hydropolitics is one of cooperation and negotiation. The author feels that in arid and politically volatile regions, especially, a water dispute may often act as a destabilizing factor. But, it is often the dispute itself, and the interdependence of the parties vis-à-vis the water source, that leads them to a cooperative venture. This chapter by Dinar gives an analysis of international rivers with a focus on the manner by which conflict turns into cooperation, ultimately culminating in an agreement. Considering the patterns of treaty making and the substance of past agreements may provide countries currently in dispute over shared water bodies, with a guide for a successful cooperative venture.

The main factor for cooperation between two countries is trust. One of the ways to build trust and share information is 'technical co-operation'. On the technical level, cooperation between South Africa and Lesotho is remarkable, where both the countries prioritized the enhancement of certain benefits not related to water in the negotiations of the project. **Mirumachi**, in "River Development and Bilateral Cooperation: Lesotho Highlands Water Project Case Study" explores the case study of the Lesotho Highlands Water Project (LHWP), one of the few successful inter-basin water transfer schemes in the world. There have been attempts to transfer water in the Zambezi River, Okavango River and Mekong River. However, none of them has achieved the same level of success as this Southern

African case. The author higlights the historical background of the LHWP for it encompasses the political, economic and social links between the two basin states, and also demonstrates how South Africa and Lesotho prioritized the enhancement of certain benefits not related to water in the negotiations of the project.

Although, water scarcity is cited as one of the reasons for conflict among different stakeholders (users or countries), yet at times, it is not just the scarcity of water that is the problem but the mismanagement of water (bad governance and poor policies dealing with water have been blamed for problems these days). This is best illustrated by Weber in **"The Political Economy of Water Supply in Suva, Fiji Islands"**. Weber analyses the situation of water supply in the Pacific Island region. The very heterogeneous nature of the islands (e.g. low lying tiny atoll and limestone islands with insufficient surface water and small groundwater lenses; bigger, high islands of volcanic origin) causes a great number of problems related to the water supply in both rural and urban population. The chapter discusses various causes of water scarcity in the Pacific Island region with a focus on the political economy of water supply in Suva, Fiji Islands. This makes an interesting case study because, with a population of less than 100,000, Suva is the largest city in the Pacific Island region, with an annual rainfall of more than 3000 mm making it one of the 'wettest' capitals on the globe. Despite this abundance of water for many years the administration of Suva is facing severe problems in supplying sufficient water to all its inhabitants. Many residential areas especially in the hilly suburbs are facing severe scarcity of water. This may be attributed to population growth, insufficient infrastructure and mismanagement of water supply system.

So far, we have discussed conflicts/cooperation among different countries but with growing needs of water in the country itself – there are multiple and competing users, such as human consumption (daily needs– drinking, cooking, bathing, etc.), agriculture (to meet the food needs of the growing population), and industry to name a few. **Molen**, in "Disputes over water, natural resources and human security in Bangladesh: Towards a conflict analysis framework" explores the nature of the conflicts, which are related to the use, ownership and management of water in particular and natural resources in general. The author in the chapter describes that water problems are not only due to resource scarcity, or unequal distribution of resources, but can also be related to seasonal abundance of water resources. This chapter brings together a combination of the analytical concepts used in the literature on conflicts and competition over natural resources, and applies these concepts to water-related conflict in Bangladesh. It suggests that social, political, economic and governance factors which contribute to vulnerability, marginalization and conflict should receive particular attention in any conflict analysis related to natural resources. The next few paragraphs will discuss some other reasons for conflicts over water (and may be how they can be resolved).

Water has a socio-economic and cultural significance, thus all the solutions cannot be technology related. In fact, it is imperative to consider cultural and social values in resolving conflict where differences in cultures exist between different countries. This is illustrated by **Wang and Byrne** in "Enhanced Sustainable River Basin Management through Conflict Resolution: A Comparative Case Study between U.S. and South Korea", in which they present an interesting comparison between two U.S. river basins (Delaware and the Susquehanna) and two South Korean river basins (the Han and Nakdong Rivers). The authors have compared river basin management practices in both the U.S. and South Korea to understand how multi-stakeholder conflicts are resolved under widely varying cultural and regulatory contexts. The cross-cultural analysis enables an evaluation of conflict resolution as a means of producing socio-politically acceptable, economically sound, technologically feasible and environmentally viable delivery of water. It is underscored that conflict resolution, in its widely varying forms, is vital in order to achieve sustainability in river basin management, despite great differences in actors, policies and culture.

Gleditsch et al. in "Conflicts in Shared River Basins", look at a measure of willingness derived from the literature on resource and environmental conflict; the sharing of a river that might give rise to upstream-downstream conflicts over scarce water. The authors have analyzed international conflict through the framework of opportunity and willingness. Opportunity has mainly been operationalized as physical proximity and willingness has been measured in a number of ways, and remains a somewhat more elusive concept. Using a new and much more extensive dataset on boundary length for the entire Correlates of War period, the authors find very different results. Gleditsh et al. study the relationship with shared rivers and water scarcity as measures of neo-Malthusian factors in willingness over a 110-year period. The results indicate that the neo-Malthusian factors are significant although not dramatic in their effects.

Another way of cooperation among nations sharing water resources is formation of the Institutions/Commissions such as International Joint Commission between Canada and the United States, Danube and Rhine River Commission in Europe, more recently the Mekong River Commission in Asia. Canada and the United States are in an envious and unique position of sharing the world's most valuable and strategic Great Lakes system. The Great Lakes, a treasure that constitutes about 20% of the entire world's accessible freshwater, are lumped together between the two countries and indeed they often provide the natural boundary between the two countries. Although the Lakes can be cut in half on political maps, but they act as one entity in reality, and thus it is impossible for one country to make use of their share of the resource without affecting the other's interest. To solve these problems, Canada and the US signed the Boundary Water Treaties way back in late 1800's and then later formed an International Joint Commission (IJC). The chapter by **Grover and Khan** describes how the IJC and its committees have helped solve the problems between two countries peacefully. The working of these IJC committees to solve technical, scientific and policy issues, is touted as a model for other countries to follow. It is being suggested that may be a similar institution or even IJC can be used to solve the issues of North (Arctic). Although, it is easy to suggest a similar model institution for all shared watersheds/river basins but it may not be easy to form such institutions in the environment of hostility and lack of trust. Lack of funding of such institutions is also a fundamental obstacle in developing countries.

China, one of the thirstiest countries in the world, is challenged by severe issues of water shortage, natural water disaster and water pollution. In "Water Conflicts in China", **Hu** describes that water conflicts exist both nationally between users, communities and administrative regions, and also involve the relevant international watercourses. According to the author, in order to resolve these problems, a national water rights system dealing with the property right on water resources, the human right to water and the environmental right to water (water for the environment) should be established and implemented, and a practical foreign policy concerning international watercourses should be outlined.

International river basins in Southeast Asia are at different stages of development ranging from basins at a fully developed state by the riparian countries to basins having opportunities for further development. Most of the existing treaties and agreements do not deal with the problems and programs of development from the perspective of a comprehensive basin-wide approach, but rather relate in general context to specific river sections and project areas and to boundary water reaches. One of the long-term collaborative efforts of water use in transboundary river basin is the shared thrust of the riparian countries in the Lower Mekong Basin to plan and implement basin wide development of water resources since 1957, even though the region has passed through political changes in basin countries and seen a number of conflicts among member states. **Babel et al.** discuss the establishment of the Mekong River Commission (MRC) following the signing of "Agreement on Cooperation for the Sustainable Development of the Mekong River Basin" in 1995 by the Governments of the four riparian countries as a major step in broadening the scope of cooperation in all fields of basin development and resource management, river navigation, flood control, fisheries, agriculture, power production and environmental protection; and how it can be successful by involving upstream countries such as China, which is not a member of MRC.

"The Hydropolitics of Cooperation: South Africa during the Cold War" by **Turton** focuses on Southern Africa, specially the four countries–South Africa, Zimbabwe, Botswana and Namibia. All of these countries are highly water stressed, with water scarcity limitations posing a strategic threat to their future economic development potential. These four countries also share two transboundary river basins–the Orange and Limpopo–along with a range of other transboundary rivers. A critical assessment of the South African case study in particular, reveals a remarkable record of cooperation, to the extent that the management of transboundary rivers can be considered a sufficiently focused issue area to drive international relations between the four states. This history of water cooperation in the face of endemic armed conflict has resulted in the evolution of robust institutions that floated like islands of peace in a sea of seething unrest. The chapter discusses how these institutions are now poised to play a significant role in post-conflict reconstruction and regional integration in the Southern African Development Community (SADC).

Water is a finite and very scarce resource in Botswana, due to the semi-arid climatic condition of the country, characterized by recurrence of droughts and very high evaporation rates in surface water resources. As a result, water demand management and conservation measures are introduced in order to reduce pressure for the development of new water resources. The major water-consuming sectors are settlement, mining and energy, livestock, irrigation and forestry, and wildlife. This is further exacerbated by the erratic rains and unpredictable climatic conditions. The forecast has shown that water consumption will rise from 116.9 million m^3 in 1990 to reach 323.5 million m^3 by the year 2020, an increase of 177% whereas water development and supplies have limitations. **Umoh and Kumar** in, "Competition For Limited Water Resources In Botswana" describe that in Botswana, it is evident that water demand by far outweighs available water supply, thus a more realistic water demand management is essential. It can be concluded from the author's discussions that, the major objective of water demand management in Botswana is to reduce the need for continuing expansion of conventional water supply systems such as boreholes and dams. The aim is to meet water demand on a sustainable basis, which calls for a three-pronged strategy: reduction of water consumption; increasing water supply from non-conventional sources and reducing losses of conventional sources. This implies that the demand management measures should include: water restriction and policing; water rationing; loss reduction programs and public education and awareness campaigns. The authors have also given some recommendations for the government which will increase the attraction of water demand management measures.

Conflicts over water allocation in the Niger River Basin of West Africa have arisen between agricultural, urban and environmental uses. The frequent droughts of the Sahelian region of the sub-continent exacerbate conflict between competing agricultural and urban demands. Lakes created by dam construction displace people living upstream of the dam, while those living downstream find the change in river regime disadvantageous. Floods no longer reach lands where crops depended on moisture left in the soil when water subsided. Dams interfere with fish migration. Kainji lake for example displaced 50,000 people who were resettled in specially built villages. Many of these people found housing and sites of the villages unsatisfactory. These aggravate conflicts. "Water Development, Use And Conflicts In The Niger River Basin Of West Africa" by **Umoh and Kumar** discusses the enduring conflict in the management of waters in the River Niger, a river which traverses four independent West African countries, often with conflicting interest on developmental needs. The authors have identified three potential elements of conflicts in the River Basin: geography, environmental change and resource dependence.

The next chapter in the book by **Hussainy and Kumar** "Water An Essential Commodity and Yet A Potential Source of Conflict With Special Reference To Australia" deals with the aspects of water as a basic commodity and yet a potential source of conflict among neighbors, they may be interstate or international persons sharing the same potential source. The activities and the outcomes impinge on the users of the same resource downstream. The impact of the upstream uses on the quantity and quality of water available for the downstream users would be a potential source of conflict. The authors have used Australia as a case study to make their point. The final section of the chapter, focuses on a Hollywood film producer who has selected water as a scare commodity as his theme for the next production. Indian born Hollywood filmmaker, Shekhar Kapur while traveling from Mumbai Airport to his hotel in an upper class suburb observed that on the surface roads were wide and smooth but under the overpasses, India's most populous slums (called *Dharavi*) existed. Kapur also noticed that the only freshwater supply they had is the water that seeps out of cracks in the pipes that carry water to steel-and-glass blocks in the downtown Mumbai. He concluded, **"Water will soon be the World's most valuable commodity, and places like '*Dharavi*' will have none**. He decided to undertake this subject as the theme for his movie project, called, '*Paani*', a Hindi word meaning water. Media productions like these can have remarkable success in teaching the general community (as suggested by Umoh and Kumar) those social issues, which were impossible tasks for many governments.

It can thus be safely concluded that water, which is the basic requirement for life, flows freely through different countries and have been a source for both conflict and cooperation. Maybe the need is to treat problems at the watershed level instead of using political boundaries to solve the issues and also build the capacity of the riparian countries sharing the watershed/ river basins on how to cooperate and solve the problems. The emphasis then will be to find solutions in a collaborative fashion instead of being combative, and stress should be on conflict avoidance and prevention.

REFERENCES

Buckles, D. (eds). 1999. Cultivating Peace, Conflict and Collaboration in Natural Resource Management, IDRC/World Bank.

Chevalier, J. and Buckles, D. 1995. A Land without Gods: Process Theory, Maldevelopment and the Mexican Nahuas. Zed Books, London, UK.

Dinar, A. 2001. Reforming Ourselves Rather than Our Water Resources: Politics of Water Scarcity at Local, National and International Levels. Gutachten im Auftrag des Bundesministeriums für wirtschaftliche Zusammenarbeit und Entwicklung Bonn, October 2001.

Elhance and Arun, P. 2000. Hydropolitics: Grounds for Despair, Reasons for Hope, International Negotiation 5: 201–222. Kluwer Law International, the Netherlands.

Elhance and Arun, P., Winter 1992-93. Swords and Ploughshares. Geography and hydropolitics in The Bulletin of the Program in Arms Control, Disarmament, and International Security, Geography and Hydropolitics. VII (2). University of Illinois at Urbana-Champaign, USA.

Gupta, A.D. 2005. Transboundary Water: An Overview Presentation at Transboundary Workshop AIT.

Homer-Dixon, T. and Blitt, J. 1998. Ecoviolence: Links among Environment, Population, and Security. Rowman & Littlefield, Lanham, MD, USA.

Homer-Dixon, T. 1999. Environment, scarcity, and violence. Princeton University Press, Princeton, USA.

Nathan, 2001. The Four Horsemen of the Apocalypse: The structural cause of violence in Africa Track Two. 10 (2) August 2001, CCR, Cape Town, South Africa.

Peet, R. and Watts, M. 1996. Liberation Ecologies: Environment, Development and Social Movements. Routledge, London, UK.

Postel, S. and Wolf, A. 2001. Dehydrating Conflict. Foreign Policy, September/October, 60-67.

Sherk, George William, Patricia Wouters and Samantha Rochford. Water Wars in the Near Future? Reconciling Competing Claims for the World's Diminishing Freshwater Resources–The Challenge of the Next Millennium, The Centre for Energy, Petroleum and Mineral Law and Policy, University of Dundee, Vol. 3, No. 2.

Wolf, A.T. 1998. Conflict and cooperation along international waterways Water Policy 1 251±265.

Wolf, A.T. 1999. Criteria for equitable allocations: The Heart of International Water Conflict, Natural Resources Forum, 23 (1), 3: 30 (http://www.transboundarywaters.orst.edu/publications/allocations) accessed on April 12, 2006.

2

Water Wars? Conflict, Cooperation, and Negotiation over Transboundary Water

Shlomi Dinar

Department of International Relations and Geography
Florida International University, 3000 N.E. 151st Street, AC I 323A
North Miami, FL 33181
dinars@fiu.edu

The Wars of the Next Century will be about Water[1]

So goes a grim 21st century prediction that has been echoed by such well-respected institutions as the World Bank. As demand for freshwater rises, together with the population, water scarcity already figures on the national security agenda of many countries, especially in the Middle East, North Africa, and Central Asia. While it is true that water disputes have taken a military turn on at least eight occasions since 1918, thousands of water agreements have been concluded, the oldest of them dating back some 1,200 years (Wolf and Hamner, 2000). Despite the sensationalist appeal of the "water-wars" thesis, the history of hydropolitics (i.e. the politics of water), has been rather one of cooperation and negotiation.

The above does not mean to suggest that conflict over water does not take place. Indeed, disputes over shared rivers transpire since water is crucial for basic survival, irreplaceable, transcends international borders, and scarce. Rather, for the same exact reasons, water has largely become a foundation for collaboration among states. Surely, water conflicts are of great concern in more volatile regions, and among otherwise unfriendly nations, where a water dispute acts as a destabilizing factor. While such

---

[1]The prediction, made by Ismail Serageldin, Vice President of the World Bank at the time, was quoted in Crosette (1995).

disputes have never turned violent, the urgency for cooperation in these regions is greatest.

This chapter will consider these two interrelated observations but provide several examples to demonstrate that disputes over shared rivers, have rarely, if at all, resulted in violence. In fact, history has demonstrated that states engaged in disputes over water have mostly entered into international water agreements. As such, this chapter will consider how disputes often turn into opportunities for cooperation and how lessons learned from these occasions may be used to foster future collaboration, especially in regions where a water dispute may escalate political tensions.

Water, Conflict and War

The claim that violent conflict over international shared rivers is a fait accompli resonates due to its melodramatic appeal. In its most acute form, the hypothesis reasons that the next war will be about water. Given that water is critical for daily life, non-substitutable, and often transboundary in nature, it follows that states will take up arms to defend access to a shared river. As water becomes increasingly scarce, such as in the Middle East, a water war is more likely. While an intereting claim, the water-wars thesis has largely been limited to the popular press and policy-orientated journals (Cooley, 1984; Starr, 1991). In reality, the last time water has played the main role in instigating a war was 4,500 years ago. This was the case of two city states, Lagash and Umma, battling over rights to exploit boundary channels along the Tigris River in modern day Iraq.

Since 1918 water has played a role only in initiating armed – or near armed – skirmishes between states. There have been only seven such recorded incidents. Not surprisingly, the majority of these episodes has taken place in the Middle East where water is scarce and the political atmosphere tense. Perhaps the most celebrated case is that between Israel and Syria between March 1965 and July 1966, whereby shots were fired between the two states over the 'all-Arab' plan to divert the headwaters of the Jordan River. Often cited as the cause of the Six-Day War between Israel and her Arab neighbors, this case only demonstrates that shots were exchanged between Syria and Israel over one country's respective water project. The war started in June 1967, one year after shots over Syria's water installations long ceased (Wolf and Hamner, 2000).

To argue that conflict is an anomaly in the realm of hydropolitics would be a distortion of reality. Political disputes regularly arise over shared rivers. In fact, this is the case in regions known for having limited water availability and historically tense relations among the riparians but also in regions with relatively ample water resources and amicable relations among the riparians.

Surely, water disputes in regions with historically tense relations between the states, and scarce water resources, have tended to be more volatile, and the above example alludes to this. In such regions, like the Middle East, water is evolving to an issue of high politics and the probability of water related violence is increasing (Gleick, 1993). As Mathews argues, environmental decline occasionally leads to conflict, especially when scarce water resources must be shared (Mathews, 1989; Homer-Dixon, 1999). Three other examples are cited below.

After signing a peace agreement with Israel in 1979, Egyptian President Anwar Sadat announced that the only issue that would prompt Egypt to declare war again would be water. Sadat's threats were directed at Ethiopia, where the majority of Egypt's Nile waters originate. Egypt claims that it is wholeheartedly dependent on the Nile waters and would not tolerate any attempts by its upstream neighbor to challenge its water use. In the early 1990s, King Hussein of Jordan issued similar war-like declarations targeted at Israel who shares the Jordan River with the Hashemite Kingdom. In 1975, Iraq reportedly mobilized troops on its border with Syria in response to Damascus's unwillingness to release additional water on the Euphrates River (Wolf and Hamner, 2000). Despite these three interesting anecdotes and 'close calls', none ended in actual war, nor was gunfire ever exchanged.

By the accounts of the water-war thesis, the Indus River, should have long been grounds for a bloody showdown between its riparians—Pakistan and India. Since India's partition in 1947, the two countries' relations have been especially tense. In fact, the water dispute that commenced in the late 1950s could have, quite easily, brought the two states to a water war. Despite this prospect, war never took place. In fact, in 1960, under the auspices of the World Bank, Pakistan and India concluded the Indus Waters Treaty. Albeit not an agreement that embodied joint utilization of the river, the treaty divided the waters among the riparians. In essence, India would utilize the eastern tributaries of the Indus Basin and Pakistan the western tributaries. This agreement has survived two wars (1965 and 1970) between the riparians. Still the lingering political tensions between the two countries make the water issue an important one. In fact, Pakistan and India are currently working out their differences on the latter's decision to build a hydroelectric dam on the Chenab River, one of the tributaries governed by the 1960 Treaty. Pakistan has argued that India's actions will affect its allocations guaranteed by the Treaty.

Understanding Conflict Over Water
Conflict may take place for several reasons, including: 1) the level of scarcity, mismanagement, or misallocation of water resources among the riparians, 2) the degree of interdependence among the states regarding the common water resource they share, 3) the geographic and historic criteria of water ownership vis-à-vis states, 4) the relative power of the parties,

5) whether a protracted conflict underlies the water dispute, and 6) an ambiguous property right allocation regime (Gleick, 1993; Dinar, 2002).

Scarcity

Considering the level of scarcity, Falkenmark has argued that environmental stress results when the population grows large in relation to the water supply derived from the global water cycle. In consequence, conflicts may easily arise when users are competing for a limited resource to supply the domestic, industrial and agricultural sectors (Falkenmark, 1992). The same author has also argued that 1,000 cubic meters of water per capita per year constitutes the minimum amount necessary for an adequate quality of life in a moderately developed country (1986). When water availability drops below this figure, scarcity problems grow intense. As water scarcity becomes more acute, violent conflict becomes more probable.

Nations may also suffer from scarcity in energy, flood prevention facilities, or pollution control, and may be increasingly apt to resort to the unilateral exploitation of an international river. Naturally, this may also provoke conflict between the riparians.

While scarcity is an important element for explaining conflict, Elhance (1999) has cautioned that by itself scarcity of natural resources does not necessarily lead to interstate conflict. Rather, it is when such a resource is rightly or wrongly perceived as being overexploited or degraded by others, at a cost to oneself, that states may become prone to conflict. Geography may be partly to blame here.

Geography

With regard to the interdependence of the parties vis-à-vis the common resource and the geographical criteria of water ownership, some scholars have argued that certain geographical configurations of rivers may be more prone to disputes. This is the case especially in upstream-downstream situations in comparison to cases where, for example, the river straddles the border (Gottmann, 1951; Falkenmark, 1990; Just and Netanyahu, 1998; Toset et al., 2000; Amery and Wolf, 2000). Dinar (2006; 2007) has labeled these configurations as "through-border" and "border-creator", respectively. The through-border configuration constitutes a river that flows from one country into another, crossing the border only once. The border-creator river constitutes a river that flows in its entirety along the border separating the countries. A geographically asymmetric relationship exists in the through-border configuration. Such a geographical advantage has its benefits. The upstream state can pollute causing a unidirectional externality in the downstream direction. Similarly, the upstream state may impound water in its territory to the detriment of the downstream state. In short, the assumption is that upstream country A can harm downstream country B's part of the river but not vice versa.[2] A geographically symmetric relationship exists in the border-creator configuration. In this instance, any state that engages in a harmful activity may hurt itself as well as its neighbor. Also, harm can be reciprocated (Dinar, 2006). It is by no means certain that conflict in the exploitation of border-creator rivers can be avoided (Falkenmark, 1986), but the geography of border-creator rivers helps by simplifying both retaliation and reciprocity (Dinar, 2007).

LeMarquand has utilized slightly different labels to describe similar river configurations. According to LeMarquand (1977), "successive rivers" (i.e., the upstream/downstream configuration) and "contiguous" (i.e., rivers where the river forms some part of the border between the two states) produce different incentives or disincentives for cooperation (1977). According to LeMarquand, when the river is contiguous, there is significant incentive for cooperation. The incentive to attain such cooperation is to avoid the "tragedy of the commons" (LeMarquand, 1977). Alternatively, cooperation finds no incentive when the upstream country uses the river's water to the detriment of the downstream country and that country has no reciprocal power over the upstream country (LeMarquand, 1977).

Relative Power

In considering the relative power of the parties, Lowi has argued that conflict is most likely when the most powerful country in the river basin is located in the upstream position. According to Lowi, the interest of the hegemonic state along a river is often a prerequisite to cooperation. But cooperation is more likely to ensue if the hegemon is located in a strategically inferior position – downstream – and if the hegemon's relationship to the water resource is that of critical need. Therefore, river basins with upstream hegemons evincing an ongoing water dispute between the riparians are the least likely to see an end to the conflict. According to Lowi, the case of the Euphrates-Tigris Basin fits this criterion. Upstream Turkey is also the hegemon in comparison to midstream Syria and downstream Iraq. A comprehensive water agreement has not been forthcoming in the basin largely due to Turkey's intransigence. In short, Turkey's upstream and hegemonic position will continue to dissuade her from giving up the status quo she benefits from (Lowi, 1993).

In contrast to Lowi, Homer-Dixon has argued that violent conflict[3] between upstream and downstream riparians is likely only when the

---

[2] This is a proto-typical case. Yet there are instances where dams built downstream cause inundation or environmental damage upstream.

[3] Homer-Dixon (1999) actually considers the outbreak of war.

downstream state fears that the upstream state will use water as a form of leverage and believes it has the military power to rectify the situation (Homer-Dixon, 1999). Therefore, the situation is most dangerous when, a) the downstream country is highly dependent on the water for its national wellbeing, b) the upstream country is threatening to restrict substantially the river's flow downstream, and c) the downstream country believes it is militarily stronger than the upstream country (Homer-Dixon, 1999).

Fredrick Frey and Thomas Naff also estimate how states size up to one another through an assessment of the riparian's geographical position along the river, projectable power or brute ability to impose its will on its rivals, and need for water (Frey, 1993; Naff, 1994). Although not developed beyond its illustrative purposes, the authors' *"power matrix model"* attempts to measure the intensity of a water conflict in a given river basin based on assigning a weighted value for each of the above three factors. With this model, a given power calculation for each riparian and the potential for conflict intensity in the region is subsequently surmised. In applying their model to the Jordan, Euphrates, and Nile River Basins, the authors conclude that the greatest potential for conflict exists when a lower riparian is more powerful than the upstream riparian and perceives its water needs to be deliberately frustrated. In fact, conflict is least likely when the upstream riparian is most powerful since the asymmetry in power inhibits conflict potential. Finally, in a river basin with relative power symmetry among the riparians but with asymmetry in geographic position and need for water, the potential for conflict is moderate. Note that the first and second assessments are directly in opposition to Lowi's assessments yet in agreement with Homer-Dixon's conclusions[4].

Protracted Conflict

An existing protracted conflict between river riparians could also intensify the particular water conflict. That is, when a water dispute unfolds within the context of a more comprehensive political conflict (such as over identity, territory, or religion) the former cannot be effectively isolated from the latter. Lowi has utilized the example of the Israeli-Palestinian conflict, and the larger Arab-Israeli conflict, to argue that without the resolution of the broader political dispute between the parties (status of Jerusalem, settlements, refugees, territory, and borders) an agreement on water issues will be difficult to achieve. In this particular view, water is relegated to the status of a low-politics matter while the aforementioned issues are considered high-politics matters (Lowi, 1993).

---

[4] Homer-Dixon (1999) discusses the case of the Euphrates River to demonstrate the small chance for violent conflict when the upstream state is stronger.

Elhance (1999) comes to a related inference and argues that domestic politics has a robust effect on water issues between river riparians. Frey (1993) seconds this view and argues that: "a nation's goals in trans-national water relations are usually the result of internal power processes, which may produce a set of goals that does not display the coherence, transitivity or 'rationality' assumed in many analyses of transparent national interest." Since water is considered a national security concern in some reaches of the world, it is understandable how a natural resource may be embroiled in nationalistic and identity issues. According to Elhance (1999), this is especially salient between Hindu India and Islamic Bangladesh along the Ganges. Despite recent democratization, Bangladesh remains vulnerable to Islamic fundamentalism and other domestic political factions that accuse their government of compromising the nation's sovereignty and national interest if they pursue negotiations with India. Such posturing often curtails cooperation and undercuts the would-be benefits.

Another example from Southern Asia is that of Nepal-India hydropolitics. In the future the parties may seek to transcend their limited agreements for exploiting their shared waters. Nonetheless, a sector of the Nepalese society regards some of the water agreements, already in effect with India, to provide that country with greater benefits at the expense of Nepal. As such, desires for pursuing additional cooperation with India have been tainted with domestic skepticism that has culminated in the incorporation of a clause in the country's constitution which requires any treaty pertaining to the exploitation of Nepal's natural resources to be ratified by a two-thirds majority of the National Assembly. (Verghese, 1996; Shrestha and Singh, 1996; Gyawali, 2000). Future cooperation among the parties will, therefore, be subject to strong domestic scrutiny and final approval.

International Water Law

Conflict over water also takes place because property rights are not clearly defined and this is largely due to vague legal principles. While the history of international water law is too rich to discuss here, the following discussion focuses on the most recent legal document dealing with international freshwater: the 1997 Convention on the Law of the Non-Navigational Uses of International Watercourses. The Convention, which was adopted by the UN General Assembly, provides a general framework agreement containing numerous articles developed for use by states in resolving their common water disputes. However, it was not ratified by a sufficient number of states, and so never entered into force. As of August 2002, only 12 out of the 35 countries needed for the Convention to be adopted have ratified it. The deadline for ratification has long since passed.

Despite the Convention's uncertain status, three of its Articles in particular, should be noted. The principle of equitable and reasonable utilization, Article 5, is an attempt to foster a compromise between two notable and opposed principles: absolute territorial sovereignty and absolute territorial integrity. The former extreme principle states that an upstream state can essentially do what it wants regardless of harm to the downstream state. Conversely, the latter principle states that the downstream state has a right not to be harmed by the upstream state. The principle of equitable and reasonable utilization establishes that a state both has a right to utilize its waters in an equitable and reasonable manner and at the same time the duty to cooperate in the river's protection and development. In other words, a state has a right to an equitable and reasonable share in the beneficial uses of the waters of the basin, yet that the state should not use these waters in such a way as to unreasonably interfere with the legitimate interests of other states (Bilder, 1976). Article 6 provides a non-exhaustive list of how equitable utilization may be determined. The list includes factors such as "effects of the use or uses of the watercourses in one state on other watercourse states", and "existing and potential uses of the watercourse". These factors are also equal to one another in weight. That is, no factor has priority over another. Article 5 is balanced by Article 7, which imposes an obligation not to cause significant harm. According to Article 7, states are obliged to undertake all necessary measures to ensure that their utilization of a shared watercourse does not significantly harm another riparian state.

International scholars have argued that Article 5 takes precedent over Article 7, even subsumes it in its broad principles of equitable and reasonable utilization (Dellapenna, 2001; McCaffrey, 2001). However, such a verdict has been of no consequence to states. The emphasis only suggests increased support for reconciling the various interests of river basin states in the development of their shared waters (Wouters, 1997). It does not say which state has the property right or which use by one state subordinates a different use by another state.

To be fair, although the 1997 Convention has stirred some controversy among states, which may favor one Article over another, it is important to note that it does not attempt to provide countries with specific guidelines for dispute resolution. Rather it attempts to codify customary law in the most general terms. As an umbrella agreement it does not pretend to replace individual agreements negotiated between countries over specific disputes. In a way, once countries agree to a specific formula, codified in an international water agreement, they have agreed on the compromise principle of equitable and reasonable utilization.

It is through existing agreements, therefore, that we may better detect how states go about reconciling conflicting interests in developing water resources or solving transboundary pollution problems. Property rights are essentially negotiated, and water treaty observations have made clear the ability of states to develop systems of property rights and liability rules in the absence of an overarching international body. As such, cooperation and negotiation are scrutinized next.

Cooperation and Negotiation

Cooperation over a shared river is fostered largely for the same reasons that conflict ensues. As Elhance has argued, the hydrology of an international river basin links all the riparian states, requiring them to share a complex network of environmental, economic, political, and security interdependencies. Therein lies the potential for interstate conflict as well as opportunities for cooperation (Elhance, 1999).

Compared to the small number of military episodes involving freshwater, history has recorded over 3,500 water agreements—one of the earliest dating back to 805 AD. As research on conflict and cooperation over water continues, more agreements are identified and catalogued. Thus, despite the controversy that international legal principles have wrought, states have been able to come to agreement and compromise.

Many of the earlier agreements signed are, for the most part, about navigational issues. And while navigational matters continue to be the source of dispute between states, concerns such as water scarcity, pollution, flood control, and hydropower have become prominent. Naturally, this is because populations have increased over time and states need to satiate their agricultural, industrial, and domestic needs.

As Deudney (1991) has argued, resource scarcity based on environmental degradation tends to encourage joint efforts to halt such degradation. Environmental disparities modify the meaning of ecological interdependence whereby "states and groups of states will try to seek alliances as they seek to exploit, or to escape, these disparities" (Brock, 1992). As Dokken claims, in some cases such environmental scarcities and environmental problems may be considered the starting points for cooperation (Dokken, 1997). In short, unable to unilaterally utilize the river in an efficient manner, states have turned to their fellow riparians to exploit the river in unison.

More generally, given the inherent inefficiency of deploying armaments against a fellow river riparian in the name of exploiting a river, water disputes have rarely turned violent and cooperation more often the desired strategy. Similarly, in river basins where power asymmetries exist between the riparians and exploitation of the weaker party seems likely, the stronger side may realize that it can do better by giving a sense of equality to the weaker side rather than by taking what it wants by force (Zartman and Rubin, 2000).

Though upstream states may obviously be in a position not to cooperate, they do not always exploit their strategic location on a river or their aggregate power to the detriment of the downstream state. Nor are powerful downstream states more inclined to dictate a given water regime with a weaker upstream state according to that state's sole desires. In fact, cooperation may come about relatively effectively and history has shown that quite clearly. Understanding such lessons and considering how other riparians have solved their water disputes may be instrumental to states in the midst of an on-going water conflict. This is especially the case for states that are situated in otherwise precarious regions where a dispute over a shared river can aggravate the larger political situation.

Foreign Policy Considerations and Linkage

While states are more likely to cooperate if unilateralism fails to sustain a satisfying outcome, cooperation may also come about due to foreign policy and linkage considerations. An example can best illustrate the point. On the Colorado River, the United States is at the same time the hegemonic and upstream state. In the early 1960s, Mexico protested that the water entering its territory was highly saline and demanded the United States rectify the situation. The United States should have had no immediate economic incentive to cooperate with Mexico. However, and contrary, to the predictions of Lowi's (1993) argument the United States not only entered into an agreement with Mexico but also paid the costs of desalinating the waters of the Colorado, via a desalting plant, that flow into the territory of its southern neighbor (This final settlement was reached in 1973). In fact, the costs of removing the salt from the Colorado River water provided to Mexico were considered uneconomical for the United States. However, not only did the United States not want to be considered a belligerent bully by its southern neighbor and the rest of Latin America, by rejecting cooperation, but also considered cooperation on the water issue a form of gaining cooperation and support on other fronts, such as migration (LeMarquand, 1977).

For one, the United States did not want to taint its regional and international image—a foreign policy consideration. Second, the United States believed that cooperation on water issues with Mexico would allow it to 'cash in' such goodwill on other matters at a later point in time — linkage.

Reciprocity

Considerations of reciprocity may also dampen a country's desire to utilize its geographical position to act strategically. Countries that share more than one river may be upstream on some rivers yet downstream on others. As such, countries may not wish to exploit their strategic location on the first river to the detriment of the other state, setting a precedent for the other state to act in the same manner on the second river where it is more strategically located. Utton (1988) has argued that although the Colorado River may flow from the United States into Mexico, several rivers flow from Mexico north to the United States, and the United States requires Mexico's help in maintaining their water quality.

A nation may follow a similar strategy when it shares several rivers, each with a different neighbor. It will not want its strategic behavior on one river, shared with one country, to affect its hydropolitical relations with another country on a different river. This has often been Syria's dilemma regarding the Yarmouk and the Euphrates Rivers. On the Yarmouk/Jordan River Basin, Syria is upstream of Jordan and Israel while on the Euphrates it is downstream of Turkey. Any strategic behavior employed on the Yarmouk could weaken its position with Turkey on the more important Euphrates River (Elhance, 1999).

Side-payments and Compensation

Beyond a general need for cooperation, a subsequent motivation for the upstream country (whether it be the hegemon or not) to conclude an agreement with the downstream country on projects that provide benefits to both countries (such as hydropower and flood control) is related to side-payments. That is, regulation of the river will generally provide external benefits downstream for which the upstream country will not be compensated unless an agreement is negotiated (LeMarquand, 1977). Similarly, if the downstream state perceives that the upstream riparian will go forth with a regulation project and is not interested in being compensated for the downstream benefits it thereby creates, the downstream state will not elect to sign an agreement since it will receive benefits at no cost to itself.

Weaker, smaller, and poorer upstream states can be even more handsomely rewarded by more powerful and larger downstream states desiring to exploit a river basin. Where regulation of the river for flood control and hydropower purposes is sought and the majority of the facilities need to be built upstream, upstream states can take advantage of the situation. An upstream state may agree to cooperate in exchange for some kind of compensation, whether side-payments or in-kind through projects that will be largely funded by the downstream country. To assume therefore, as Lowi (1993) does, that a more powerful country acts as a malign and coercive hegemon is problematic. Along these lines, it is also puzzling how a state might use brute force to coerce a weaker state to cooperate when the stronger state depends on the weaker state to honor an agreement. In reality, consent by the weaker upstream state is a prerequisite to a cooperative agreement. Benefits in the upstream direction need to follow to consummate the deal.

In cases, where negative externalities, like pollution, are concerned a polluting upstream state may not necessarily have the same motivation to cooperate as in, for example, utilizing the river for hydropower or flood-control purposes. In the through-border situation the externality is unidirectional and largely affects the downstream state. Abating the pollution would mean that the upstream state would incur the costs but reap no benefits. As such, its decision not to abate may be deemed most economical and its geographic advantage provides the upstream state with the necessary disincentives. Despite this seemingly deadlocked situation, side-payments may again be used by the downstream state to provide the needed incentive to abate the pollution upstream.

In a recent study, Dinar has found that of the entire set of agreements governing rivers that fall into the classification of through-border, a large percentage embody side-payments from the downstream to the upstream state (2007). That is, the great majority of recorded agreements that pertain to such issues as hydropower, pollution, and flood control, from 1864 to 2002 institute side-payments as the basis for solving the dispute.[5] As suggested before, these solutions to property right disputes embody agreements between riparians in both traditionally unstable and peaceful regions. Thus, similar compromises and settlements endure despite the diversity among regions.

The following are selected examples, illustrating the three instances outlined above where side-payments are instituted in an agreement.

The case of the 1961 Columbia River Agreement embodies the first instance noted whereby the upstream state chooses to cooperate not only out of a desire to exploit the river but also to gain compensation from the downstream state. Indeed, Canada (the upstream state) was reluctant, in the first place, to go ahead with its projects (construction of dams for flood-control and hydropower generation) unless it was assured of receiving some compensation for the unrealized benefits it was to send downstream

---

[5] Water-quantity agreements are not included in this count. Ten such agreements were identified but only two embodied side-payments, illustrating that utilizing side-payments to solve water allocation disputes is not too common (Dinar, 2007). This phenomenon is not out of line with some other findings and claims about the notion of payment for water allocations (Wolf, 1999; McCaffrey, 2001). Regarding the phenomenon of compensation and payment for water, legal scholars have argued that "modern international law does not accept the notion that seems to underlie such a claim for compensation, namely, that a state owns the waters of an international watercourse that are, for the most part, situated in its territory and is free to do with them as it pleases regardless of the consequences for other riparian states;" "On the contrary, upper riparians are under an obligation not to prevent such waters from flowing to a lower riparian country" (McCaffrey, 2001).

to the United States (LeMarquand, 1976). Barrett notes that the United States believed that Canada would want to develop the Columbia River on its side of the border anyway, and so felt that it did not need to compensate Canada for constructing the project. When Canada threatened to construct an alternative project on a different river which would provide the United States with no benefits, the United States heeded the threat as a credible one and Canada was able to secure a more attractive compensation deal (Barrett, 1994).

The case of the 1974 Chukha Hydroelectric Agreement on the Wangchhu River embodies the second instance noted whereby a weaker, poorer and smaller upstream state gains compensation from a stronger and richer downstream state for cooperating in the exploitation of the river. In this case the hydropower installations were built mainly in upstream Bhutan and downstream India financed the project. In addition, the agreement also stipulated that all of the surplus energy that Bhutan was not using would be sold to India. To this day, power export to India accounts for a handsome share of Bhutan's domestic revenue (Bandyopadhyay, 2002).

The case of the 1976 Rhine Chlorides Agreement embodies the third instance noted whereby a polluting upstream state receives compensation from the downstream state encouraging it to abate the pollution. In this instance, the Netherlands is situated downstream and was the sole victim from chloride emissions originating in upstream France, and to a lesser degree in Germany and Switzerland. For many years the Dutch attempted to persuade the upstream states to abate their salt emissions. But it was not until a compromise that stipulated the Dutch contribution to the costs of abating pollution upstream, that the deal was consummated (LeMarquand, 1977; Barrett, 2003). When faced with the salinity issue and the French refusal to underwrite all the abatement costs, the Netherlands had little choice but to contribute to the costs and at a higher level than either of the two major polluters, France and Germany (LeMarquand, 1977).

Concluding Remarks

As Wolf and Hamner have noted in a survey of hundreds of non-navigational water treaties: "...the more valuable lesson of international water is as a resource whose characteristics tend to induce cooperation." (Wolf and Hamner, 2000). Indeed the history of hydropolitics is one of cooperation rather than water-wars. If the past is any indication of the future, then the wars of the coming centuries will not be about water. This is not to say that conflict does not ensue among states sharing an international river. In fact, such disputes are most alarming in regions characterized by hostile relations between the riparians where the shared river may act as a destabilizing element. However, conflict rarely, if ever, erupts into a violent episode. Rather, it is conflict that provides impetus for cooperation, and cooperation is almost always codified in international treaties. It is, therefore, not surprising that the list of recorded international agreements is growing. In addition, learning from such agreements may be instrumental in promoting cooperation in otherwise volatile regions.

The same variables that bring about conflict may also bring about cooperation. Scarcity is chief among them. While scarcity (in water quantity, hydropower, flood-control, and water quality) may force states to act unilaterally to satiate their respective needs, it may, at the same time, facilitate a cooperative venture. Acting unilaterally to utilize a river may either be inefficient or impossible. Therefore, exploiting the river in unison may be the better alternative.

There are other variables that may exacerbate inter-state hydropolitics, such as the geographical position of a state, brute power differences among the riparians, and domestic and regional politics. Surely, tense political relations between the riparians aggravate the water conflict and even make it difficult to resolve. But on the whole, the aforementioned variables are not deterministic. For example, upstream hegemons are not destined to act in an uncooperative fashion. At the same time, downstream hegemons are not, by default, malign. They must take into account the desires and needs of the weaker upstream state. The efficiency of jointly utilizing the river depends on such recognition. In short, a state's incentives not to cooperate are not predetermined nor could the incentives to cooperate not be created.

States are motivated to cooperate not only because the unilateral option for exploiting a river is untenable but also because their need to exploit a river does not operate in a political vacuum. A state may be concerned with how a non-cooperative act may be perceived internationally or regionally (foreign policy considerations). A state may likewise be building a reservoir of goodwill to be 'cashed in' at a later time with regards to other issues (linkage). A state sharing other rivers with another state may also not want its uncooperative behavior to be reciprocated by that state on the rivers where it occupies a disadvantaged position (reciprocity). Finally, cooperation may be motivated by compensation. An upstream state, for example, may realize that the potential benefits it will create in the downstream direction, in the event it exploits the river, will go unrewarded in the absence of some kind of an agreement (side-payments and compensation). This same remedy may also apply to pollution problems where upstream states may require some monetary incentive to abate pollution in favor of the downstream state.

Among the aforementioned elements, side-payments seem to be clearly articulated in an agreement. In fact, research has shown that compensation embodies the compromise to the property right conflict between upstream and downstream states on issues such as pollution abatement, flood-control, and hydropower (Dinar, 2006; 2007). Therefore, while international water law is confined to vague principles, states have been able to agree on specific compromise solutions throughout history. This phenomenon is true across different continents and between both developing and developed countries. The policy relevance of such findings may be striking for all cases but especially in regions where water can, nonetheless, exacerbate the overall political tensions and a water agreement may alleviate those strains. Side-payments frequently occur to offset an asymmetric geographical relationship between upstream and downstream states, and are commonly regarded as an appropriate instrumentality for fostering cooperation along a given through-border river. In such situations, compromise expressed as compensation, may also serve as recognition of the downstream benefits created by works upstream.

Among the regions susceptible to instability due to an ongoing water dispute, the Aral Sea Basin seems to be ripe for this type of solution. In this basin Tajikistan, Kyrgyzstan, Uzbekistan, Kazakhstan, and Turkmenistan share the Syr Darya and Amu Darya Rivers. Kyrgyzstan and Tajikistan are upstream while Uzebekistan, Kazakhstan, and Turkmenistan are downstream. While the hydropolitics of Central Asia dates back to the Soviet era, tensions over these two rivers began after independence was achieved in the early 1990s. In essence, the five republics agreed to continue the water utilization regime, which was in place during Soviet times. The cotton producing downstream states would utilize the majority of the water while Kyrgyzstan and Tajikistan would release the water in the spring and summer from their Soviet built dams. This would mean that the upstream countries would not be able to release water in the winter for the production of hydroelectricity.

In exchange for this service and as compensation for the foregone uses, Kazakhstan, Turkmenistan and Uzbekistan have continued to sell coal and fuel to Tajikistan and Kyrgyzstan. Yet unlike the type of system that existed in Soviet times, the coal and fuel downstream states were now providing upstream states was made available only at market prices. In short, these resources were no longer provided at no – or little – charge, although they were extracted and produced relatively cheaply. This has caused much consternation upstream. A second qualm of upstream states has been their claim that they are receiving no monetary compensation from downstream states to upkeep the reservoirs as they did in Soviet times.

Despite the merits of linking and trading water and fuel, the barter regime has not worked. Both upstream and downstream states have fallen short on their commitments largely due to the problems noted above. As such the current regime is neither efficient nor sustainable. Upstream states have threatened to stop water deliveries if the current water regime is not altered. Downstream states have warned Kyrgyzstan and Kazakhstan not to *up the ante*. Needless to say all parties realize that a change in the current regime is needed, though some states may be more enthusiastic than others about how exactly to go about this shift.

In fact, changing the barter system currently utilized to a side-payment regime may be a way out of the impasse. First and foremost, upstream states should be compensated for the benefits they are foregoing upstream — production of hydroelectricity in the winter — in favor of those they are creating downstream — water for cotton growing in the spring and summer (Dinar, 2005). Kyrgyzstan has already hinted that it will be willing to forego hydroelectric production in the winter if fair monetary compensation is forthcoming (ICG, 2002). Downstream states should also be prepared to compensate Kyrgyzstan and Tajikistan for reservoir upkeep since they too benefit from the reservoirs.

This monetary option should be embraced by downstream states. First, these countries constitute some of the largest cotton exporting countries, and a small percentage of cotton revenues may be allocated to upstream states for the benefits they are creating downstream. In addition, downstream states would actually be guaranteeing themselves an uninterrupted flow of water since upstream states will be fairly and appropriately compensated for not producing hydropower in the winter and sticking to the agreed water release schedule. Second, by providing monetary compensation for the reservoirs they too are utilizing, downstream states would then have a right to demand proper upkeep. For their part, upstream states will have an incentive and obligation to care for the reservoirs, given the compensation they are receiving (Dinar, 2005).

REFERENCES

Amery, H. and Wolf, A. 2000. Water, Geography and Peace in the Middle East: An Introduction, *In:* Water in the Middle East: A Geography of Peace, H. Amery and A. Wolf (Eds), University of Texas Press, Austin, USA.

Bandyopadhyay, J. 2002. Water Management in the Ganges - Brahmaputra Basin: Emerging Challenges for the 21st Century, In: Conflict Management of Water Resources, A. Chatterji, S. Arlosoroff and G. Guha (Eds.), Ashgate, Hampshire, U.K.

Barrett, S. 1994. Conflict and Cooperation in Managing International Water Resources, Policy Research Working Paper, N 1303, The World Bank, Washington, D.C., USA.

Barrett, S. 2003. Environment and Statecraft: The Strategy of Environmental Treaty-Making, Oxford University Press, Oxford, UK.

Bilder, R. 1976. The Settlement of International Environmental Disputes, University of Wisconsin Sea Grant College Program, Technical Report, N 231.

Brock, L. 1992. Security Through Defending the Environment: An Illusion?" *In:* New Agendas for Peace Research: Conflict and Security Reexamined, E. Boulding (Ed.), Boulder Lynne Rienner, CO, USA.

Cooley, J. 1984. The War Over Water, Foreign Policy, N 52.

Crossete, B. 1995. Severe Water Crisis Ahead for Poorest Nations in the Next Two Decades, The New York Times, August 10, Section 1.

Dellapenna, J. 2001. The Customary International Law of Transboundary Fresh Waters, International Journal of Global Environmental Issues, 1 (3) (4).

Deudney, D. 1991. Environment and Security, Bulletin of Atomic Scientists, 47(3).

Dinar, S. 2002. Water, Security, Conflict, and Cooperation, SAIS Review, 22 (2).

Dinar, S. 2005. Treaty Principles and Patterns: Selected International Water Agreements as Lessons for the Resolution of the Central Asia Water Dispute, *In:* Transboundary Water Resources: Strategies for Regional Security and Ecological Stability, H. Vogtmann and N. Dobretsov (Eds.), Springer, Dordrecht, the Netherlands.

Dinar, S. 2006. Assessing Side - Payment and Cost-Sharing Patterns in International Water Agreements: The Geographic and Economic Connection, Political Geography, 25 (4).

Dinar, S. 2007. International Water Treaties: Negotiation and Cooperation along Transboundary Rivers, Routledge, London, UK.

Dokken, K. 1997. Environmental Conflict and International Integration, *In:* Conflict and the Environment, N.P. Gleditsch (Ed), Kluwer Academic Publishers, Dordrecht, the Netherlands.

Elhance, A. 1999. Hydropolitics in the 3rd World: Conflict and Cooperation in International River Basins, United States Institute of Peace Press, Washington, D.C., USA.

Falkenmark, M. 1986. Fresh waters as a factor in strategic policy and action, *In:* Global Resources and International Conflict: Environmental Action in Strategic Policy and Option, A. Westing (Ed.), Oxford University Press, Oxford, UK.

Falkenmark, M. 1986. Fresh Water: Time for a Modified Approach, Ambio, 15 (4).

Falkenmark, M. 1992. Water Scarcity Generates Environmental Stress and Potential Conflicts, *In:* Water, Development, and the Environment, W. James and J. Niemczynowicz (Eds.), Lewis Publishers, Boca Raton, FL, USA.

Falkenmark, M. 1990. Global Water Issues Confronting Humanity. Journal of Peace Research, 27(2).

Frey, F. 1993. The Political Context of Conflict and Cooperation over International River Basins. Water International, 18 (1).

Gleick, P. 1993. Water and Conflict: Fresh Water Resources and International Security, International Security, 18 (1).

Gottman, J. 1951. Geography and International Relations, World Politics, 3 (2).

Gyawali, D. 2000. Nepali-India Water Resource Relations, *In:* Power and Negotiation, I.W. Zartman and J. Rubin (Eds.), The University of Michigan Press, Ann Arbor, MI, USA.

Homer-Dixon, T. 1999. Environment, Scarcity, and Violence, Princeton University Press, Princeton, NJ, USA.

International Crisis Group. 2002. Central Asia: Water and Conflict, Asia Report, N 34.

Just, R. and Netanyahu, S. 1998. International Water Resource Conflicts: Experience and Potential, *In:* Conflict and Cooperation on Trans-Boundary Water Resources, R. Just and S. Netanyahu (Eds.), Kluwer Academic Publishers, Boston, USA.

LeMarquand, D. 1977. International Rivers: The Politics of Cooperation, Westwater Research Center, Vancouver, Canada.

LeMarquand, D. 1976. Politics of International River Basin Cooperation and Management, Natural Resources Journal, 16.

Lowi, M. 1993. Water and Power: The Politics of a Scarce Resource in the Jordan River Basin. Cambridge University Press, Cambridge, UK.

Mathews, J. 1989. Redefining Security, Foreign Affairs, 68 (2).

McCaffrey, S. 2001. The Law of International Watercourses: Non-Navigational Uses, Oxford University Press, Oxford, UK.

Naff, T. 1994. Conflict and Water Use in the Middle East, *In:* Water in the Arab World: Perspectives and Prognoses, P. Rogers and P. Lydon (Eds.). Harvard University Press, Cambridge, USA.

Naff, T. and Frey, F. 1985. Water: An Emerging Issue in the Middle East, *In:* Changing Patterns of Power in the Middle East, T. Naff and M. Wolfgang. (Eds.) The Annals of the American Academy of Political and Social Science, 482.

Shrestha, H.M. and Singh, M.L. 1996. The Ganges-Brahmaputra System: A Nepalese Perspective in the Context of Regional Cooperation, *In:* Asian International Waters: from Ganges-Brahmaputra to Mekong, A. Biswas and T. Hashimoto (Eds.), Oxford University Press, Bombay, India.

Starr, J. 1991. Water Wars, Foreign Policy, 82.

Toset, H.P.W., Gleditsch, N.P. and Hegre, H. 2000. Shared Rivers and Interstate Conflict, Political Geography, 19 (8).

Utton, A. 1988. Problems and Successes of International Water Agreements: The Example of the United States and Mexico, *In:* International Environmental Diplomacy, J.E. Carol (Ed.), Cambridge University Press, Cambridge, UK.

Verghese, B.G. 1996. Towards an Eastern Himalayan Rivers Concord, *In:* Asian International Waters: From Ganges-Brahmaputra to Mekong, A. Biswas and T. Hashimoto (Eds.), Oxford University Press, Bombay, India.

Wolf, A. 1999. Criteria for Equitable Allocations: The Heart of International Water Conflict, Natural Resources Forum, 23 (1).

Wolf, A. and Hamner, J. 2000. Trends in Transboundary Water Disputes and Dispute Resolution, *In:* Water for Peace in the Middle East and Southern Africa, Green Cross International, Geneva.

Wouters, P. (Ed.) 1997. International Water Law: Selected Writings of Professor Charles B. Bourne, Kluwer Law International, London, UK.

3

Conflicts in Shared River Basins[1]

Nils Petter Gleditsch[*], Håvard Hegre and
Hans Petter Wollebæk Toset

International Peace Research Institute Oslo (PRIO),
Norwegian University of Science and Technology, and University of Oslo
* Corresponding author, at PRIO, Hausmanns gate 7, 0186 Oslo, Norway.
E-mail address: nilspg@prio.no (N.P. Gleditsch).

WATER AND HUMAN SECURITY

Water is an essential commodity for human existence. Water is used for consumption, for maintaining public health, for agriculture, for industry, and for transportation. Serious scarcities of water will affect virtually every aspect of human life. Given the importance of water, it is not surprising that water is expected to be among the commodities which people will be especially concerned to preserve, even to the point of fighting. For a country heavily dependent on river water for its economic development, the threat of having its water supply severely constrained by an upstream user may seem threatening indeed.

Indeed, the idea of 'water wars' has become part of the political rhetoric. As early as 1967, just before the Six-Day War between Israel and its Arab neighbors, Prime Minister Levi Eshkol declared that "water is a question of survival for Israel", and therefore "Israel will use all means necessary to secure that the water continues to flow" (Biliouri, 1997). In the mid-1980s,

[1] This chapter is a revised version of Hans Petter Wollebæk Toset, Nils Petter Gleditsch and Håvard Hegre, 2000. 'Shared Rivers and Interstate Conflict', *Political Geography* 19(6): 971-996. We thank the editor of *Political Geography* and the journal's publisher Elsevier for permission to reprint material from this article. We acknowledge the financial support of Research Council of Norway for our research. Detailed credits can be found in the first note of Toset et al. (2000) as well as our subsequent articles. The data used for analysis can be downloaded from www.prio.no/cscw/datasets. The first author can be contacted at nilspg@prio.no.

US intelligence services are said to have "estimated that there were at least ten places in the world where war could break out over dwindling shared water" (Starr, 1991). In 1991, the then Crown Prince of Jordan is reported to have said that the 1967 war "was brought on very largely over water related matters" and that unless there was an interstate water agreement in the Middle East by 2000 "countries in the region will be forced into conflict" (cited from Irani, 1991). At the Habitat Conference in Istanbul in 1996, the Secretary-General of the Conference was reported to have told the participants that "the scarcity of water is replacing oil as a flashpoint for conflict between nations in an increasingly urbanized world" (cited from Lonergan, 1997). Hilde Frafjord Johnson, then Norwegian Minister of Development and Human Rights, argued that in many countries water shortages could develop into major conflicts (interview in *Dagbladet*, 20 August 1998). The media have reported numerous similar statements by politicians, spokespersons for international organizations, and others.

More tempered fears have been expressed by, for instance, Elhance (1999), who argues that an unequal distribution of freshwater does not in itself necessarily lead to acute interstate conflict, but that "severe scarcities of *an essential, non-substitutable, and shared resource*" like freshwater may make states prone to conflict. Dupont (1998), who views environmental security as an important issue in the Asia-Pacific region, believes that "environmental issues are unlikely to be the primary cause of a major conflict between states." He takes a middle ground in concluding that water wars are less likely between countries "with shared values and generally cooperative relations." Trolldalen (1992) asserts that "Competition for both quality and quantity of shared water at a local level often leads to international water conflicts", but he does not specify that such conflicts are necessarily violent. Homer-Dixon (1999) argues that although there have been a number of resources wars over non-renewable resources like oil and minerals, there are few examples of wars over renewable resources. Among the renewables, Homer-Dixon believes that water is the most likely candidate for stimulating international war. However, wars over river water are likely to take place only under special conditions such as high dependence on water in a downstream country, a history of antagonism between the two countries, and so on.

RESOURCES AND CONFLICTS

Most conflicts are over scarce resources of one kind or another, at least if territory is counted as a scarce resource. Holsti (1991) concluded that among interstate wars in the period 1648-1989, territory was by far the single most important issue category. In the first period such issues figured in about half the wars, declining to about one-third in the post-World War

II period. In a reanalysis, Vasquez (1993; 1995) found that between 79% and 93% of all wars over Holsti's five time-periods involved territory-related issues. Huth (1996), in a study of territorial disputes in 1950-90, characterized this issue as "one of the enduring features of international politics." This relationship holds for interstate as well as intrastate conflict. Wallensteen and Sollenberg (2000) found that among the 110 conflicts in 73 countries in the post-Cold War period (1989-99)—the vast majority of which were intrastate conflicts—over half were over territory, the rest over government. The territorial explanation for war is also consistent with the finding that wars occur most frequently between neighbors (Bremer, 1992) and between proximate countries (Gleditsch, 1995). There is still some dispute whether wars between neighbors occur mainly because they fight over territory or because they generate disagreements in their day-to-day interaction, or because they are more easily available for fights—but Vasquez (1995) has presented a strong case for the territorial explanation.

But what is it about territory that makes it worth fighting for? Territory can be a symbol of self-determination and national identity, but it can also be a proxy for tangible resources found on the territory. Such resources are *strategic raw materials, sources of energy, food,* and—emerging strongly in the public debate—*access to freshwater resources.*

The hypothesis that scarce resources create conflict is closely related to the debate about environmental conflict (Gleditsch, 1997). Many social scientists, such as Bächler et al. (1996) and Homer-Dixon (1991, 1999), have posited that environmental degradation depletes the stock of scarce resources, so that resource conflicts are generated or exacerbated in new areas and for resources that previously were plentiful. Others, in a new wave of writing on economic motives for war (Collier, 2000; de Soysa, 2000), maintain that resource abundance may provide a motive for loot-seeking, a factor in many violent civil conflicts in Africa.

In this chapter we do not address directly either the environmental issue or the question of loot-seeking behavior in conflict. We simply set out to study how the supply of a specific resource, water, may be associated with violent conflict between states, on its own or in interaction with other factors.

WATER–A RENEWABLE RESOURCE?[2]

Most writing on water conflict focuses on water scarcity. But is there a scarcity of water? The total amount of water on earth has been calculated

---

[2] In addition to specific sources cited in the text, this section relies on Beaumont (1997), Falkenmark (1990), Shiklomanov (1993), and *Aschehoug og Gyldendals Store Norske Leksikon* (1989). There are many definitional problems, including whether water availability should be measured as precipitation or runoff, but they are not critical here.

at 1650 mill, km^3, or more than 0.25 km^3 per person. Unlike non-renewable resources such as petroleum, water is rarely consumed in the sense of becoming unavailable for use by human beings. Rather, it circulates in a never-ending hydrological cycle, which regularly cleans and desalinates the water by evaporation and precipitation.

But there is also bad news: 97% of the water is unusable for human consumption as saltwater and only 3% is freshwater. A total of 87% of the freshwater is not directly available, either because it is locked up in icecaps or in deep aquifers, or because it is polluted. Water has to be regarded as a finite and fixed resource, and the rise of the global population has progressively reduced the world run-off per capita, from 40,000 m^3 per person in 1800 to 6840 m^3 in 1995, estimated to fall further to 4692 m^3 by 2025. Water resources are enormously skewed geographically. North America has an annual run-off of 17,000 m^3 per capita per year, while Africa has 6000, and Egypt has 50. Less than 1% of all the world's usable freshwater is found in the Middle East or North Africa, although this region contains 5% of the world's population. Many countries with lower water availability today, particularly in Africa, also have high rates of population growth, so that their water shortages may be exacerbated in the future. Increasing standards of living may lead to greater demands for water. In a study for the International Water Management Institute, Seckler, Molden and Barker (1998) estimate that "slightly more than one billion people live in arid regions that will face absolute water scarcity by 2025." In particular, they see groundwater depletion and pollution "as a major threat to food security in the coming century." The existing demand for water exceeded the renewable supply for half a dozen countries in the late 1980s and more countries will move into this category. Lundqvist (1998) fears impending 'hydrocide', where pollution and heavy water withdrawals will cause disease, ecosystem disturbance, and societal disorder. Many countries are highly dependent on water that originates outside their border—over 90% in the case of countries like Egypt, Hungary, and Mauritania (Gleick, 1993a).

Very little water is needed for vital human life processes. A person perhaps requires 1-2 m^3 of drinking water per year, a minuscule amount compared to what is available even in many arid countries. More water is required for the transportation of urban or industrial wastes. Counting these uses, as well as normal inefficiencies and losses, the annual freshwater needed for urban life has been estimated at about 250 m^3 per capita, a much higher figure but still low compared even to the projected availability of 4692 m^3 in 2025. Even though urban and industrial uses of water can be quite wasteful, few countries would experience severe water shortages if this were the end of the story. Although there are many problems of interpretation and measurement, the overall message seems clear: there is no scarcity of water for the globe as a whole. However, many areas have water shortages relative to their present needs, and this problem may increase unless changes are made in the patterns of supply or consumption. Securing adequate and plentiful water for human objectives is a political and economic issue rather than one of absolute physical constraints.

The question of water as a lootable resource has not been given comparable attention in the water conflict literature. However, an abundance of water can stimulate transportation on the waterways, facilitate the disposal of waste, and generate other exploitable resources, like fishing and hydroelectric power. Once again, it seems probable that other factors than the absolute amount of water available will decide whether such exploitation follows a violent or a peaceful track.

WATER CONFLICTS

Based in part on the optimistic and the pessimistic aspects of the resource situation, two different scenarios may be outlined. The *conflict scenario* foresees growing and increasingly serious water scarcities in a number of countries. "Where water is scarce, competition for limited supplies can lead nations to see access to water as a matter of national security" (Gleick, 1993a). The current trends in population and development will make water "an increasingly salient element of interstate politics, including violent conflict" (Gleick, 1993a). In order to identify potential trouble areas we need to look to "rivers, lakes, and water aquifers shared by two or more nations" (Gleick, 1993a). While flowing in its natural course, a trans-boundary river does not necessarily follow state boundaries. To overcome food scarcity and poverty, an upstream country might be tempted to use the water to increase its biomass production in agriculture and forestry. Such use (or misuse) can affect the quantity and quality of water sent to the downstream neighbor. In particular, there is a "serious risk of international conflict", especially in the Middle East and Africa, "between upstream and downstream countries" (Falkenmark, 1990).[3]

The *cooperation scenario,* while freely admitting the possibility of conflict, denies its inevitability (Kukk and Deese, 1996; Lowi, 1995; Rogers, 1996)[4]. As noted above, all countries have access to a sufficient amount of drinking water. Irrigation makes the most severe demands on freshwater supplies, and even a small concentration of salt progressively worsens the soil quality, requiring ever more water per hectare. In Egypt, it takes about 3000 tons of water to produce a ton of wheat. Most of this water is wasted in evaporation. What goes up, must come down, but not necessarily in the

---

[3] A number of other examples are cited by Wolf (1999b, p. 243).
[4] All the figures in this paragraph are from Beaumont (1997).

same area. When viewed in a global perspective, producing more food in areas where water was more plentiful might conserve water. This would require less self-sufficiency and more international trade, but such policies are controversial for a variety of reasons.

The cooperation scenario further points to the possibility of cooperative arrangements for sharing river resources between the upstream and downstream countries, including treaties and joint river administrations. Such arrangements have been in force on the Danube[5] and the Rhine for decades. Even the Mekong River Basin has had a UN-sponsored committee in operation since 1957, but the long and numerous Indo-China wars prevented the body from making much progress (Dupont, 1998). In April 1995, Cambodia, Laos, Thailand, and Vietnam did, however, sign an agreement establishing the Mekong River Commission, whose mandate calls for cooperation in the management and conservation of water resources in the river basin, including irrigation, hydropower, navigation, flood control, fisheries, timber floating, and tourism. In fact, the Mekong River provides a better example of how an ideologically driven armed conflict prevents cooperation in the development of shared water resources than an example of how a shared river generates conflict among the riparian states.[6] Another example of how water shortage may lead to cooperation rather than armed conflict is provided by the example of the island state of Singapore, which relies on neighboring Malaysia for about 50% of its water needs. Singapore's water is supplied under two agreements, signed in 1961-62, which expire in 2011 and 2061 respectively (Dupont, 1998). While increased use of water in both countries may lead to haggling over the terms of these agreements when they run out, armed conflict does not seem very likely. As part of NAFTA, the US and Mexico have established a Border Environmental Cooperation Commission, which among other things deals with transboundary river issues over a border exceeding 3000 km (Milich and Varady, 1999).

The cooperation scenario also points to more realistic pricing as a way of regulating the use of water. Traditionally, water has been perceived as a public good, to be consumed or polluted at will. As Falkenmark and Lundqvist (1998) argue, "... most people tend to take water for granted". Increasingly, the use of water comes under public and private regulation, which permits realistic pricing. As Beaumont (1997) points out, in restaurants people are willing to pay up to US $1000 per m^3 for brand-

---

[5] The Danube is by a wide margin the river shared by most countries (19). Next on the list are the Congo and the Niger (shared by 11) (Wolf et al., 1999, p. 424).

[6] The members have also engaged in talks with China and Myanmar (Chou, 1998, p. 5). However, China, which is upstream to all the other riparian states (the Mekong originates in Tibet), has not signed the 1995 agreement. She has ambitious hydroelectric plans that may influence conditions downstream adversely (Dupont, 1998, p. 72).

named bottled waters that are not very different in quality from the water obtained from the public supply. The cooperative scenario, then, argues that the same conflict-regulation processes as other scarce resources, including the judicious use of market mechanisms, may peacefully regulate conflict over water resources. Of course, human affairs are not always conducted wisely, so violent conflict could still occur.

The incentives for cooperation may depend on the nature of the water dispute. Wallensteen and Swain (1997) argue that in rivers where there is a *quality* problem, such as in the Rhine or the Colorado River, there is a strong incentive for cooperation among the riparian states. In the Nile and Ganges, characterized by a *quantity* problem, the incentives for cooperation are less obvious. In these cases, negotiations tend to be bilateral, and military threats and boycotts routinely become part of the bargaining behavior.

Previous Studies

Apart from the studies of territory and war, there are relatively few large-n studies of the relationship between resource scarcity and interstate conflict. Tir and Diehl (1998) have summarized the literature on population pressure and interstate conflict, and have tested the relationship between conflict and population density and growth over the period 1930-89. They concluded that with appropriate control variables, population growth did appear to be moderately related to interstate conflict, but that population density had no effect. In neo-Malthusian thinking, involving environmental pessimism, population pressure plays a major role in increasing resource scarcity (Ehrlich and Ehrlich, 1972; Homer-Dixon, 1999). Cornucopian thinkers, on the other hand, who are more inclined to technological optimism, argue that population growth is outpaced by human innovation and therefore has little effect on resource scarcity (Lomborg, 2001, Ch. 13).[7]

Summarizing a large number of case studies of internal armed conflicts that he has carried out with various associates, Homer-Dixon concludes that "environmental scarcity has often spurred violence in the past" (1999) and that "in coming decades the world will probably see a steady increase in the incidence of violent conflict caused, at least in part, by environmental scarcity" (1999). However, he has also made it clear that at this stage he cannot identify any clear "causal effect", and that his work is limited to establishing "causal mechanisms" (Schwarz, Deligiannis and Homer-Dixon, 2001).

In view of the many alarming public statements regarding water and conflict, there is surprisingly little relevant systematic research on this

---

[7] For a debate starkly contrasting the two views, see Myers & Simon (1994). A interpretative survey of the debate is found in Ohlsson (1998). For more recent work on demographic factors in conflict, see Brunborg & Urdal (2005).

issue. Work by Falkenmark (1990), Gleick (1993a, 1993b), and others is valuable in clarifying the mechanisms by which conflict could occur, and by mapping the potential locations. However, these authors have not demonstrated that problems of water-sharing have actually played an important role in escalating conflicts to war.

Many authors have pointed to the Middle East as a particularly likely location for a 'water war'. Water played an important role when Israel in March, May, and August 1965, as well as in July 1966, attacked the water diversion works of Syria, Jordan, and Lebanon with tanks and aircraft. This project, named the Headwater Diversion Plan, would have channeled the Hasbani River in Lebanon and Banias River in Syria, two of the sources of the Jordan River, around Lake Tiberias through Syria to the Yarmouk River where the water would have been regulated by a Jordanian dam at Mukheiba (Naff and Matson, 1984). It has also been argued that these trends towards competitive utilization of the water in the Jordan River system played a key role in the Six-Day War in 1967. In that war Israel destroyed a Jordanian dam on the Yarmouk, the most important tributary to the Jordan River. Regardless of the role of the water, Israel, by conquering the West Bank and the Golan Heights from Syria, improved its hydrostrategic position through control of the Upper Jordan River. The occupation of the Golan Heights made it impossible for the Arab states to divert the Jordan headwaters. The 1969 ceasefire lines gave Israel control of half the length of the Yarmouk River, compared to 10 km before the war. During the summer of 1969 Israel also bombed the East Ghor Canal, today's King Abdallah Canal, the most vulnerable target among Jordanian water facilities (Naff and Matson, 1984). Although such conflicts over shared water resources appear to be zero-sum games, it seems far-fetched to argue that water is the main or even a very important general reason for war in the Middle East. After King Hussein gave up his claim to the West Bank in 1988, this dispute became detached from other strategic interests and could therefore be regarded as a genuine water conflict (Libiszewski, 1995). Nevertheless, the basic issues of nationalism and control of land territory seem vastly more important factors in most of the disputes in the Middle East. Wolf (1999b) says categorically that "the only problem with these theories is a complete lack of evidence" and that "water was neither a cause nor a goal of any Arab-Israeli warfare."

The conflict scenario could be defended on the grounds that the future is likely to be different from the past. Gleditsch (1998) has criticized the widespread tendency in studies of environmental security to refer to future crises as empirical evidence. A convincing scenario that argues that the future is going to be different from the past requires a clear specification of theoretical mechanisms. Citing a 1979 statement by Anwar Sadat that "the only matter that could take Egypt to war again is water" (Gleick,

1993a) is not equivalent to showing why and when such threats are likely to be carried out. However, Gleick (1993a) does not rely only on the future: "History is replete with examples of competition and disputes over shared water resources", he argues. He goes on to say that he will "describe ways in which water resources have historically been the objective of interstate politics, including violent conflict." His examples of water-induced conflict, apart from the Six-Day War, are verbal conflicts between states, threats of violence, and water-related violence in ongoing wars, rather than a historical argument that conflicts over scarce water played an important role in the outbreak of the war. Gleick (1998) identifies in detail 54 historical and ongoing disputes and conflicts over freshwater resources. In most of these disputes, water is an instrument of war or a strategic target, rather than a scarce resource at the root of the dispute.

In this chapter, we conduct a multivariate study of the effect of shared rivers on international conflict.

SHARED RIVERS

Based on the literature on water shortages, we expect countries with shared rivers to have more armed conflict and militarized disputes than other dyads. We could have formulated a hypothesis in terms of shared water resources generally, but we have data only on shared rivers. We do not necessarily expect the relationship between shared rivers and conflict to be very strong, or to dominate other factors of conflict. But even if the cooperative scenario is often correct, there should be some cases where foolish policies have prevailed and led to conflict. Thus our first hypothesis is:

H1: Everything else being equal, countries that share a river have more dyadic conflict behavior

Boundaries between two countries may vary from a few kilometers to several thousand. Thus, it is not surprising that dyads may share more than one river. We assume that an additional shared river increases the likelihood that some conflict issue will arise. Of course, having more shared rivers might also lead to more cooperation, if the cooperation scenario is correct. However, our hypotheses focus on the conflict aspect. Thus the next hypothesis:

H2: The more shared rivers between two countries, the higher the probability of conflict behavior between them

Water shortage is not the only mechanism by which rivers may generate conflict. Conflict may also be generated by the use of a river for navigation purposes. For instance, the use of the Mekong river for transportation through Vietnam into Phnom Penh led Cambodia to oppose a development aid-funded bridge project in Vietnam unless the bridge was high enough to accommodate the ships (Chou, 1998). Another potential source of conflict derives from the use of rivers for international boundaries. While in a sense they are obvious candidates for this role, rivers are somewhat devious boundaries because they do not stay in the same place. They erode the landscape, dig new outlets, etc. The legal boundary usually follows the *Thalweg* (i.e. the line following the deepest part of a river), but its location may change over time. An island that belongs to one country may eventually end up on the other side of this line. A dispute about the boundaries in the Ussuri River occasioned an armed confrontation between China and the Soviet Union in 1969. Finally, shared rivers may generate conflicts over pollution (Shmueli, 1999), although it is not generally suggested that by itself this is sufficient to lead to armed conflict.

Rivers may run along the border, as does the Congo, which separates the former French and Belgian colonies. They may also run from one country into another, as when the Nile crosses from Sudan into Egypt. We shall refer to the first type as a *river boundary* relationship, to the second as an *upstream/downstream* relationship. Figure 1 illustrates the two, plus a mixed type. The three types of conflict issues that we have mentioned impinge in slightly different ways on these two relationships: *sharing water resources* can create problems in both types of relationships, but much more seriously in the upstream/downstream relationship. In a river boundary situation, the country on the left bank may divert water, but the

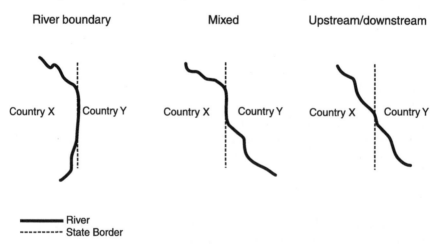

Fig. 1. Different forms of shared rivers in dyads. Each shared river is classified as one of three types: river boundary; mixed; upstream/downstream. A dyadic relationship may be classified as one of these three or as having several rivers with different types. In order to count as a river boundary type or a mixed type, the river has to run along the boundary of the dyad for more than 10 km.

country on the right bank has an obvious form of retaliation: to divert water to its own side of the river. *Navigation problems* and *transborder pollution* may occur in both cases, while *boundary problems* would occur only in the river boundary situation. It is not obvious how this adds up, but given the focus on water shortages in the literature, we expect the upstream/downstream relationship to have a higher conflict potential. We anticipate that the incentives for cooperation will be higher for countries sharing a river boundary than in an upstream/downstream situation characterized by zero-sum relations and lack of confidence. Our next hypothesis is therefore:

H3: Among countries with shared rivers, upstream/downstream situations have more dyadic conflict behavior

We also want to test a common assumption in the neo-Malthusian literature, that resource and environmental issues are becoming more important conflict factors. We therefore formulate a hypothesis to the effect that to the extent there is a resource scarcity problem linked to shared water resources, it has probably become more serious over time, because of population growth and increasing consumption. Moreover, we would expect water issues to emerge as more important as a factor in global conflict after the end of the Cold War, when the world is no longer locked into a tightly bipolar confrontation between East and West reinforced by mutual nuclear deterrence.

Our fourth hypothesis is:

H4: The relationship between shared rivers and conflict is accentuated over time

Finally, we want to test more directly a hypothesis about water sharing, the main mechanism in the presumed causal chain from shared rivers to conflict behavior. We follow the dominant view of water conflict in positing that water scarcity might be related to conflict:

H5: Everything else being equal, two contiguous countries with water scarcity are more likely to have conflict behavior

Finally, we will investigate the interaction between water scarcity and river borders:

H6: Water scarcity increases the extent to which river-sharing is associated with dyadic conflict behavior

OTHER FACTORS

The phrase 'everything else being equal' implies a series of hypotheses about interstate conflict. Multivariate studies such as Bremer (1992), Oneal and Russett (1999), and Hegre (2000) indicate which factors increase the probability of interstate conflict behavior between two countries A and B: one or both are major powers, the two are about equally powerful, neither country is a democracy, the two are not allied to one another, they have a history of violence, neither country is rich, they trade little with one another,[8] and A is already at war with a third country C which is contiguous or allied with B. In this study, we control for regime type, economic development, major power, peace history, and alliances. In addition, contiguity is a selection factor here: our analyses are for contiguous countries only.

For some of these controls, a multivariate framework is essential. For instance, allied countries and rich countries tend to cluster geographically and allied dyads therefore appear to fight each other more frequently if one does not control for contiguity.

We study the years 1880-1992, a period for which data are widely available for all control variables, but also report results for a shorter time period.

DATA

Shared Rivers

The data on shared rivers are taken from Toset (1998). He, in turn, started from the contiguity dataset of the Correlates of War (COW) Project (Gochman, 1991). This dataset distinguishes between land contiguity through the main territory of the states, land contiguity through dependent territory, and contiguity by sea in three categories if the countries are separated by less than 150 miles of water. The latter form of contiguity is less relevant here, since by definition two countries cannot share a river if they are separated by sea. We exclude contiguity of dependent territories, since we assume that most of the conflict-generating effect of having dependencies is picked up by the great power variable.

In order to identify shared rivers, Toset (1998) relied heavily on a UN register (CNRET, 1978) which claims to include all rivers 10 km long or longer for all the world except Asia. The culmination of a 20-year effort by a now defunct UN body, its final report listed 214 major shared international freshwater resources, 148 of which flow through two countries only, and the rest through three or more.[9] The number 214 is cited frequently even though it was never completely accurate and has become less so with the

---

[8] The relationship between trade and conflict remains somewhat contested; see Oneal and Russett (1999), Beck, Katz and Tucker (1998) Barbieri and Schneider (1999).

[9] For a brief description of the CNRET reports, see Swain (1997). A recent World Bank report refers to "over 245 river basins", which serve about 40% of the world's population and half its area (Salman and Boisson de Chazournes, 1998). Wolf et al. (1999) have updated the UN register to list 261 international rivers, which cover almost half the land surface of the globe. Milich and Varady (1999) put the number of shared river basins at "more than 300".

further proliferation of new states.[10] Toset supplemented the CNRET data with more detailed information for Asia and for historical boundaries from sources such as Granzow (1898), *Westermanns Atlas zur Weltgeschichte* (1956), and *The Times Atlas* (1997). The maps varied in scale and reliability and the coding was bound to involve some inaccuracy, particularly for the older dyads. An example of a particularly complex and difficult area is the desert and swamp territory between Iraq and Syria. Biswas (1990) has criticized CNRET for not consulting other projects and for not making use of the best maps. Its definition of river basin also excludes a number of international river systems from its register. For instance, CNRET recognizes only one river basin in the India-Bangladesh area, Ganges-Brahmaputra (shared by India, Bangladesh, China (Tibet), Nepal, and Bhutan), while the Indo-Bangladesh Joint River Commission identified 54 river systems divided between those countries (Swain, 1996). Nevertheless, the CNRET register provided the best starting point for a systematic empirical study.

Coded in dichotomous form, Toset's dataset contains a total of 1274 dyads with shared rivers over the 1816-1992 period, and 13,707 dyad-years with shared rivers.[11] About 80% of the contiguous dyads share rivers. Of these, in turn about 8% are coded as having a very high number of shared rivers (10 or more). The approximately 8000 river boundary dyad-years coded on the basis of CNRET (1978) were coded as having a short, medium, or long river boundary. A total of 30% were in the 'short' category (less than 100 km) and 26% were 'long'. Of the shared river dyad-years, 8% had river boundaries, while 30% were simple upstream/downstream types. Also, 13% of the shared river dyad-years had multiple border crossings and no less than 48% had more than one type.

Toset's dataset also includes the total area of the river basin, each state's share of the river basin, and which of the riparian states is upstream. These additional variables have not been used in the analysis. The upstream/downstream variable is potentially the most interesting one. Since an upstream country can restrict the supply of water, we might expect downstream countries to be more likely to initiate conflict behavior against upstream countries. However, the downstream country can restrict navigation for upstream countries, leading to precisely the opposite prediction. We do not have data on the kind of restrictions that upstream countries may have imposed on their downstream brethren or vice versa. In any case, we do not think that the COW data on the initiation of conflict behavior are reliable enough to justify such an analysis.[12]

---

[10] In particular, the fragmentation of the Soviet Union. Russia and the Ukraine (since 1991) have more shared rivers than any other dyad. Part of this increase in internationally shared rivers was offset by the reunification of Germany and the Yemen after the end of the Cold War.

[11] A dyad is a pair of states. In the dataset, information is recorded for each 'dyad-year', i.e. for each dyad for every year in the period.

[12] This problem is discussed in Gleditsch and Hegre (1997).

A weakness of the Contiguity dataset is that it does not have clear criteria for how to deal with territory occupied by another state. An example is the border area around the Golan Heights between Israel and Syria, occupied by Israel since 1967. In generating river boundaries in this area, Toset used the original borders if the occupation was not recognized by the states involved.

Conflict Behavior

The source of the main dependent variable is the Militarized Interstate Disputes (MIDs) dataset from the Correlates of War project (Gochman and Maoz, 1984; Singer and Small, 1994). The interstate war variable of the COW project has one general weakness from an analysis point of view: because there are few of them, the results may be overly dependent on a few historical events, particularly in analyses of shorter time periods. A specific weakness of using interstate wars for this particular analysis is that it may be unreasonable to expect disputes over water to escalate all the way to war. Some of the more dramatic predictions, of course, foresee that the escalation will go that far. But it seems more reasonable to hypothesize that many of the conflicts will de-escalate before they cross a threshold of 1000 battle-deaths in a single year. The MIDs—which include a range of low-level hostilities including threats to use force and displays of force—are much more numerous, but are widely assumed to be somewhat less reliably coded and with greater uncertainty than the wars about the start and end dates. They may also suffer from what might be called an 'attention bias': while a war can scarcely be hidden from public view, a militarized dispute may not catch the attention of the media and thus will not have been caught by the COW coders (Gleditsch, 1999). The harm done by the least serious MIDs (threat of force) is also frequently quite marginal and the practical significance of a relationship between shared rivers and such conflict behavior can be questioned. We choose an intermediate solution here and measure conflict behavior as the onset of a MID with at least one casualty. This should reduce attention bias considerably.

Control Variables

We measure *major power* with the standard dichotomous variable from the COW project (Small and Singer, 1982) and score the dyad-years as involving 0, 1, or 2 major powers. *Regime type* is measured by Polity III (Jaggers and Gurr, 1995). Countries with 3 or higher on the difference between institutionalized democracy and institutionalized autocracy were characterized as democratic, and the dyad-years were coded as involving 0, 1, or 2 democracies. Dyads in which at least one country was in regime transition or missing data for other reasons were coded as a separate category labeled Regime transition/NA. *Economic development* was coded by using the log of energy consumption per capita from the National Material Capabilities Dataset of the Correlates of War Project (Singer and Small, 1993, and additional data coded from UN sources taken from Gissinger and Gleditsch, 1999). Energy consumption and GDP per capita each have their problems as measures of economic development; decisive for our choice was the fact that the former variable is available for a much longer time-span. We assume that the least developed country in the dyad is the one least constrained of the two against the use of force.[13] Consequently, we use the lower log of energy consumption per capita in the dyad as the dyadic form of the development variable. Another variable from the COW project is *alliances* (Singer and Small, 1993). The COW project lists three types of alliances: defense pacts, neutrality pacts, and ententes. We excluded the neutrality pacts and merged the other two categories as our indicator of shared alliance (cf. Raknerud and Hegre, 1997). An alliance concluded in one year is coded as in effect from the next year. An alliance ended in a particular year is coded as a non-alliance from that same year. An alliance formed and ended in the same year is not coded at all. To control for temporal dependence between units of analysis, we added the variable *Peace history* to the model. The variable was defined as $-\exp\{(-\text{years in peace})/a\}$, where 4.329 was chosen as a value for a. 'Years in peace" is either the number of years since the two states were on opposite sides of a militarized dispute or since the youngest state gained independence.[14] The function models the log odds of a militarized dispute in the dyad as high just after a previous dispute/independence then decreasing at a constant rate. The variable ranges from –1 to 0. For this value for a, the impact of Peace history on the log odds of conflict has a half-life of exactly 3 years.[15] Finally, *freshwater availability* was taken from a dataset generated by Hauge and Ellingsen (1998) in a study of civil war. They coded low, medium, and high availability on the basis of data from *World Resources* (WRI, 1986-95). We have used 10,000 m^3 per capita per year

---

[13] See Dixon (1994) on this 'weak-link assumption" in the context of testing the democratic peace.

[14] See Raknerud and Hegre (1997) for a discussion of the problems with temporal dependence in a similar context. The value 4.1 for a is the value that maximizes the log likelihood in Model 1 (log likelihood= –1043.27). We chose the value 4.329 since this corresponds to an integer half-life (3 years) with insignificant loss of goodness-of-fit.

[15] We also estimated all models using the 'cubic splines" correction for temporal dependence proposed by Beck et al. (1998). This method applies a smoothing function to the variable counting years of continuous peace in the dyad before the year of observation and the probability of conflict instead of the decaying function used above. We used S-Plus to estimate this model. However, this estimation yielded a lower log likelihood value than the estimations reported above (–1053.98 as compared to –1043.30). Since it is both more parsimonious and fits the data better, we prefer the decaying-function correction. The choice between the two has only minor consequences for the other estimates of the model.

as the cutoff point for a dichotomous measure of freshwater availability, and have coded dyad-years as having 0, 1, or 2 countries with low availability. This variable includes water from rivers as well as ground water. Water in rivers with their sources from outside the border is excluded for all countries, except Africa and South America. This variable was only used for analyzing data from the most recent period. Even so, it is problematic that it is only available for a single year. The variable is also insensitive to seasonal variations in rainfall and to national differences in the use of water.[16] However, given the importance of water scarcity to theories of water conflict, we do need to include such a variable.

RESULTS

Table 1 reports results from the estimation of bivariate logistic regressions for the shared river and water availability variables as well as for the

Table 1. Bivariate analyses of outbreak of militarized interstate disputes, all contiguous dyads, 1880-1992

Variable	Parameter estimate	Standard error	Odds ratio	Relative risk*	N
Shared river vs. no shared river	1.3***	0.23	3.7	3.2	13,899
One democracy vs. two democracies	0.79***	0.21	2.2	2.1	13,899
Zero democracies vs. two democracies	0.28*	0.21	1.3	1.3	13,899
Transition/missing regime data vs. two democracies	0.64***	0.26	1.9	1.8	13,899
One or two major powers vs. no major powers	−0.49***	0.17	0.61	0.63	13,899
Energy consumption per capita	−0.11***	0.033	0.90[b]	0.90	11,665
Shared alliance vs. no alliance	−0.59***	0.13	0.55	0.57	13,899
Peace history	−3.6***	0.19	36.6[c]	13.2	13,899
Two countries with low water availability vs. zero countries with water scarcity[d]	1.6***	0.60	4.8	4.0	3,069
One country with low water availability vs. zero countries with water scarcity[d]	1.5***	0.63	4.4	3.8	3,069

*p < 0.10; ** p < 0.05; ***p < 0.01. All p-values refer to one-sided tests.

[a] The relative risk is calculated on the assumption that the baseline probability of dispute is 0.05. If the baseline probability is lower, the relative risk is closer to the corresponding odds ratio.

[b] For energy consumption per capita the odds ratio refers to the difference between a dyad with lower energy consumption which is e=2.1 times larger than another dyad (i.e. the odds ratio for dyads with one unit's difference on the ln(energy consumption per capita).

[c] For peace history the odds ratio refers to the difference between a dyad with several decades of peaceful existence with one with a peace history close to zero years.

[d] The figures for water availability were available for the 1980-92 period only.

---

[16] We are grateful to Phil Steinberg for this point.

control variables. We report the parameter estimates, but have also calculated the odds ratios and the relative risks of conflict. The *odds ratio* is the ratio of the estimated odds of conflict for one group to a reference group, where the odds is the probability of conflict in the group divided by the probability of no conflict:

$$\text{odds ratio} = \frac{p(\text{war})_1 / p(\text{not war})_1}{p(\text{war})_0 / p(\text{not war})_0}$$

where the subscripts refer to the two groups we compare. For the dichotomous variables, group '0' refers to the dyad without the relevant characteristic. The *relative risk* is the probability of conflict when a risk factor was present divided by the probability of conflict when it is not present:

$$\text{relative risk} = \frac{p(\text{war})_1}{p(\text{war})_0}$$

Since the probability of war is dependent on other variables in the model, we assume that $p(\text{war})_0 = 0.05$.

The relationship between odds ratio and relative risk depends on the two probabilities of conflict (see Agresti, 1990):

$$\text{odds ratio} = \text{relative risk} \left(\frac{1 - p(\text{war})_1}{1 - p(\text{war})_0} \right)$$

When the baseline probabilities are very low and the odds ratio moderate, the relative risk is roughly equal to the odds ratio, as can be seen from Table 1.

The bivariate analyses show preliminary support for Hypotheses 1 and 5: countries that share a river have more dyadic conflict behavior, and dyads where one or two countries have low water availability also have more dyadic conflict behavior. For the control variables, the results are generally in line with previous studies. One exception is the major power variable. Most studies (e.g. Bremer, 1992; Raknerud and Hegre, 1997) find major powers to be more often engaged in militarized conflicts than minor powers, whereas we find that dyads including one or two major powers have less conflict. This discrepancy may reflect that we study only contiguous dyads, where the major powers' ability to wage war over long distances is less relevant. At the same time, most of the dyads containing major powers consist of only one of these. In such dyads, the major power is much more powerful than the other state. Many studies have found power preponderance in a dyad to reduce the probability of war (e.g. Oneal and Russett, 1999).

All these computations are for contiguous dyads only. At the bivariate level, the impact of Contiguity is an order of magnitude higher than the impact (in the contiguous dyads) of Shared river or any of the other variables in Table 1. Bremer (1992) finds contiguous dyads to be more than five times as likely between contiguous dyads than between non-contiguous dyads. Raknerud and Hegre (1997; p. 397) estimated a contiguous dyad to be from 13 to 40 times more war-prone than a non-contiguous dyad (the estimated relative risk varies with the number of states in the international system).

In Table 2, we test Hypotheses 1 and 2 in two multivariate models. Model 1 investigates the relationship between the dichotomous indicator 'Shared river' or not and militarized disputes with at least one casualty.[17] Model 2 studies the relationship between 'Number of shared rivers' and

Table 2. Logistic regression for the outbreak of interstate militarized disputes, all contiguous dyads, 1880-1992

Variable	Model 1		Model 2	
	Parameter estimate	Standard error estimate	Parameter estimate	Standard error estimate
Constant	−5.3***	0.31	−4.8***	0.23
Shared river				
No	ref.			
Yes	0.87***	0.24		
Number of shared rivers			0.063***	0.016
Regime type				
Two democracies	ref.		ref.	
One democracy	0.10	0.23	0.11	0.24
Zero democracies	−0.07	0.24	−0.14	0.30
Regime transition/NA	−0.070	0.30	−0.050	0.19
Major powers in dyad				
None	ref.		ref.	
One or two	−0.62***	0.19	−0.74***	0.18
Economic development	−0.054*	0.042	−0.064*	0.042
Alliance				
No	ref.		ref.	
Yes	−0.43***	0.15	−0.47***	0.15
Peace history	−3.5***	0.21	−3.5***	0.21
Number of dyad-years	11,476		11,476	
Log likelihood	−1043.30		−1044.33	

*$p < 0.10$; **$p < 0.05$; ***$p < 0.001$. All p-values refer to one-sided tests.
'ref.' signifies that this category is the reference category for the categorical variable.

---

[17] We have also tested Model 1 using all MIDs and all MIDs with at least 25 casualties. The 'Shared river' variable was significant also using these dependent variables, although less significant than for the MID-with-one-casualty variable. This is not surprising given that the full set of MIDs contains much more noise (e.g. attention bias and non-systematic errors) and the set of MIDs with 25 casualties contains fewer conflict onsets.

dispute. All the control variables listed above except water availability have been included in the models.

Compared to the bivariate results, the parameter estimate for 'Shared rivers' (Model 1) is slightly smaller - corresponding to an odds ratio of 2.4. However, the relationship is still highly significant, both statistically and substantially. Dyads that share a river are 2.4 times more likely to be involved in militarized disputes than other contiguous dyads.[18] The estimated effect of this variable is stronger than any of the control variables.

In the multivariate model, the regime type variables are all insignificant. This may be due to the inclusion of an economic development variable (cf. Hegre, 2000). According to the estimates in Model 1, a dyad with 1.16 as the lower log of energy consumption per capita (corresponding to Austria in 1985) is 23% less likely to be involved in a militarized conflict than one with lower log of energy consumption per capita of −3.76 (Mali in 1986).[19] As in the bivariate analyses, we find the existence of major powers in the dyad and shared alliances to roughly halve the risk of conflict.

Hypothesis 2 is tested in Model 2 in Table 1. The positive parameter estimate for 'Number of shared rivers' is consistent with the hypothesis: the more rivers two countries share, the higher the likelihood of conflict. However, the more refined measure does not add much explanatory power over the simple dichotomous 'Shared river': the parameter estimate is small, and the log likelihood for Model 2 is lower than that for Model 1.

An analysis for a shorter and more recent period (1980-92) is reported in Table 3, Model 3. This analysis gives very similar results. The parameter estimate is identical to the longer period. The estimate for the standard error is larger as expected when reducing the number of observations from 11,476 to 2747, such that the statistical significance of the variable is considerably lower in Model 3 than in Model 1.

Hypothesis 3 is tested in Model 4. 'Upstream/downstream' is estimated to be the most conflict-prone type of shared river, 'Mixed Boundary' is the second most conflict-prone, and 'River boundary only' the third. This is in line with the hypothesis. However, the differences between the types are not statistically significant.

Hypothesis 4 suggested that shared rivers should be more likely to lead to conflict in the later period than in the shorter period. This can be tested

---

[18] Throughout, we use 'more likely' as a generic term. Here, it refers to the odds ratio. For the baseline probability in this dataset the odds ratio is roughly equal to the relative risk.

[19] The dyadic development variable ranges from −9 (Afghanistan in the 1920s) to nearly 3 (Qatar and the United Arab Emirates in the early 1980s), with 80% of the observations between −3.76 and 1.16.

Table 3. Logistic regression for the outbreak of interstate militarized disputes, all contiguous dyads, 1880-92

Variable	Model 3		Model 4	
	Parameter estimate	Standard error estimate	Parameter estimate	Standard error estimate
Constant	–6.1***	0.77	–6.1***	0.77
Shared river				
No	ref.			
Yes	0.87***	0.48		
Type of shared rivers				
None			ref.	
River boundary only			0.82*	0.54
Mixed boundary			1.02**	0.59
Upstream/downstream			1.24**	0.60
Several categories			0.74	0.53
Regime type				
Two democracies	ref.		ref.	
One democracy	0.57	0.60	0.52	0.60
Zero democracies	0.34	0.59	0.27	0.60
Regime transition/NA	–0.93	1.2	–0.99	1.2
Major powers in dyad				
None	ref.		ref.	
One or two	–0.27	0.48	–0.28	0.49
Economic development	–0.13*	0.097	–0.12	0.098
Alliance				
No	ref.		ref.	
Yes	–0.34	0.31	–0.27	0.33
Peace history	–3.5***	0.46	–3.6***	0.47
Number of dyad-years	2747		2747	
Log likelihood	–225.48		–224.75	

*$p < 0.10$; ** $p < 0.05$; *** $p < 0.001$. All p-values refer to one-sided tests.
'ref.' signifies that this category is the reference category for the categorical variable.

by comparing the results in Table 3 with those in Table 2. The parameter estimate is the same for both periods, so we have no basis for concluding that the effect of shared rivers is increasing over time.[20]

The control variables are less significant in Models 3 and 4 than in the first two models. Economic development remains weakly significant in Model 3, but becomes insignificant in Model 4.

Hypotheses 5 and 6 are tested in Table 4. Hypothesis 5 receives clear support in Model 5: dyads where at least one of the states has low freshwater availability are considerably more likely to get into serious militarized conflicts than states with ample supply of water. Dyads with water scarcity

---

[20] As alternative tests of Hypothesis 4, we tried adding indicator variables for various periods and adding a linear time variable to Model 1. This led to severe problems of collinearity and showed no signs of support for the hypothesis.

are estimated to have approximately four times higher risk of conflict than dyads without.[21]

In Model 6, we have added an interaction term between 'Shared rivers' and 'Low water availability' to test Hypothesis 6. The positive and significant estimate for the interaction term may indicate that low water availability increases the risk of militarized conflicts over rivers. The improvement in log likelihood from Model 5 is 1.48. The likelihood ratio test P-value is 0.085, implying that the interaction term somewhat improves the goodness-of-fit of the model. However, the estimation suffers from serious problems of collinearity, warning us not to draw to firm conclusions from these results.

DISCUSSION

Our empirical results provide some support for the idea that shared river dyads have a higher frequency of dispute outbreaks than other contiguous dyads. Although the effect of a shared river is much lower than that of contiguity itself, it is of the same order of magnitude as the standard variables used to account for interstate conflict, such as regime type, economic development, great power status, and alliances. Clearly the analysis can be refined in various ways, for instance, by developing better measures of different types of shared rivers, by distinguishing between major and minor rivers, and by including a more complete set of control variables.

One potentially important factor that we have not controlled for is the length of the border separating the two states in the dyad. The longer the border, the more opportunities for conflict and potential contentious issues for conflicts. At the same time, it is more likely that states with a long shared border share a river than states with a short shared border. There is therefore a danger that the shared river variable might act as a proxy for border length. However, the 'Number of rivers' variable should then be even more highly correlated with the length of the shared border. If the relationship between shared rivers and conflicts was indeed spurious, we would expect this variable to be even more closely related to conflict than the 'Shared river' variable. This is not the case.

Of course, a border crossed by several rivers might have greater permeability than a border through a desert or over inhospitable mountains. The permeability would permit interaction, which in turn could generate conflict. Borders without rivers are also likely to be found in sparsely

---

[21] We also estimated models where we distinguished between one state with low water availability and two states with low water availability. We found no difference between these two categories, nor did the division lead to a better goodness-of-fit.

Table 4. Logistic regression for the outbreak of interstate militarized disputes, all contiguous dyads, 1880-92

Variable	Model 5		Model 6	
	Parameter estimate	Standard error estimate	Parameter estimate	Standard error estimate
Constant	–7.5***	0.98	–5.4***	1.2
Shared river				
No	ref.		ref.	
Yes	0.98***	0.49	–1.5	1.3
Water availability				
Neither state low	ref.		ref.	
One or both states low	1.5***	0.62	–0.80	1.1
Shared river and water availability				
Shared river and one or two states with low water availability			2.7**	1.4
Other combinations			refs.	
Regime type				
Two democracies	ref.		ref.	
One democracy	0.58	0.61	0.59	0.61
Zero democracies	0.18	0.59	0.20	0.59
Regime transition/NA	–0.99	1.2	–0.95	1.2
Major powers in dyad				
None	ref.		ref.	
One or two	–0.26	0.48	–0.26	0.48
Economic development	–0.11	0.094	–0.11	0.094
Alliance				
No	ref.		ref.	
Yes	–0.16	0.33	–0.13	0.32
Peace history	–3.5***	0.46	–3.5***	0.46
Number of dyad-years	2747		2747	
Log likelihood	–220.73		–219.24	

*$p < 0.10$; ** $p < 0.05$; *** $p < 0.001$. All p-values refer to one-sided tests.
'ref.' signifies that this category is the reference category for the categorical variable.

populated regions where there is a smaller chance that frictions develop. It would be useful if future studies of the relationship between shared rivers and conflict could include the length of the border as a control variable.

We asked at the outset what it is about territory that makes it worth fighting for. Our tentative conclusion is that there is something to shared rivers as a source of conflict. Whether this 'something' is mainly water scarcity is not clear. The results indicate that low availability of water in both countries in the dyad is significantly related to disputes. However, the results for the type of shared river variable indicate that it is not only water scarcity and the potential for serious upstream/downstream conflict that threatens the peace in shared river dyads. For the river boundary dyads in particular, it may rather be a question of friction over navigation, pollution, fishing rights, or territorial issues. Better data for the type of shared river might put us in a position to answer such questions more precisely.

We have no data on the actual issues involved in the COW disputes in the shared river dyads and in other dyads. The MID dataset includes a dispute-type variable which allows distinguishing between territorial conflicts and other conflicts, but not between different types of territorial conflict. The Issues Correlates of War project (Hensel 1998, www.icow.org) has collected data on freshwater claims, but so far these data are limited to a few regions. Aaron Wolf and associates have created a Water Events Database with comprehensive data on cooperative and hostile acts. So far, however, this dataset has only been used in bivariate analysis at the river basin level and a dyad-year database has not yet been created. Conceivably, the shared river dyads might be feuding over entirely different issues. If we had data which showed that to be the case, we would need to look for a third variable which would account for the statistical relationship between shared rivers and disputes. If no such variable could be found, we should still suspect that the shared rivers in many cases were at the root of the problem and that the other issues were mostly rhetorical. On the other hand, if it could be demonstrated by issue coding of the disputes that shared river dyads do indeed feud over their joint water resources, the case for considering shared rivers as a causal factor in conflict would be strengthened.

Wolf (1999b) has studied crisis behavior on the basis of the International Crisis Behavior dataset (Brecher and Wilkenfeld, 1997). The ICB project identified 412 crises for the period 1918-94. Through a search of the text files for the dataset, Wolf found four disputes where 'water was at least partially a cause'. His own research added three more cases.[22] In three out of these seven crises, not a single shot was fired. None of the others were violent enough to qualify as wars. Wolf (1999b) argues that the last and only water war was the conflict between the Sumerian city states of Lagash and Umma, which occurred 4500 years ago. However, the lack of clean freshwater may lead to political instability and acute small-scale violence.

Wolf feels that the history of water dispute resolution is much more impressive. He cites studies from FAO (1978, 1984) which have identified more than 3600 treaties relating to international water resources between 805 and 1984, most of which concern aspects of navigation. Since 1814, around 300 treaties have been concluded about non-navigational issues relating to international water resources. The UN General Assembly in

---

[22] The freshwater dispute database is described in Wolf (1999a). Data from this project are available on http://terra.geo.orst.edu/users/tfdd/

1997 passed an international convention on the non-navigational uses of international waterways (McCaffrey, 1998). And The World Bank and other international agencies are actively promoting the development of water regimes and joint exploitation of international rivers (Krishna, 1998).

As noted earlier, one possible objection to our work is that the relationship between shared rivers and conflict may be spurious. Countries with long land boundaries are more likely to share international rivers and may also have more conflict because of other shared resources or for other reasons. Furlong and Gleditsch (2003) published a measure of the length of the land boundaries that made it possible to test this hypothesis. Furlong, Gleditsch and Hegre (2006) showed that the relationship between shared rivers and conflicts is not eliminated when controlling for the length of the land boundary. Although the length of the land boundary is significantly related to conflict, it is this variable – and not shared rivers – that becomes insignificant in a multivariate analysis, Gleditsch *et al.* (2006) have confirmed the basic findings of this article with an improved data set for shared rivers.

We conclude, then, that the sharing of international rivers does seem to be associated with conflict between nations, as well as with activities directed at conflict prevention. At this stage we do not have much solid evidence for saying that sharing a river provides a major source of armed conflict, or that water scarcity is the only or even the main issue in whatever such conflicts do occur.

REFERENCES

Agresti, A. 1990. Categorical Data Analysis. Wiley, New York: USA.

Aschehoug og Gyldendals Store Norske Leksikon 1989. Kunnskapsforlaget. Oslo: Norway.

Bächler, G., Böge, V., Klötzli, S., Libiszewski, S., and Spillmann, K.R. 1996. Kriegsursache Umweltzerztörung. Ökologische Konflikte in der Dritten Welt und Wege ihrer friedlichen Bearbeituing (Environmental destruction as a cause of war. Ecological conflicts in the third world and peaceful ways of resolving them). Zurich: Megger (in German) (three volumes).

Barbieri, K. and Schneider, G. 1999. Trade and conflict [special issue]. Journal of Peace Research, 36 (4).

Beaumont, P. 1997. Water and armed conflict in the Middle East—fantasy or reality? *In:* Conflict and the Environment N.P. Gleditsch, (Ed.), pp. 355-374. Dordrecht: Kluwer Academic, The Netherlands.

Beck, N., Katz, J.N. and Tucker, R. 1998. Taking time seriously: Time-series-cross-section analysis with a binary dependent variable. American Journal of Political Science, 42 (4) 1260-1288.

Biliouri, D. 1997. Environmental issues as potential threats to security. Paper presented to the 38th Annual Convention of the International Studies Association, Toronto, 18-21 March.

Biswas, A. K. 1990. Objectives and concepts of environmentally-sound water management. *In* Environmentally-Sound Water Management N. C. Thanh and A. K. Biswas, (eds.), pp. 30-58. Delhi: Oxford University Press.

Brecher, M. and Wilkenfeld, J. 1997 A Study of Crisis. Ann Arbor, MI: University of Michigan Press, USA

Bremer, S. 1992. Dangerous dyads: conditions affecting the likelihood of interstate war, 1816-1965. Journal of Conflict Resolution, 36 (2), 309-341.

Brunborg, H. & Urdal, H. (guest editors) 2005 Special Issue on 'Demography of Conflict and Violence', Journal of Peace Research 42(4), July.

Chou, M. T. 1998. Institutional conflicts in the Mekong Basin. Paper presented to the Workshop on Water and Human Security in Southeast Asia and Oceania, Australian National University, 16-18 November.

CNRET 1978. Register of International Rivers. Oxford: Pergamon, for Centre of Natural Resources, Energy, and Transport of the Department of Economic and Social Affairs of the United Nations.

Collier, P. 2000. Doing well out of war: an economic perspective. *In* Greed and Grievance. Economic Agendas in Civil Wars, M. Berdal and D.M. Malone (Eds.), pp. 91-112. Boulder, CO, and London: Lynne Rienner.

de Soysa, I. 2000 The resource curse: are civil wars driven by rapacity or paucity? *In* Greed and Grievance. Economic Agendas in Civil Wars M. Berdal and D.M. Malone (Eds.), pp. 113-135. Boulder, CO and London: Lynne Rienner.

Dixon, W.J. 1994. Democracy and the peaceful settlement of international conflict: an empirical analysis. American Political Science Review, 88 (1), 1-17.

Dupont, A. 1998. The Environment and Security in Pacific Asia. Oxford: Oxford University Press, for International Institute for Strategic Studies (Adelphi paper 319), UK.

Ehrlich, P.R. and Ehrlich, A.H. 1972. Population, Resources, Environment. Issues in Human Ecology. (2nd ed.). San Francisco, CA: Freeman, USA.

Elhance, A.P. 1999. Hydropolitics in the Third World. Conflict and Cooperation in International River Basins. Washington, DC: United States Institute of Peace Press.

Falkenmark, M. 1990. Global water issues facing humanity. Journal of Peace Research, 27 (2), 177-190.

Falkenmark, M. and Lundqvist, J. 1998. Towards water security; political determination and human adaptation crucial. Natural Resources Forum, 21 (1), 37-51.

FAO 1978. Systematic index of international water resources treaties, declarations, acts and cases, by basin, 1. Legislative study (15). Rome: Food and Agricultural Organization.

FAO 1984. Systematic index of international water resources treaties, declarations, acts and cases, by basin, 2. Legislative study (34). Rome: Food and Agricultural Organization.

Furlong, K. and Gleditsch, N.P. 2003, The Boundary Dataset, Conflict Management and Peace Science 20(1): 93-117, [Data on http://www.prio.no/cscw/datasets].

Furlong, K., Gleditsch, N.P. and Hegre, H. 2006, Geographic Opportunities and Neomalthusian Willingness: Boundaries, Shared Rivers, and Conflict, International Interactions 32 (1): 79-108.

Gissinger, R. and Gleditsch, N.P. 1999 Globalization and conflict. Welfare, distribution, and political unrest. Journal of World-Systems Research, 5 (2), 274-300 http://csf.colorado.edu/wsystems/jwsr.htm.

Gleditsch, N.P. 1995. Geography, democracy, and peace. International Interactions, 20 (4), 297-323.

Gleditsch, N.P. 1997 Conflict and the Environment. Dordrecht: Kluwer Academic, The Netherlands.

Gleditsch, N.P. 1998. Armed conflict and the environment: a critique of the literature. Journal of Peace Research, 35 (4), 363-380 (Revised version in P.F. Diehl and N.P. Gleditsch (Eds.), Environmental conflict (Chapter 12), Boulder, CO: Westview, 251-272).

Gleditsch, N. P. 1999. 'Do Open Windows Encourage Conflicts?', Statsvetenskaplig tidskrift 102(3), 333-349.

Gleditsch, N.P. and Hegre, H. 1997. Peace and democracy: three levels of analyses. Journal of Conflict Resolution, 41 (2), 283-310.

Gleditsch, N.P., Furlong, K., Hegre, H., Lacina, B. and Owen, T. 2006. Conflicts over shared rivers: Resource scarcity or fuzzy boundaries?', Political Geography 25 (4): 361-382.

Gleick, P. H. 1993a. Water and conflict: fresh water resources and international security. International Security, 18 (1), 79-112.

Gleick, P. H. 1993b. Water in Crisis, a Guide to the World's Fresh Water Resources. New York and Oxford: Oxford University Press, for Pacific Institute for Studies in Development, Environment, and Security and Stockholm Environment Institute.

Gleick, P. H. 1998. The world's water 1998-1999: the biennial report on freshwater resources. Washington, DC: Island Press, USA.

Gochman, C.S. 1991. Interstate metrics: conceptualizing, operationalizing, and measuring the geographic proximity of states since the Congress of Vienna. International Interactions, 17 (1), 93-112.

Gochman, C. and Maoz, Z. 1984. Militarized interstate disputes, 1816-1976: procedures, patterns, and insights. Journal of Conflict Resolution, 28 (4), 585-616.

Granzow, F. C.1898. Geogrqfisk lexikon. Copenhagen: Det Nordiske Forlag.

Hauge, W. and Ellingsen, T. 1998. Beyond environmental security: causal pathways to conflict. Journal of Peace Research, 35 (3), 299-317 (Revised version in P.F. Diehl and N.P. Gleditsch (Eds.), Environmental conflict (Chapter 3). Boulder, CO: Westview, 36-57.

Hegre, H. 2000. Development and the liberal peace: what does it take to be a trading state? Journal of Peace Research, 37 (1), 5-30.

Hensel, P.R. 1998. Reliability and validity issues in the issue correlates of war (ICOW) project. Paper presented at the 39th Annual Convention of the International Studies Association, Minneapolis, MI, 18-21 March.

Holsti, K.J. 1991. Peace and War. Armed Conflicts and International Order 1648-1989. Cambridge: Cambridge University Press, Cambridge, UK.

Homer-Dixon, T.F. 1991. On the threshold: environmental changes as causes of acute conflict. International Security, 16 (2), 76-116 (Reprinted in 1995 in S. Lynn-Jones and S. E. Miller (Eds.) Global dangers. Changing dimensions of international security (pp. 43-83). Cambridge, MA, and London: MIT Press), USA.

Homer-Dixon, T.F. 1999. Environment, Scarcity, and Violence. Princeton, NJ: Princeton University Press, USA.

Huth, P.K. 1996. Standing your ground. Territorial disputes and international conflict. Ann Arbor, MI: University of Michigan Press, USA.

Irani, R. 1991. Water wars. New Statesman and Society, 4 (149), 24-25.

Jaggers, K. and Gurr, T.R. 1995. Tracking democracy's third wave with the Polity III data. Journal of Peace Research, 32 (4), 469-482.

Krishna, R. 1998. The evolution and context of the bank policy for projects of International waterways. In International Watercourses. Enhancing Cooperation and Managing Conflict. M.A. Salman and L. Boisson De Chazoumes (Eds.), Proceedings of a World Bank Seminar pp. 31-43. Washington, DC: World Bank (World Bank Technical Paper 414). USA.

Kukk, C. and Deese, D.A. 1996. At the water's edge, regional conflict and cooperation over fresh water. UCLA Journal of International Law and Foreign Affairs, 1 (1), 21-65.

Libiszewski, S. 1995. Water disputes in the Jordan basin region and their role in the resolution of the Arab-Israeli conflict, ENCOP Occasional Paper (13). http://www.fsk.ethz.ch/encop/13/enl3-ch6.htm.

Lomborg, B. 2001. The skeptical environmentalist. Measuring the real state of the world, Cambridge: Cambridge University Press, Cambridge, UK.

Lonergan, S. 1997. Water resources and conflict: examples from the Middle East. *In* Conflict and the Environment, N.P. Gleditsch (Ed.), pp. 375-384. Dordrecht: Kluwer Academic, Cambridge, UK.

Lowi, M. L. 1995. Rivers of conflict, rivers of peace. Journal of International Affairs, 49 (1), 123-144.

Lundqvist, J. 1998. Avert looming hydrocide. Ambio, 27 (6), 428-433.

McCaffrey, S. 1998. The UN convention on the law of the non-navigational uses of international watercourses: prospects and pitfalls. *In* International Watercourses. Enhancing Cooperation and Managing Conflict. M.A. Salman and L. Boisson De Chazournes (Eds), Proceedings of a World Bank Seminar (pp.17-28). Washington, DC: World Bank (World Bank Technical Paper 414) USA.

Milich, L. and Varady, R.G. 1999. Openness, sustainability, and public participation: new designs for transboundary river basin institutions. Journal of Environment and Development, 8 (3), 258-306.

Myers, N. and Simon, J. 1994. Scarcity or Abundance? A Debate on the Environment. New York & London. Norton, USA.

Naff, T. and Matson, R.C. 1984. Water in the Middle East. Conflict or cooperation? Boulder, CO: Westview, USA.

Ohlsson, L. 1998. Environment, scarcity, and conflict. A study of Malthusian concerns. PhD dissertation, Department of Peace and Development Research, Göteborg University, Sweden.

Oneal, J.R. and Russett, B. 1999. Assessing the liberal peace debate with alternative specifications: trade still reduces conflict. Journal of Peace Research, 36 (4), 423-442.

Rogers, K.S. 1996. Pre-empting violent conflict: learning from environmental cooperation. *In* Conflict and the Environment N.P. Gleditsch (Ed.) pp. 503-518. Dordrecht: Kluwer Academic, The Netherlands.

Raknerud, A. and Hegre, H. 1997. The hazard of war: reassessing the evidence for the democratic peace. Journal of Peace Research, 34 (4), 385-404.

Salman, M.A. and Boisson De Chazournes, L. 1998. International watercourses. Enhancing cooperation and managing conflict. Proceedings of a World Bank Seminar. Washington, DC: World Bank (World Bank Technical Paper 414). USA.

Schwarz, D.M., Deligiannis, T. and Homer-Dixon, T.F. 2001. The environment and violent conflict. *In* Environmental Conflict. P.F. Diehl and N.P. Gleditsch (Eds.), Boulder, CO: Westview USA (Chapter 13), 273-294.

Seckler, D., Molden, D. and Barker, R. 1998. Water scarcity in the twenty-first century. Colombo: International Water Management Institute (IWMI Water Brief 1) Sri Lanka (Reprinted in 1999 in Water Resources Development, 15(1-2), 29-42.

Shiklomanov, I.A. 1993. World fresh water resources. *In* Water in Crisis, a Guide to the World's Fresh Water Resources P.H. Gleick (Ed.), pp. 13-24. New York and Oxford: Oxford University Press, for Pacific Institute for Studies in Development, Environment, and Security and Stockholm Environment Institute.

Shmueli, D.F. 1999. Water quality in international river basins. Political Geography, 18 (4), 437-476.

Singer, J.D. and Small, M.F. 1993. Correlates of war project: national material capabilities data, 1816-1985. Ann Arbor, MI: Inter-university Consortium for Political and Social Research (ICPSR 09903). USA.

Singer, J.D. and Small, M.F. 1994. Correlates of war project: international and civil war data 1816- 1992. Ann Arbor, MI: Inter-university Consortium for Political and Social Research (ICPSR 09905). USA.

Small, M.F. and Singer, J.D. 1982. Resort to Arms. Iinternational and Civil Wars, 1816-1980. Beverly Hills, CA: Sage. USA.

Starr, J. R. 1991. Water wars. Foreign Policy (82), 17-36.

Swain, A. 1996. Displacing the conflict: Environmental destruction in Bangladesh and ethnic conflict in India. Journal of Peace Research, 33 (2), 189-204.

Swain, A. 1997. Sharing international rivers: A regional approach. *In* Conflict and the Environment N.P. Gleditsch (Ed.) pp. 403-416. Dordrecht: Kluwer Academic, The Netherlands.

The Times Atlas 1997. The Atlas of the World. (7th ed). London: HarperCollins.

Tir, J. and Diehl, P.F. 1998. Demographic pressure and interstate conflict. Journal of Peace Research, 35 (3), 319-340 (Revised version in P.F. Diehl and N.P. Gleditsch (Eds.), Environmental Conflict (Chapter 4). Boulder, CO: Westview, 58-83.

Toset, H. P. W. 1998. Kampen om elvevannet—en kvantitativ analyse av elvegrensenes betydning for lands involvering i internasjonale militariserte disputter 1816-1992 (The struggle over river water. A quantitative analysis of the significance of river boundaries for the involvement of countries in international militarized disputes). Thesis for the Cand.polit. degree, NTNU, March.

Trolldalen, J. M. 1992. International environmental conflict resolution. The role of the United Nations. Oslo and Washington, DC: World Foundation for Environment and Development. USA.

Vasquez, J. A. 1993. The War Puzzle. Cambridge: Cambridge University Press, Cambridge, UK.

Vasquez, J.A. 1995. Why do neigbors fight? Proximity, interaction or territoriality? Journal of Peace Research, 32 (3), 277-293.

Wallensteen, P. and Swain, A. 1997. International fresh water resources: conflict or cooperation?. *In* Comprehensive Assessment of the Freshwater Rresources of the World. // Stockholm: Stockholm Environment Institute.

Wallensteen, P. and Sollenberg, M. 2000. Armed conflict 1989-99. Journal of Peace Research, 37 (5), 651-665.

Westermanns Atlas zur Weltgeschichte. Teil 1: Vorzeit und Altertum. Braunschweig: Georg Westermann.

Wolf, A. T. 1999a. The transboundary freshwater dispute database project. Water International, 24 (2), 160-163.

Wolf, A. T. 1999b. 'Water wars' and water reality: conflict and cooperation along international waterways. *In* Environmental Change, Adaptation, and Human Security, Slangarman (Ed.), pp. 251-265. Dordrecht: Kluwer Academic, The Netherlands.

Wolf, A. T., Natharius, J.A., Danielson, J.J., Ward, B.S. and Pender, J.K. 1999. International river basins of the world. Water Resources Development, 15 (4), 387-427.

WRI (Annual) World Resources. New York and Oxford: Oxford University Press, for World Resources Institute.

Asia

4

Water Conflicts in China

Desheng Hu

Economist, Professor of Law, Zhengzhou University, P. R. China

China, one of the thirstiest countries in the world, is challenged by severe issues of water shortage, natural water disaster and water pollution. Water conflicts not only exist nationally between users, communities and administrative regions, but also involve the relevant international watercourses. In order to resolve these problems, a national water rights system dealing with the property right on water resources, the human right to water and the environmental right to water (water for the environment) should be established and implemented, and a practical foreign policy concerning international watercourses should be outlined.

INTRODUCTION: THE NATIONAL CONTEXT OF CHINA

Water Resources and Water Uses

According to a report by the Ministry of Water Resources (MWR) of the People's Republic of China (PRC) (MWR, 2004), water resources in China are relatively abundant in terms of both type and total quantity (with regard to quantity, the total renewable water resources is about 2,812 billion m^3), but there is a serious scarcity in a per capita sense with merely about 2,200 m^3. Moreover, its availability varies dramatically according to seasons and across regions, and most rivers are polluted. In order to lessen the water shortage problem and tackle the uneven distribution of water resources for meeting the increasing water demands, 86,000 reservoirs with an overall storage capacity of about 471 billion m^3 have been built, inter-basin water transfer projects with a combined capacity of transferring 15.3 million m^3 have been constructed, and the famous South-to-North Water Transfer Project is under construction. Under Article 4 of the Water

Law of the PRC (2002 Revision) (hereinafter Water Law 2002), the categories of beneficial water use in China include water uses for livelihood, for production and business, and for the eco-environment. The statistics (MWR, 2004) suggest that, in 2003, domestic use accounted for 11.9%, production and business use for 86.6% (22.1% for industry use and 64.5% for agriculture use) and eco-environment use for 1.5%.

National Water Regime

Under Article 9 of the Constitution of the PRC (2004 Revision), the State has right over all the water systems. Further, Article 3 of the Water Law 2002 provides that "[w]ater resource is owned by the State". At present, there are four laws directly related to water resources enacted by the National People's Congress (the top legislator) or its Standing Committee, these comprise the Water Law 2002, the Law for Water Pollution Control of the PRC (1996 Revision), the Flood Control Law of the PRC, and the Law on Water and Soil Conservation of the PRC. Each of these contains provision(s) that deal(s) with the relevant water conflicts, which will be discussed in detail below. There are also legal documents issued by the MWR and other governmental organizations that handle water conflicts, such as the Rules on Prevention and Handling of Inter-Province Water Conflicts (2004) issued by the MWR.

WATER CONFLICTS, PRESENT GOVERNANCE AND ISSUES

An Overview of Water Conflicts

Water conflicts in China exist between different water use categories, water users, administrative regions as well as communities, and also involve international watercourses concerning China. In 2000, there were 6,037 domestic water conflicts excluding those involving eco-environment uses, with 5,697 of them were being mediated by the competent department of water administration or basin administration authority (MWR, 2001); in 2001, there were 6,338 conflicts, with 5,595 being mediated (MWR, 2002); and in 2002, there were 9,972 conflicts, with 9,349 being settled by negotiation or mediation (MWR, 2003). Alongside China's rapid social and economic development, the number of water conflicts arising from water pollution and hydropower development has increased in recent years (Jing, 2005). Under the present governance, law and policy as well as social customs play a role through a wide range of organizations, such as the government at various levels or its relevant working departments, the court of justice and the community. However, as will be discussed below, many domestic problems relating to water conflicts were not, or have not been, settled in a sustainable, equitable, reasonable and prompt way, and, on the one hand, as more and more water conflicts occur, this situation has lead to the utilization of water resources in an unsustainable way. On the other hand, the law and policy dealing with present and/or potential conflicts concerning international watercourses are insufficiently clear due to their impractical nature.

Water Conflicts between Water Use Categories

As mentioned above, the water uses for livelihood, for production and business, and for the eco-environment are the three beneficial water uses in accordance with the law. However, in some regions or sometimes, on the one hand, the amount of water available cannot meet these three water uses simultaneously. On the other hand, water use in one category may have an adverse effect on water use in other categories. Therefore, water conflicts arise between these three categories. Those that have attracted special attention include basic water needs for livelihood and/or water requirements for the eco-environment.

Regarding the environment requirement for water, for decades, the Huanghe River, the **Heihe** River, the Tarim River, the Taihu Lake, the Nansihu Lake and the Baiyangdian Lake have all suffered a shortage of minimum appropriate water for survival. The Baiyangdian Lake is discussed as an example (Wen, 2003; Jin, 2003). Located in the Hebei Province and with a surface area of 366 km^2, the Baiyangdian Lake is the biggest freshwater lake as well as the largest wetland in the North China Plain. It is composed of 143 lakes that are connected by over 3,700 channels, and is rich in ecological resources and biodiversity. There are 47 types of hydraulic plants, 54 types of fish, and 192 types of birds, including 187 types of National I and II Protected Types of Birds on this lake. All of these have formed excellent ecological links. There are few historical records of the dry lake phenomenon, which is defined as a water level of less than 6.5 m. However, since 1960's, because of the regional drought climate combined with the increasing water demands from the growing population and developing industries, the water input into the lake has decreased greatly, and the dry lake phenomenon has occurred many times, including for a continuous five-year period from 1983 to July 1988. Fortunately, because of profuse rainfall in August 1988 the lake was restored. In order to prevent the dry lake phenomena from happening again, a special management institution – the Baiyangdian Lake Management Section–was established to do the integrated management of the lake, and ecological water input action has increased 10 times, with a total of 800 million m^3 flowing in from the reservoirs lying upstream of the basin during the time period from 1992 to December 2003. However, continuous drought in the basin in the years of lower precipitation, especially in 2002 and 2003, resulted in the water level of the lake remaining below 6.5 m and only a little water existing in the channels from late August 2003. No water could be taken from other parts of the basin to input into the lake, and thus the dry lake phenomenon occurred. In order to maintain the eco-environment of the lake, the Haihe River Water Resources Commission and the Water Resources Bureau of Hebei Province decided to channel water into the lake through an inter basin water transfer, and the "Implementation Project on the Emergency Ecological Water Input by Diverting Water from the Yuecheng Reservoir to the Baiyangdian Lake" was discussed and approved on December 16, 2003. The Project started on December 28, 2003, and water began to be diverted from the Yuecheng Reservoir on February 16, 2004. On March 1st, 2004, the diverted water began to reach the lake. By June 29, 2004, the transfer was completed, with 390 million m³ of water having being diverted from the Yuecheng Reservoir and 160 million m³ of water having been input into the lake.

It is a common phenomenon that production and business water use often has an adverse impact on the livelihood and eco-environmental water uses, not only because the former category needs a certain quantity of water but also because production and business activities usually worsen the quality of the water resources. Therefore, almost every country has promulgated the relevant standards to prevent water resources from being polluted, and China is no exception. However, the initiatives to increase profit sometimes lead to enterprises contravening these standards and polluting the water resources, in which case water conflicts arise. When the existing mechanism can not operate effectively to stop illegal pollution, or takes a long time to enforce, harsh conflicts may arise. The Maxi River water conflicts can be cited as an example of such a conflict (Wu, 2002).

The Maxi River runs through Shengze Town (Suzhou City, Jiangshu Province) and then Jiaxing City (Zhejiang Province). Since the 1990's, the printing and dyeing industry has developed rapidly in Shengze Town. However, the wastewater discharged by the industry seriously polluted the river, and, furthermore, this pollution not only severely damaged the fishery in Jiaxing City but also decreased the quality of the water for livelihood use. According to the monitoring by the Taihu Basin Authority (TBA), the water quality in the river permanently fell into only the IV-V level standards, and thus was relatively poor. Therefore, a water conflict arose. In accordance with the Law for Water Pollution Control: (a) any unit that has caused a water pollution hazard shall have the responsibility to eliminate and compensate the unit or individual who has suffered direct losses; (b) the environmental protection department or the navigation administration office of the communications department, at the request of the parties, may settle the dispute over compensation liability or amount;

when any party refuses to accept the settlement decision, he may take legal action to the relevant court of justice; however, any party may take legal action directly. Although the residents and units in Jiaxing City employed every possible legal method for remedy, no radical measures were implemented. On May 14, 2001, with State Environmental Protection Administration (SEPA) mediation, an agreement to settle the conflict was concluded between the governments of the two provinces. According to this, the Suzhou Government agreed to ensure that the water flow in the river when leaving its borders reach the required legal standard by the end of 2003, and to pay one million CNY to the Jiaxing Government in compensation. However, the residents of Jiaxing City considered this agreement unfair, since they had to suffer damage for a further three years. On November 22, 2001, they sunk boats in order to create a dam to block the river and prevent the polluted water from flowing into their territory. The conflict became complicated, since Article 22 of the Flood Control Law prohibits any construction of buildings or structures that may block flood passage, within the management areas of river channels and lakes. Under this emergency, the MWR and SEPA both sent working groups to Jiaxing City to mediate the conflict, and, later, a new agreement was reached. Under the new agreement, the Jiangsu Government should immediately (a) order all enterprises that were discharging the waster water illegally to cease production, and also punish them in accordance with the law; and (b) close down all hidden pipes for discharging waste water. The Zhejiang Government, meanwhile, should immediately organize the dismantling of the dam. According to the monitoring by the TBA, the water quality in the river reached the 3rd-4th level standards on January 11, 2002. Since that time, through fear that the river being maybe cut off again, the printing and dyeing industry in Shengze Town no longer discharges waste water illegally.

Another example is the Shi River water conflict concerning soil erosion (The Water and Soil Conservation Office of Nanping City 2002). In the upstream of the Shi River, there are several companies engaged in mining ore. In 2002, due to the soil erosion caused by the mining activities of these companies, the riverbed rose, which endangered the riverbank and caused flooding which submerged some agricultural land that was utilized by others. This caused conflict between the farmers and the mining company. According to Article 39 of the Law on Water and Soil Conservation: (a) any individual or unit that causes damage through soil erosion has a duty to remove the damage and compensate those who have directly suffered the damage; and (b) any dispute over the liability or amount of compensation may, upon the request by the party(ies), be dealt with by the department of water administration; when the party is dissatisfied with the decision thus made, he may ask the relevant court of justice for a verdict; however, the party may directly go to the court. Unfortunately, the mining companies refused to adopt measures to stop the erosion, and the relevant authorities were unable to deal with the conflict in due process and promptly. In order to restore the river at a lower cost, the peasants who suffered the damage began to dig up a road, which was necessary for the companies' transportation. Hence, the conflict became more serious. In this situation, the Water and Soil Conservation Office of Shunchang County immediately took action about the conflict. Under the mediation of the authority, the mining companies agreed to make remedy and promised to adopt the relevant measures for soil conservation. Finally, the companies built a wall 34 m long to prevent the waste from entering the river, cleaned up the existing waste stones of 3,200 m³, repaired a 110 m length of the damaged riverbank, and cleared away 980 m³ of sludge.

Water Conflicts between Users

Water users may fall into different water use catalogues. As discussed above, water conflicts may arise between users belonging to different water use categories, on the one hand, while on the other hand, water conflicts may also occur between users within the same category. This phenomenon is more common during periods of drought in rural areas. For example, in the spring and summer of 2001, Xinyang City of Henan Province experienced its most serious drought in the last 60 years, which led to about 418,000 hectares of cultivated land being left unharvested, and more than 900,000 people and 240 larger livestock lacked drinking water. Nearly 4,000 water conflicts arose because of this situation. The court in the city mediated over 3,000 water conflicts, judged more than 600 water conflicts, and interdicted over 30 water conflicts from evolving into fighting (Zhang et al., 2001).

During late July 2003, due to less rainfall and a high evaporation rate, Jinyun County of Zhejing Province suffered a serious drought, and there was insufficient water for irrigation. In only a few days, seven water conflicts occurred involving violence, and five people were injured. In order to safeguard the social order, the security bureau of the county engaged in preventing and mediating the water conflicts (Wang, 2003).

Water Conflicts between Communities

Due to of insufficient water resources, water conflicts often arise between different communities over obtaining enough available water in an appropriate quantity. Regarding how to resolve this type of water conflict, Article 57 of the Water Law 2002 provides that, "[a]ny water dispute between units or individuals shall be resolved through negotiation. Where a party is unwilling to have the dispute resolved through negotiation, or a substantial agreement can not be concluded through negotiation, the parties concerned may request the local people's government at or above the county level or the department authorized by such a government to mediate the water dispute, or may directly initiate a civil legal action in the people's court. Where the local people's government at or above the county level or the department authorized by such a government can not mediate the water dispute successfully, the parties concerned may initiate a civil legal action in the people's court. Pending the settlement of a water dispute, no party may unilaterally alter the existing water regime". The water conflict between Zhongcun and Kaiyang is given below as a case study to demonstrate this (Zhang and Zhong, 2001).

In Meiyuan Village in Oubei Township, Yongjia County, Zhejiang Province, there exist several communities, including Zhongcun and Kaiyang. Due to the lack of proper and convenient sources of drinking water, after obtaining the approval of the government, Zhongcun raised funds and built a dam one km away from its location in the Longtan Stream and constructed a pipe to bring water to Zhongcun. Since then, the community with a population of about 230 and the primary school, which is located in the community with 200 teachers and students, has been able to enjoy high quality drinking water in peace. This situation continued until January 1996, when Kaiyang, which is 2 km far away to Zhongcun, built another dam as its drinking water source 200 m upstream. The previous drinking water source for Kaiyang had been a pond of an area of about 666 m^2 and a depth of over 10 m, before the area was impounded and sold for real estate development by its inhabitants. However, this meant that Kaiyang no longer had a proper drinking water source, so the community built the dam in the Longtan Stream, and diverted water for drinking use. The dam built by Kaiyang has been adversely influencing the quantity of the water that the dam built by Zhongcun can collect. Particularly in the dry season, the dam built by Zhongcun can no longer collect water. Therefore, this situation has led to endless water conflicts between the two communities. In order to resolve this dispute, the two communities frequently asked both the township and the county governments to settle the dispute. However, the governments did not deal with this dispute. Subsequently, in November 1998, Zhongcun initiated a civil action against Kaiyang in the court, but the verdict was that the dispute be settled by the competent department of water administration of the county. On December 1, 1999, Zhongcun submitted an appeal, requesting the government of the county to settle the dispute, but without any result. On March 3, 2000, Zhongcun initiated an administrative action against the government of the county for failing to deal with the dispute. On August 10, 2000, the Wenzhou Medium Court issued a judgment that charged the government of the county with settling the dispute within two months after the judgment became valid. The government appealed, but the judgment has been upheld.

Another case is the Mongjia Pond water conflict (Li 2004), a water conflict that had continued for over one hundred years. Zhangzhuang Village lies three km south of Minggang Town, Pingqiao District, Xinyang City, Henan Province. The village contains two communities, Mongzhuang and Xizhangzhuang. Between these two communities is located a large lake of about 6.6 hectares called Mongjia Pond. Since the Qing Dynasty, there has been a water conflict between these two communities, often involving violence, over how to allocate the water in the lake. In 2001, due to a serious drought, large scale fighting occurred. In order to settle this conflict once and for all, the Government of Pingqiao District organized a special working group to deal with this dispute. Under the mediation of the working group, the two communities finally reached an agreement to settle the conflict. According to this agreement, Mongzhuang should supply a piece of land, on which the two communities could build a new pond named Zhangjia Pond, for which a certain part of water sources for the Mongjia Pond should be utilized. Since this agreement was implemented, the two communities have experienced no further water conflicts between them.

Water Conflicts between Administrative Regions

Watercourses do not recognize the political borders of administrative regions. When a watercourse flows through/alongside more than one administrative region, on the one hand, there is a potential for water conflict(s) between these regions. On the other hand, when an inter-basin water diversion programme that crosses two or more administrative regions is implemented, the possibility for water conflict remains. In China, there are four levels of local authority, namely, provincial, prefecture, county and township, and therefore four levels of administrative region exist. Due to the fact that almost every river flows through/alongside more than one administrative region, many water conflicts have arisen between administrative regions, and will continue to do so. This category of water conflicts arises due to allocation of water between upstream and down-stream or between the right and left bank, or due to when and how to utilize water resources, or due to matters concerning maintaining the appropriate flow of rivers, etc. Some of the conflicts, mostly those that arose recently, have been resolved, while the others, mainly those that came into existence before October 1949, are still pending. The most famous ones include: the Tonmin River water conflict between Sichuan and Guizhou Province; the Danjiang-Jingziguan water conflict between

Hubei and Henan Province; the Ning River water conflict between Hebei Province and Tianjin City; the Dayankeng water conflict between Zhejiang and Fujian Province; the Huolin River Basin water conflict between Jilin Province and Inner Mongolia Autonomous Region; the Jindi River water conflict between Henan and Shandong Province; and the Zhanghe River water conflict between Shānxi, Hebei and Henan Province. Water Law 2002 makes a provision in Article 56, regarding how to deal with this type of water conflict, although it is far from sufficient. According to this article, this type of conflict shall be resolved through negotiation. However, when a substantial agreement cannot be concluded through negotiation, it shall be adjudicated by the common people's government at the next higher level of the administration regions concerned, and the adjudication shall be obeyed and executed by all parties concerned. Pending the settlement of a water conflict, no party may, in the absence of an agreement concluded by the parties concerned or an approval from their common people's government at the next higher level, construct any project for draining, blocking, drawing, diverting or storing water within a reasonable area on either side of the boundary between the administrative regions concerned, or unilaterally alter the existing water regime.

The conflicts that arose in 2003 in the Huanghe River basin, where a water shortage always exists, and issues concerning how much, when and how the water resources should be utilized are all important, and may be taken as examples. (i) Water conflicts between upstream and downstream. For instance, in the summer of that year, eight incidents that violated the annual water allocation plan of the Huanghe River arose in different provinces. During the last 10 days of June, the water users in Inner Mongolia Autonomous Region suddenly diverted an amount of water that exceeded the quota allocated to the region on a large scale. On June 29, the section flow at Toudaoguai was only 45 m^3/s, meaning that a zero flow would soon appear in the downstream of the Huanghe River, leading to the water users in the downstream provinces having no water to divert, and the downstream of the river suffering a complete lack of water. In order to deal with this incident, the Huanghe River Water Resources Commission (HRWRC) immediately adopted measures to supervise the diversion in the region. However, this failed to stop the illegal diversion, and on July 1 the section flow at Toudaoguai decreased to only 15 m^3/s. In these circumstances, the commission telegraphed both the secretary of the Committee of the Communist Party of China and the chairman of the Government of the region. Only then did the illegal diversion gradually stop. (ii) Water conflicts between the right and left bank. For example, on June 17, the section flow at Tongguan suddenly decreased, and was only 39 m^3/s on June 20. This must have been caused by the illegal diversion of water by water users in either Shānxi or Shǎnxi Province, since the Huanghe River forms the border of these two provinces, and it is difficult to judge which side was diverting the water illegally. In order to prevent a zero flow from appearing downstream, the HRWRC ordered the two provinces to close all diversion gates, and also sent staff to supervise the implementation. (iii) Water conflicts concerning when and how to utilize water resources. In order to operate machinery and produce more electricity, from April 30, the Liujiaxia Hydropower Station gave away an amount of water that exceeded the quota provided by the water allocation plan during this period. However, this water was wasted as the downstream did not need much water at this time. Therefore, on May 5, the HRWRC ordered the station to cease giving away water in accordance with the principle of "the electricity allocation plan succumbs to the water allocation plan".

The Beitou River water conflict was a conflict at the prefecture level that was settled by the Government of Shǎnxi Province (Qin, 2005). Jiajiayuan Village, with a population of almost a thousand, is a mountain village in Shǎnxi Province. Before 1980, it was under the jurisdiction of Linggao Commune, Pucheng County, Weinan City. Due to the lack of access to drinking water, an abstracting-water project was built with the approval of the government in 1970. This project could divert water from the Beitou River, and then transfer water to Jiajiayuan Village, and, therefore, the drinking water problem was resolved. In December 1971, the Linggao Reservoir was built on the river, which meant water abstraction from the reservoir rather than from the river. However, in 1980, due to the adjustment of the administrative region, Jiajiayuan Village came under the administration of Gaolouhe Township, Yintai Distict, Tongchun City. Therefore, from then on, the village and the reservoir came to belong to different administrative regions at the prefecture level. In August 1983 and August 1995, while the village maintained the abstracting-water project, the reservoir administration tried to prevent this on many occasions. In addition, the reservoir always prohibited water from being abstracted from the reservoir. All these factors made it very difficult for the village to get drinking water. Therefore, a water conflict arose between the different regions. In 2002, the village obtained funding from the State to reconstruct the abstracting-water project and its pipes. However, due to the prevention by the reservoir, the construction could not be implemented in due course. In early 2004, the Government of Tongchun City applied to the Government of Shǎnxi Province to judge the water conflicts in accordance with Article 56 of the Water Law 2002. On September 29, 2004, the Government of Shǎnxi Province issued a judgment with the following opinion: the Management Unit of the reservoir shall allow the village to abstract water from the reservoir within the amount allocated to the village under the design of the reservoir, i.e., 2,430 m^3 per day. Finally, the water conflict, which lasted for over 30 years, was resolved completely.

However, complicated water conflicts are often difficult to settle merely through a decision issued by the government, and the Zhanghe River water conflict between Shǎnxi, Hebei and Henan Province is discussed here as an example (Jing, 2002). In the upstream of Zhanghe River, although the natural conditions are abominable, there is a considerable population. As a result, issues concerning the shortage of water resources (per capita 400 m³) and cultivated land (per capita 1,300-2,000 m²) exist alongside the river. Due to the lack of integrated planning and management as well as overexploitation, the river flow has decreased. In order to reclaim more land for cultivation and more water for irrigation purposes, a water conflict involving six county regions from three provinces arose in the 1950's. Sometimes the conflict developed into fights with guns and canons. Since the 1980's, over 30 battles have occurred because of the water conflict, with explosions in Hongqiqu (Henan Province), and Dayuequ and Baishanqu (Hebei Province). During the Spring Festival of 1999, fighting in Gucheng Village (Henan Province) and Huanglongkou Village (Hebei Province) left nearly 100 villagers wounded, houses damaged, the facilities for household and production destroyed, and a direct loss of 8 million CNY. In order to resolve this conflict, the State Council and its relevant working departments, such as the MWR and the Ministry of Public Security, issued many documents concerning the allocation of water resources. However, these documents were not executed effectively, due to both subjective and impersonal reasons, until March 1999. Also, the Zhanghe River Upstream Administration Bureau was established in 1992 to manage a 108 km length of river upstream and the key waterworks for four irrigation areas in an integrated way. During the two years following the fighting in 1999, the central government and relevant local governments and their working departments, especially those in charge of water administration and public security, strictly punished the individuals and units engaging in the water conflict, and invested in many waterworks in order to enforce the law and documents issued by the governments. Further, water rights transfer is encouraged by the government. However, the water conflicts still remain due to the allocation and management of water resources. Fortunately, there has been no severe fighting concerning water conflicts.

Water Conflicts in International Watercourses

There are over 112 Chinese watercourses that connect with over 40 international watercourses. The international watercourses with which China is involved share the following three characteristics: (a) in most circumstances, China is an upstream country; (b) these international watercourses flow through arid, poor regions; and (c) most of these international watercourses are still under initial natural circumstances with lower or the lowest development (Chen, 2002). The latter two characteristics have made a commendable record in the realm of water-sharing. However, with the economic and social development of the riparian states during the last two decades, especially in recent years, there is the potential for an increasing number of water conflicts arising between China and other watercourse states in many fields. To date, the Chinese government has signed many agreements with other states. For example, on the exchange of information, agreements have been reached with Viet Nam (December 2000), India (April 2002), the Mekong River Commission (April 2002). In addition, in September 2001, the Cooperation Agreement on the Utilization and Protection of International Watercourses was signed between Kazakhstan and China. This was the first inter-governmental agreement on the comprehensive utilization and protection of international watercourses that China concluded with a foreign country. As to the Lancang-Mekong River, several agreements have been concluded concerning navigation, the provision of information, the building of hydroelectric power stations, etc. Plans for the comprehensive utilization of the waters of the Ertis (Irtysh) River and Heilong River have been completed by Russia and China. An Associated Committee on Boundary Water between Mongolia and China has been established and operates effectively. A negotiation mechanism has also been established to deal with cooperation about the hydrology of the boundary river between North Korea and China. It can be predicted that, in the development and utilization of international watercourses, China and other watercourses states will have to cooperate in order to deal with many water related conflicts.

RECOMMENDATIONS AND CONCLUSIONS

Water conflicts in China appear in the exploitation, utilization, saving and protection of water resources, as well as in the prevention and control of water disasters. The root cause of water conflicts is that there are insufficient water resources to meet the water demands of human beings for survival, social and economic development and of the environment to remain in a healthy state in different areas or regions simultaneously. To resolve water conflicts, sustainable integrated water resources management, within which water law should play a vital role, should be implemented or enforced. In more detail: (i) a national water resources strategy planning and relevant basin or region plans, as well as a water allocation plan, should be formulated in accordance with the principle of sustainable development; (ii) integrated water resources management should be able to ensure both that the multiple functions of water resources are reasonably balanced and that the competing water needs are taken into proper consideration, so that the economic, social and environmental values of water resources can co-exist equitably in harmony; (iii) on the subject of priority, with the principle of water conservation should be followed in the order that the water is first available for basic human needs, followed by water for the fulfilment of other human rights (in China, water for irrigating land allocated to peasants for living on shall be included), the environmental requirement for water, and finally water for production and business; (iv) as to water law, a rights approach system, which should clearly define the human right to water, the property right of water resources and the environmental right to water, is required; (v) concerning international watercourses, all water conflicts should be settled in accordance with the relevant bilateral or multilateral international treaties to which China is a contracting party, wherever lack of concrete provisions, the principles of international law recognized in the UN Charter, including the Five Principles of Peaceful Coexistence, should be applied; (vi) in dealing with water conflicts, particularly domestic ones, the relevant authorities should perform their duties completely and handle the matters concerning water conflicts in due process and as soon as possible, on the one hand. On the other hand, coordination between different authorities should be taken into consideration where appropriate, since sometimes a conflict is caused by more than one element or involves two or more aspects; (vii) public participation, particularly stakeholder participation, shall be encouraged and promoted. The process of public participation may not only ensure that public/stakeholders exercize their lawful rights, but also help them to acknowledge the rights of others, understand the conflict, be aware of water scarcity, and realize how and why the law and policy make provisions for this element but not that one. As a direct result, a decision can be made or an agreement concluded fairly, which can then be executed and enforced. Further, potential water conflicts that may be caused due to a lack of knowledge of the law and policy, or the rights of others etc., may be avoided; and (viii) the judicial method is not always the best solution to water conflicts, as shown by the statistics about China and further illustrated by the Gabcikovo-Nagymaros case between Hungary and Slovakia, which, at the time of writing, has not yet been settled, even though the International Court of Justice gave its verdict in 1997.

REFERENCES

Chen, L. 2002. Address at the 2002 National Water Conference on International Cooperation, http://www.mwr.gov.cn/tbtj/020421b.htm

Jin, H. 2003. The Baiyangdian is Definitely Dry, China Water Resources News, November 4th, 2003.

Jing, Z. 2002. An Investigation Report on the Settlement of Water Conflict in Upstream of the Zhanghe River, China Water Resources News, http://shuizheng.chinawater.com.cn/ssjf/20021021/200210170090.htm, October 21, 2002.

Jing, Z. 2005. Address at the 2005 National Water Administration Conference, http://www.mwr.gov.cn/bzzs/20050225/50708.asp

Li, D. 2004. Focusing on a One-Hundred Year Water Conflict: How Central China Get Rid of Water Difficulty, International Online, http://gb.chinabroadcast.cn/3821/2004/07/18/301@235601.htm, July 18, 2004.

MWR 2001. Water Statistics Gazette 2000, http://www.cws.net.cn/gazette/tongjigb/index.html.

MWR 2002. Water Statistics Gazette 2001, http://www.mwr.gov.cn/yzndsj/2001sltjgb.htm.

MWR 2003. Water Statistics Gazette 2002, http://www.mwr.gov.cn/gonggao/030613.doc.

MWR 2004. Annual Report 2003, Beijing.

Qin, Y. 2005. A Judgment Resolved a Thirty-Year Water Dispute, China Water Resources News, May 26, 2005.

Wang, L. 2003. More Water Conflicts Arising, Policeman Strengthening Mediation Work, Lishui Daily, July 30, 2003.

The Water and Soil Conservation Office of Nanping City (2002) Dealing with Dispute Concerning Water and Soil Erosion, Stabilizing Social Order in Countryside, http://www.swcc.org.cn/page1_view.aspid=5144.

Wen, Z. 2003. The Ecological Function of the Baiyangdian Wetland and its Protection, Journal of Xingtai University, 18 (4) 30-32.

Wu, J. 2002. An Introduction to Modern Water Resources Management, China WaterPower Press, Beijing, China.

Zhang, H. and Zhong X. 2001. Zhejina Higher Court on Water Conflict: Peasant Has Defeated County Government, People's Daily-Huadong News, March 22, 2001.

Zhang, H., Wang, T. and Li, M. 2001. Judging Water Conflict Case ASAP and Making Donation, Xinyang Court Does Best for Mitigating Disaster, Legal Daily, September 10, 2001.

5

Disputes over Water, Natural Resources and Human Security in Bangladesh: Towards a Conflict Analysis Framework

Irna van der Molen[1] and Atiq Rahman[2]
[1]University of Twente, The Netherlands
In cooperation with
[2]Bangladesh Centre for Advanced Studies (BCAS) , Bangladesh

INTRODUCTION

Why is conflict analysis of conflicts related to natural resources so important? What makes it different from other conflicts? Let us start by a hypothetical situation and assume a situation of community outrage over an incidental land-dispute (over irrigated land) or between individuals from two different ethnic groups. This community outrage may be orchestrated and serve as a strategic goal, if this takes place in a context where ethnic identity is heavily polarized and politicized. The question then is to what extent the issue at stake is access to land and water; or whether the issue at stake is ethnic strife fuelled by identity politics. For the individuals involved, it can be seen as a simple conflict over access to, or ownership of land. For the larger community, it is likely to reflect or symbolize the tensions between two ethnic groups, and for the political leaders, it serves as a means in their political struggles and their desire to generate popular support. Several cases will be used to illustrate that existing typologies (of scarcity, of property, and of conflict related to natural resources) can be useful, but that one should recognize these are simple stereotypes of the more complex and fluid reality.

For the purpose of this chapter, conflict will be defined as "a struggle by individuals or collectivities over incompatible values and purposes, or over claims to status, power, and valued resources, in which the aims of the conflicting parties are to assert their values or claims over those of others" (Goodhand and Hulme, 1999)[1]. As indicated by Bennett et al. (2001) and Richards (2005), conflict is not necessarily always negative. Positive conflict "highlights incompatible goals or objectives, thus focusing attention on something that needs to change for the benefit of all concerned" (Bennett et al. 2001). It can also be perceived as a process through which society is adapting to a new political, economic or physical environment.

Violent conflict is usually related to a variety of causes, and can seldom be related to water resources only. Therefore, in this chapter, we will not only refer to freshwater resources, but also to wetlands and dry lands, vegetation, forests, biodiversity, mineral resources, aquatic resources, and coastal resources, simply because the interrelation between fresh-water resources and these other resources is too important to ignore.

The remainder of this chapter is divided into five sections. The first section starts with an academic debate on the relation between environmental scarcity, environmental degradation and conflict. This is followed by a framework for analysis of water-related conflict. The third section discusses the impact of climate change on human security and the likelihood of climate change-induced conflict in Bangladesh. We then apply the framework for analysis from section three, to case studies at local and sub-regional in Bangladesh; and to inter-state conflict and cooperation between Bangladesh and India over the Farakka Barrage. The final section uses the findings to reflect on the environment-conflict debate, and on the framework for conflict analysis in relation to natural resource management.

DEBATE ON ENVIRONMENT AND CONFLICT

Some authors claim that the next war(s) will be fought over water; and that climate change, sea level rise, changes in temperature and precipitation, and the expected impacts of increased frequency and intensity of cyclones on already vulnerable ecosystems is likely to contribute to more (armed) conflict. The hypothesis that (increasing) environmental scarcity can be considered as one of the key factors contributing to violent conflict is, however, contested by several authors (Barnett, Gleditsch), and not yet supported by research on cross-boundary conflict. Within the literature on

---

[1] J. Goodhand and D. Hulme, 'From Wars to Complex Political Emergencies: Understanding Conflict and Peace-building in the New World Disorder', *Third World Quarterly*, Vol. 20 (1999), no. 1, p. 24. Bennett et al. (2001) use a similar definition, when they state "conflict emerges when 'the interests of two or more parties clash and at least one of the parties seeks to assert its interests at the expense of another party's interests" (FAO cited in Bennett et al., 2001, p. 366).

environment, human security and conflict, one can witness a vivid academic debate over the contribution of environmental scarcity (or competition over natural resources) to violent conflict.

A number of theories has been used for the explanation of conflict, such as frustration-aggression theories, group-identity theories and structural theories. Frustration-aggression theories can be described as "psychological theories of individual behavior to explain civil strife, revolutions, insurgencies, strikes, riots, coups" (Homer-Dixon, 1999), often through a combination of relative-deprivation, grievance and increased opportunity for particular groups to act violently. Group-identity theories, on the other hand, are based in social psychological theories of group behavior which explain inter-group conflicts are related to nationalism, ethnicity and religion. Finally, structural theories are grounded in micro-economics and game theory, based on rational choice theory; and look at the opportunity structure (summarized from Homer-Dixon, 1999). For an overview and explanation of these, see Homer-Dixon (1999).

Richards (2005) also provides an overview of theories, but concentrates primarily on theories which explain 'new wars' (post-Cold War). He identifies three streams and labels these as: (a) 'Malthus with guns', which explains conflict from competition over natural resources; the Homer-Dixon school; (b) New Barbarism, which explains conflict from 'ancient hatreds' between different ethnic and cultural groups[2]; and (c) 'Greed, not grievance' school, which emphasizes economic considerations and opportunity for rent-seeking behavior in particular in countries with mineral wealth, such as diamonds, oil, coltan[3].

Two of these theories, the 'Homer-Dixon' school and the 'greed, not grievance' theory by Collier, explicitly refer to natural resources. Whereas Homer-Dixon emphasizes natural resource scarcity, Collier tends to look at natural wealth, in particular mineral resources, such as oil, gas, diamonds, coltan, but also rain forests used for timber. The literature on environmental scarcity and conflict relation is highly contested[4]; especially the literature by Gleick (1992); Homer-Dixon (1991; 1999) and Homer-Dixon and Blitt (1998).

The framework used in this chapter is based on a combination of three interrelated factors contributing to water-related conflict: (i) environmental

---

[2] See Kaplan, 1993, 1994, and in adjusted form: S. Huntington, 1998.

[3] Le Billon (2001); Collier (2000)

[4] Several authors state that the hypothesis that environment is a major factor in explaining conflict (in the south) serves a northern agenda; that it is based on insufficient empirical evidence; and that statements which suggest that climate change could be an important cause for future violent conflict are 'highly speculative' (Barnett, 2003; Gleditsch, 1998; Richards, 2005). A comprehensive deconstruction of this particular debate has been provided by Barnett (2000 and 2003). Richards (2005) opposes the 'environment-conflict' literature by Homer-Dixon on more conceptual and methodological reasons.

scarcity, competition over natural resources and environmental degradation, (ii) poor economic conditions for, and relative deprivation of, particular groups, (iii) poor governance (Smith *in* Austen et al., 2004).

The theory by Collier, stating that natural wealth in combination with other factors, can result in violent conflict provides interesting insights. But this theory is less useful for the purpose of this study since resource abundance of water resources does not have the same features as natural wealth, which can be looted. Although inter-ethnic conflict over power and access to economic resources[5] is a fourth potential factor, this requires a separate analysis and will therefore not be elaborated upon in great detail in this chapter.

FRAMEWORK FOR CONFLICT-ANALYSIS RELATED TO NATURAL RESOURCES

An analysis of conflict related to natural resources, or more in particular, water resources, requires a sequence of steps: (i) identification of multiple issues involved; (ii) identification of the actors involved; (iii) description of history and distribution of the conflict; (iv) level and intensity of the conflict; (v) identification of the underlying political, economic, social and institutional structures which may contribute to the conflict; and (vi) prediction of the impacts of the conflict on human security (in terms of violence, vulnerability and marginalization, displacement and or famine).[6] It is in particular in step (iii) and (v) where dimensions of the conflict should be analyzed which are related to relative deprivation, grievances, structural inequalities, or group-identity.

Identification of the Issues Involved: What is the Conflict About?

Access to natural resources: Access (to natural resources) will be defined as the right to use, extract, discharge, exploit, inundate, and navigate, which can be either for consumption, production or for non-productive purposes. It is important to recognize that access to various natural resources can be temporary (seasonal) and spatially based, depending on the actors involved, the purpose of access (subsistence or profit-making), the impact of access to other groups and related to productive activities.

---

[5] See Dan Smith who states: "What are commonly called ethnic conflicts, then, are in the end conflicts over power or for access to economic resources (including environmental resources in cases not discussed here), which come to wear an ethnic mask. Ethnic difference is of central importance not as sole cause of armed conflict, but rather as an instrument of mobilisation for political leaders" (Smith, D. in: Austin et al. (eds), 2004, p. 123).

[6] Another taxonomy is described by e.g. Neefjes (2000), who suggests to look at (a) trigger mechanisms for violent conflict, displacement and/or famine; such as natural disasters, cumulative (environmental) change, accidents and warfare; (b) political-geographical factors; (c) categories of environmental change; (d) the historical dimension of 'society induced environmental transformation' (Neefjes, 2000, p. 39).

Ownership of natural resources: Previous research shows the possibility of different combinations of ownership such as traditional rights, common property, public property, and private property. Public property (including state-owned land, forestry, water bodies, canals, rivers) can be leased out on a temporary (seasonal, one year, two years) or permanent basis to local farmers, fishermen and other users, such as industries. Ownership of natural resources may be different from ownership of (often public) goods, such as infrastructure and irrigation structures, embankments. For such goods, ownership not only refers to user rights, but also to responsibilities for provision of services, goods, operation and maintenance.

The variety of activities and nature of produced goods is related to the variety of resources involved, such as land, water bodies, fishing grounds, vegetation, biodiversity, landscape and mineral resources. Different perceptions of ownership may exist simultaneously among the various stakeholders. Conflict may well be related to competing (illegal) mechanisms of 'ownership', such as encroachment of (by sedimentation) elevated areas in riverbeds.

Use of natural resources: This includes conflict over the ways, quantities and frequencies in which water and other resources (dry lands, wetlands, fishing grounds, mangroves, vegetation, mineral resources, soils, sedimentation, sand, stones, vegetation) are used, extracted, excavated, inundated, polluted, and fenced off. Conflict over the allocation of water resources to sectoral interest groups; conflicts may also arise about the use, extraction, excavation of natural resources between groups at different locations (upstream/downstream). Finally, one can observe many conflicts which relate to competition over resources for different production possibilities (agriculture versus fisheries in the same area).

Impact of human activities and development processes[7] on the use and quality of natural resources: Discharge of effluents into water resources, indiscriminate use of chemical fertilizers and pesticides; dumping of waste on land and in water, conflict over increased vulnerability due to erosion can be attributed to e.g. legal and illegal mining (mineral resources; gems or precious stones), logging, building practices on encroached land, unsustainable cultivation practices (e.g. the spatial expansion of slash and burn cultivation) or erosion-inducing fishing-methods. Conflict over the use of natural resources is often related to conflict over *management of these resources.*

Access to services, facilities, markets and infrastructure and other non-natural resources, which can provide alternative livelihood strategies, provides relief or resilience in case of extreme events, and reduces people's

---

[7] Referred to as 'society induced environmental transformation' by Neefjes, 2000, p. 39

vulnerability. This may include access to Natural Resource Conservation Projects or development programmes.

Access to, mechanisms and hierarchy of decision-making: The access to, and use of, natural resources is arranged by operational rules. One's access to decision-making, however, is formally arranged through constitutional rules while decision-making rules determine the process of decision-making. The formal mechanisms and hierarchy is, however, not always representative of the real mechanisms of decision-making. The possibility for resource-capture, exclusion or manipulation is created by access of more powerful groups to decision-making, the mechanisms of decision-making, their position in the hierarchy of decision-making.

Management /Implementation of policies, programs and decisions: This refers to the availability of standards, monitoring and enforcement of laws and regulations. Conflict can arise over (lack of) standards, monitoring of discharge, (lack of) enforcement, involvement of enforcing authorities in circumventing laws and regulations at the disadvantage of more vulnerable groups, e.g. in close cooperation with privileged and influential groups. It is important to differentiate between the impacts of natural resources depletion (and scarcity), and impacts of environmental degradation through pollution or as the result of human activities and development processes. Whereas the economic value of natural resource depletion or the cost of substitution by other resources can be estimated; the costs of biodiversity loss, health problems and human insecurity are difficult to estimate (Barnett, 2000). In the case of soil, water and biodiversity, it is not easy to make a clear distinction, since these resources are being used for both economic and ecological purposes. Even more, one of the case studies shows that these can be used even for political purposes. Environmental degradation could be mentioned as a separate issue, but given its linkage with resource depletion (less of the same resources are left which are still in good condition), it is included in the category 'use of natural resource' and 'impact of human activities and development processes'.

Identification of Actors: Who are Involved and What are their Features?

After identification of the issues, it is essential to identify the relevant parties in the conflict[8] and their position in the process. Given the importance of politics and power structures in many conflict, this dimension

---

[8] An example of a more actor-oriented approach is given by Bennett et al. (2001) who provide a typology for conflict in tropical fisheries. They identify five different types of conflict: Type I conflict: Who controls fisheries; Type II conflict: How is fisheries controlled; Type III conflict: Relations between fisher men; Type IV conflict: Relations between fisher men and other users of the aquatic environment; Type V conflict: Relations between fisheries and non-fishery issues

should be explicitly be addressed, and not incorporated in other general categories such as 'interests' of the key stakeholders.
- Description of main stakeholder groups, size of the group and internal sub-groups (gender!)
  - Causing the conflicts (or being blamed for it);
  - Affected by the conflict;
  - Who might assist in managing the conflict;
  - Who might undermine management of the conflict;
  - Identification of group representatives (Warner, 2000).
- Underlying values, interests, needs and fears of key stakeholders
- Their access to, ownership and use of natural resources
- Their access to, and control over political resources
- Their access to, and control over economic resources
- The economic benefits from their political position
- Their access to, and position in decision-making or implementation
- Their presence in, access to, and control over, geographic areas
- Relationships within and between groups, such as farmers-fishermen

History and Distribution of the Conflict or Disputes

Apart from the issues and stakeholders involved, it is important to know about the history and distribution of the conflict, as these can give information about structural causes, economic grievances, geographic characteristics or temporal characteristics. Warner (2001) refers to the following dimensions:
- The historical context of the conflict(s) including:
  - The past and predicted escalation of the conflict(s);
  - The underlying structural causes, if relevant;
  - The part played by local economic grievances;
  - Other contributing factors (e.g. demographics, environmental degradation);
  - Past efforts at conflict management and why these were ineffective
- The geographical distribution of conflicts or disputes.
- The temporal distribution of the conflict(s), e.g. seasonality, proximity to local or national elections, etc.

Warner (2000) provides an inventory of several developmental pressures that can fuel conflict over community based Natural Resource Management: (a) the introduction of productivity enhancing technologies, such as synthetic fertilizers, agricultural mechanization, permanent irrigation, joint management regimes; (b) growing awareness of the commercial value of former common property and shifts towards privatization of property rights; (c) a shift from subsistence to cash economy and (d) changes in rural employment activities, such as crop processing, manufacturing, mining activities, etc. (Warner, 2000). Other examples are given by Mbonile (summarized from Mbonile, 2005), who identified several developmental processes leading to water conflicts such as rapid population growth and migration in pursuit of environmental advantage; urbanization of river basin, with increased water consumption; land alienation and marginalization of indigenous populations; conversion of property rights from common property to private property and government policy which can intensified the shortage of water through deliberate encouragement of migration of population

There are many other examples of developmental processes which may contribute to, aggravate or amplify existing tensions and conflicts. Several authors (e.g. Klugman, 1999; Muscat 2002; Homer-Dixon, 1999) refer to root causes and triggers, and conditions for the eruption of violent conflict. Triggers are events that immediately precipitate violent conflict, which take place in a situation of already existing tensions between particular groups due to e.g. political, economic or ethnic differences and a history of polarization (see the example used in the Introduction). Accelerators are events, or developments, which worsen grievances and deprivation (Klugman, 1999; Muscat, 2002). Violence tend to erupt only in the situation where one can speak of (a) horizontal (economic, political, social) inequalities between groups, sometimes actively emphasized in identity politics; (b) mobilization of groups, often with a history of earlier hostilities; (c) absence of alternative sources for income-generation (Muscat, 2002).

Level, Intensity of Conflict

The intensity of conflict can vary widely from occasional low-level, low-intensity to high-level and high-intensity conflict, civil war or even regional wars. However, given the vast amount of literature which indicates that war over natural resources between riparian states is more an exception than a rule, this paper will only consider two intensities: non-violent conflict and violent conflict, either at community/societal, sub-national level, national level or international level. At *community/societal level*, the conflict may be explained by competing demands of (powerful) groups, with privileged access to political power and economic resources. Such competition may reflect political culture, ethnic divisions in society; antagonistic social identities, and hostile in-group/out-group discourses (Fumerton, 2005). At *state level* the conflict may be affected by state attributes, such as: institutional failure, weak state, poor governance; destabilizing processes of political or economic transition; discriminatory or repressive policies or structures; unequal relations of resource control and distribution; refugee flows (Fumerton, 2005), or controversial settlement schemes by the state (Mahaweli programme in Sri Lanka; land disputes in Chittagong

Hill Tracts, Bangladesh). At *international level,* conflict may be shaped by factors such as globalization, trade policy, spill or regional conflict between two or more countries.

Underlying Social, Political, Economic and Institutional Structures

Conflict can emerge over the issues identified in the first stage, but can also be related to underlying social structures, political structures, economic structures and institutional arrangements. Examples of social structures are: the hierarchy and interaction in social networks; participation criteria and (de facto) exclusion mechanisms to exclude particular groups from social activities such as meetings, gatherings, festivities or community development; patron-client relationships; reciprocity mechanisms which may be asymmetrical (more beneficial for members of one group than for members of another group); relationships between individuals from various ethnic, religious, class or caste background. Conflict over natural resources can be aggravated by social exclusion; by tensions over language, religion, ethnicity; or failure of dispute resolution mechanisms.

Conflict may be embedded both in social and political structures, such as conflict between two groups from the same settlement who support (and are supported by) two different political parties. Inter-personal conflict can be related to access to politicians and personal favors from politicians (jobs, promotions, protection). Resentment can emerge over the manipulation of law-enforcement through the deliberate use of political relations. Similarly, the nature of the political and decision-making processes, the exclusion of particular groups from the political and decision-making process; the involvement of politicians in extra-judicial spheres, their alliance with the informal economy or crime; are all likely to result in resentment among large parts of the population.

The economic dimension of conflict should not be ignored. Economic decline and macro-economic instability; increased poverty, unemployment, inflation, reduced food security, widening economic disparities along regional or ethnic divisions; increasing competition over natural resources; and shadow economies can all contribute to conflict. Bangladesh is rich in natural resources, such as freshwater fish, sea fish, shrimps and prawns, plant resources, mangroves, birds and animals, some mineral resources. Conflict may arise, if one of their main income generating, or food-supplying strategies is severely constrained or threatened, if the households are increasingly indebted, and alternative strategies are not available. As previous research has demonstrated: "The issue is not competition over scarce resources, but rather competition to gain dominant control over substantial income generating resources, or more equitable access to the spoils of resource extraction" (Barnett, 2003).

Finally, as mentioned before, the possibility for resource capture and manipulation of decision-making processes is embedded in *institutional failure*, which gives rent-seeking elites the opportunity to circumvent relevant decisions and regulations. Table 1 provides a number of examples of institutional failure at various levels.

Impacts of conflict on human security

Human security has been defined by the UNDP in its Human Development Report: "Job security, income security, health security, environmental security, security from crime – these are the emerging concerns of human security all over the world"[9]. Human security is severely threatened if people's resilience to cope with extreme events (epidemics, violence, cyclones, storms, major flooding) is minimal or non-existing. Coping strategies are 'a short-term response to an immediate and inhabitual decline in the access to basic needs in abnormal seasons or years' (adjusted from Davies, 1993, quoted in Chadwick, 2000). People's resilience to extreme events can be strongly affected by conflict, also by non-violent conflict, such as the denial of access of vulnerable fishermen to fishing grounds and the obstruction of their involvement in day-labor, thereby impeding their livelihood strategies.[10]

What starts as a temporary condition (vulnerability) may turn into a more permanent condition (marginalization) with sometimes devastating impacts (such as chronic food insecurity, chronic poverty, increased infant mortality, increased vulnerability to diseases, migration). For the purpose of this chapter, a distinction is made between (a) social marginalization; (b) ecological marginalization; (c) political marginalization; and (d) economic marginalization.

- Social marginalization occurs when social exclusion results in increase of poverty, health problems, infant mortality, migration, disruption of social support networks.
- Ecological marginalization occurs when lack of access to natural resources caused by unequal distribution and poor governance forces populations to migrate to ecologically fragile regions.
- Political marginalization occurs when decision-making, and manipulation of decision-making processes, operate to reinforce existing political inequalities and large parts of the population are excluded from access to the political and decision-making process.

---

[9] UNDP, *Human Development Report 1994*, Oxford University Press (New York and Oxford), 1994, pp. 24-25.

[10] The erosion of individual and group resilience is part of the vulnerability context in Sustainable Livelihoods Framework. (http://www.livelihoods.org/info/guidance_sheets_pdfs/section2.pdf)

Table 1: Institutional failure at various levels of conflict

Level	Some relevant issues	Examples of conflict/disputes	Institutional failure / poor governance
Global and international (mega level)	Climate Change, Trade, Sustainable Development and Biodiversity	Conflict over import subsidies and trade tariffs in food production, conditionalities regarding food safety, environmental standards and labor conditions; conflict between the private and public sector over the provisions of biodiversity or climate change conventions	Lack of monitoring standards, manipulation of research data at national level for (or on the request of) international organizations. Manipulation of negotiation process by powerful parties who are formally not part of the negotiation process (observer status). Stakeholder pressure of the largest industries and Development Banks in the negotiation process to avoid negative consequences
Regional/ multi-country (sub-mega level)	Regional Water Basin Management, Multinational Environmental Agreements	Conflict over dams, freshwater withdrawal for irrigation, navigation, and power from transboundary river basin, cross-boundary aquifers, disputes over the provisions of multinational environmental agreements; discharge of pollutants and waste water into rivers affecting other countries	Lack of monitoring standards, and monitoring procedures, pressure of large industries in the negotiation or implementation process to avoid negative consequences
National (macro level)	Natural Resource Management, Environment Policy; National Conservation Strategy; National Environmental Management Action Plan (NEMAP)	Disputes over the process of formulation of poverty alleviation strategies and poverty reduction strategy papers (PRSPs), disputes over the implementation of environmental policies and plans, in terms of implementation time; risk and uncertainty; potential adverse effects, and equity concerns; discharge of pollutants and waste water into rivers	Lack of institutional capacity for the enforcement or environmental regulations; political influence and alliance with criminal networks; corruption of police forces; lack of people's participation in the formulation of policies; lack of scientific education in the areas of concern
Sub-National (meso level)	Natural Resource Management, water	Disputes over the lack of voice in the formulation process or implementation	Lack of communication, consultation and participation of relevant stakeholders and local

Contd.

Table 1 Contd.

Level	Some relevant issues	Examples of conflict/disputes	Institutional Failure / poor governance
	program, infrastructure, livelihoods[11]	of environmental policies and plans, in terms of implementation time; risk and uncertainty; potential adverse effects, and equity concerns; disputes over project management; offsite environmental impacts	populations; circumvention of existing rules and regulations, non-effective law enforcement; political influence and alliances; intimidation
Community/ societal (micro level)	Social mobilization, local action for environmental natural resource management, livelihoods	Disputes over fishery grounds between private and communal land owners; disputes over lease arrangements; over land boundaries and property rights; over resource capture by local elite; and conflict over off-site environmental impacts affecting livelihoods	Lack of communication, consultation and participation of relevant population groups; circumvention of existing rules and regulations, non-effective law enforcement; political influence and alliances; intimidation

[11] For the definition of livelihoods we use the definition as given by the DfID: "A livelihood comprises the capabilities, assets and activities required for a means of living. A livelihood is sustainable when it can cope with and recover from stresses and shocks and maintain or enhance its capabilities and assets both now and in the future, while not undermining the natural resource base" (DfID, undated, www.livelihoods.org).

- Economic marginalization occurs when alternative livelihood strategies are severely constrained, access to relief measures is not provided and thereby the resilience to cope with extreme events is reduced, resulting in poverty, health problems, infant mortality, migration.

IMPACTS OF CLIMATE CHANGE ON HUMAN SECURITY

Extreme events such as floods, droughts, cyclones, storm surges, sea level rise, low river flows, river erosion, tsunamis, or forest fires, can be a potential trigger for conflict. Extreme floods can create an acute scarcity of safe drinking water, shelter and disrupt communication possibilities. The displacement of populations to already fragile or densely populated may result in aggravation of already existing tensions.

The likelihood of violent conflict further increases if the influx of new settlers is perceived as threat to the identity of the host population (Barnett, 2003). Figure 1, summarizes the most common causes of conflict related to natural resources; the resulting vulnerabilities; and the eventual impacts in terms of human security and marginalization. This framework as illustrated in Figure 1 will serve as starting point for the case studies in Bangladesh, in the following sections.

Climate Change, Vulnerability and Human Security in Bangladesh

The waters of the Ganges-Padma, the Jamuna-Brahmaputra and the Meghna River, flow towards Bangladesh, which forms one of the largest deltas in the world. Because of its geographic location vis-à-vis the Himalayas, the low-land, the large number of water bodies in the delta, and the sheer magnitude of land which is annually submerged under water due to monsoon rains and water from the Himalayas[12], Bangladesh is extremely vulnerable to climate change. Climate and regional hydrologic models suggest changes in the variability of storms, in the frequency, intensity and area of tropical disturbances, and in the frequency of droughts and flooding in particular areas.

Climate change predictions for Bangladesh

A World Bank Report[13] predicts the following critical impacts of climate change in Bangladesh:

---

[12] Estimated average of 26,000 km², with extremes of 82,000–95,000 km² (Craig et al., 2004, p. 272).

[13] World Bank, October 2000. *Bangladesh: Climate Change and Sustainable Development*. Report 21104 BD, South Asia Rural Development Team, p. 46

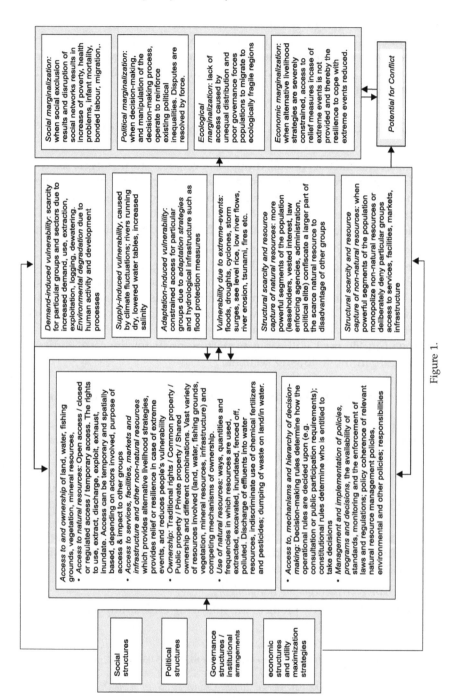

Figure 1.

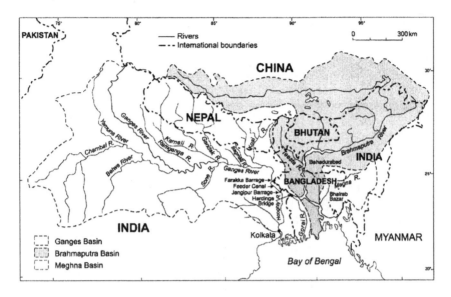

Fig. 2. The Ganges, Brahmaputra and Meghna River Basins
(Source: Monirul Qader Mirza, 2002)

- Drainage congestion problems, resulting in an increase of the period of inundation, and expansion of wetland areas. This can result in lower agricultural productivity, and human health problems due to water-borne diseases and food insecurity.

- Reduced fresh water availability due to growing demands stimulated by climate changes, population growth and economic development (demand-induced scarcity). Low river flows and increased evapo-transpiration in the dry period are expected to reduce the availability of fresh water (supply-induced scarcity). Low river flow and sea level rise stimulate saline water intrusion (in estuaries and into groundwater) in coastal zone, significantly affecting fresh water availability.

- Disturbance of Bangladesh' extreme dynamic riverine and morphological processes, in particular (i) increased river bank erosion, coastal bank erosion and bed level changes of rivers and estuaries; resulting in the loss of land and homesteads of hundreds of thousands of people; and (ii) disturbance of the balance between river sediment transport and deposition in rivers, flood plains and coastal areas; increase river bank erosion and drainage congestion/risks for flooding;

• Increased intensity of disasters such as cyclones/storm surges, floods and droughts[14].

Climate change and increased intensity of disasters
Mirza (2002) shows the potential impact of climate changes on the probability of the occurrence of floods in Bangladesh and its implications for the basin areas of the Ganges, Brahmaputra and the Meghna river:

> "[...] future changes in precipitation regime have four distinct implications. *First*, the [...] onset and withdrawal of monsoons may be delayed or advanced. *Second*, an increase in monsoon precipitation in the Ganges, Brahmaputra and Meghna basins may increase the magnitude, frequency, depth, extent and duration of floods. *Third*, timing of peaking in the major rivers may also change [...] *Fourth*, increased magnitude, depth and duration of floods will bring a dramatic change in land-use patterns in Bangladesh" (Mirza, 2002).

In order to understand the magnitude of the potential impact of increased flooding in Bangladesh, it is worth noting that the floodplains of Bangladesh cover approximately 80% of the total land area (FAO/UNDP, 1988)[15]. Another 8% of the total land surface is constituted by terrace areas, while the northern and eastern hills and the Akhaura Terrace cover another 12% of the total land area (Rahman, Chowdury and Ahmed in: Najam, 2003). These floodplains are essential for the livelihoods of its population (Rashid, 1991; Mirza, 2002; Craig et al., 2004; Rahman, Chowdury and Ahmed in: Najam, 2003), both in terms of fisheries and agriculture. The floodplains create the opportunity for fisheries, supplying ca. 80% of the daily animal protein intake (Rashid, 1991). The floodplains do not only provide opportunities for fisheries[16], but also for agriculture: approximately two-thirds (63%) of Bangladesh's cultivated area is estimated to fall into the category of floodplain (Islam, 2001).

Climate change, water resources and human security
Whereas the normal and moderate annual floods are associated with benefits, such as good agricultural output, bathing opportunities, soil fertility increase by the deposition of silt, washing out of garbage (Chadwick,

---

[14] For more detailed information on (impacts of) predicted changes in temperature and precipitation, sea level rise and subsidence, cyclones and storm surges on rivers and cross-boundary river flows, fresh and brackish water resources, geological resources, coastal morphology, mangrove forests, land resources, beaches, fisheries, aquatic ecosystems, wetlands, coral reefs, see: Agrawala, S. et al; 2003 (OECD); BCAS and DOE, undated; Huq et al. 2002; Ahmed, 2000, Choudhury et al. undated; more references bibl. OECD report.

[15] FAO/UNDP, 1988. *Land resources appraisals of Bangladesh for agricultural development. Agro-ecological regions of Bangladesh.* Report No. 2. FAO, Rome.

[16] For more detailed information on the impact of fisheries on livelihoods in Bangladesh, see Craig et al., 2004.

2000), an increased magnitude, depth and duration of floods is likely to generate severe constraints on human activity. Although they provide opportunities, floodplains also have environmental risks associated with them. As Soussan (2000) observes, "If the flood arrives earlier then expected, or excessive water accumulates and remains for longer than the crop can withstand, agricultural losses become inevitable. The extremely high floods, known as *bonna*, often breach infrastructure, disrupting communication, and threaten industry and residential areas as well as cropland" (Soussan, 2000)[17]. Soussan further points to the associated problems of extreme floods, such as riverbank erosion (resulting in permanent loss of land, property and infrastructure), and drainage problems. A classification of floods is provided by Mirza (2002), who shows not only the area inundated and the probability of occurrence, but also sketches the way in which livelihoods are affected by these floods (Table 2).

As Table 2 shows, severe floods, catastrophic floods and exceptional floods are a direct and immediate threat to human security. Due to floods, human activities are severely hampered; communication is disrupted and relief operations are required to prevent further loss of human life and it also leads to the outbreak of epidemics. For example, in 1998, Bangladesh faced an extreme situation: the rains which started early in July, continued and intensified throughout July and August. The drainage capacity was constrained due to road construction and flood control embankments, and due to simultaneous storm surges and high tides in the Bay of Bengal (Chadwick, 2000). At the end of August, 100,250 square kilometers were flooded (68% of the country). This resulted in loss of crops and livestock; poor quality drinking water; lack of paid labor, lack of fuel to cook; lack of food supplies; destruction of homestead crops and trees, and various other problems (Chadwick, 2000)[18].

Climate Change and Future War?

Several authors, politicians, leaders of international organizations and journalists have cautioned the world community that the increasing scarcity of freshwater resources might lead to national and international conflicts[20]. Three observations seem to support this assumption.

---

[17] Soussan, J.G. 2000. 'Leeds/BCAS Research into Sustainable Local Water Resources Management – Meeting Needs and resolving Conflicts' in: Clemett, A.; M.T. Chadwick and J.F.F. Barr. 2000. *People's Livelihoods at the Land-Water Interface; Emerging Perspectives on Interactions between People and the Floodplain Environment; Symposium Proceedings*; University of Leeds/University of Newcastle, UK, September 2000.

[18] Chadwick, M.T. 2000. 'Impacts of the 1998 Flood on Floodplain Livelihoods' In: Clemett, A.; Chadwick, M.T. and Barr, J.F.F. 2000. *People's Livelihoods at the Land-Water Interface: Emerging Perspectives on Interactions between People and the Floodplain Environment*; Symposium Proceedings, Dhaka, January 2000, page 27

Table 2. Flood classifications (area inundated, chances of occurrence and physical damage) in Bangladesh

Types of floods	Range of flooded area (sq.km.)	Range of per cent inundation	Probability of occurrence[19]	Physical parameters affected
			Parameters	
Normal flood [which years]	31,000	21	0.5	Hampers normal human activities Cropping pattern is adjusted with inundation May increase soil fertility Economic loss is minimum
Moderate flood [which years]	31,000 – 38,000	21–26	0.3	Hampers human activity moderately Damage limited to crops Economic loss is moderate Evacuation not necessary People take their own measures
Severe flood [which years]	38,000 – 50,000	26–34	0.10	Hampers human activities severely Damage is mainly to crops, infrastructure (roads, railways, power, telecommunications, etc.) and certain urban centers Economic loss is higher Requires evacuation Requires relief operation
Catastrophic flood [which years]	50,000 – 57,000	34–38.5	0.05	Hampers human activities very severely Extensive damage to crops of all types of lands, cultured fisheries, lives and property in both urban and rural centers, all types of infrastructure, etc. Requires extensive relief operation Very high economic loss Requires international support
Exceptional flood [which years]	> 57,000	> 38.5	0.05	Hampers human activities exceptionally Extensive damage to crops of all types of lands, cultured fisheries, lives and property in both urban and rural centers,

Contd.

Table 2 Contd.

Types of floods	Parameters			
	Range of flooded area (sq.km.)	Range of per cent inundation	Probability of occurrence[19]	Physical parameters affected
				all types of infrastructure, etc.
				Requires extensive relief operation
				Disrupts communication
				Closing of educational institutions
				Exceptional economic loss
				Usually requires international support

Source: Mirza, 2002a

[19]Probability of occurrence was calculated based on area flooded during 1954-1999.

First of all, more than 200 river systems are shared by two or more countries. Toset (2000) indicates that 'many rivers run between countries with a history of conflict, where water plays an important role in the economic life of the country'. Secondly, some countries depend for more than 80% of their renewable water resources on upstream countries such as Syria and Sudan; Turkmenistan, Egypt, Mauritania, Kuwait and Bahrain (Ragab and Prudhomme, 2002). Such dependency is expected to create potential for conflict. Finally, overexploitation of water, as shown in section one, in combination with the impacts of climate change, may well lead governments to divert major rivers, construct large dams, or tap from underground aquifers which extend beneath the neighbors territory. Ragab and Prudhomme (2002) view the potential draining of these aquifers as major potential for future conflict.

However, as indicated before, the evidence that armed conflict over water resources at international level likely to emerge is not very convincing. For example, Kliot et al. (2001) examined the nature, characteristics and shortcomings of cooperative arrangements for the management of 12 transboundary river basins[21] and they conclude that "[…] many institutions which govern the management of transboundary water resources point to the fact that in many river basins countries are able to overcome their differences and cooperate to the benefit of all" (Kliot et al, 2001).

Similar findings have been presented by Wolf (1999) and Yoffe et al. (2000). They conclude: "We found that international relations over freshwater resources are overwhelmingly cooperative and cover a wide range of issue areas, including water quantity, quality, joint management and hydropower" (Yoffe et al. 2000) and "Most of the commonly cited indicators linking freshwater to conflict proved unsupported by data. Neither spatial proximity, government type, climate, basin water stress, dams or development, nor dependence on freshwater resources in terms of agricultural or energy needs showed a significant association with conflict over freshwater resources" (Yoffe et al. 2000).

Barnett (2003) has discussed in detail the question whether, on which scale, and through which variables, climate change may contribute to violent conflict between countries. He argues that the ways in which climate change may contribute to future violent conflicts [between countries] are highly uncertain and will operate through a much more complex environmental and social process:

> "It may be possible that as climate change retards growth in climate sensitive economies this may lead to greater frustration

---

[20] For several quotes reflecting the assumption that water shortages are likely to develop into violent conflict, see Toset et al., 2000, pp. 972-973.

[21] The Mekong, Indus, Ganges, Nile, Jordan, Danube, Elbe, Rio Grande and Colorado, Rio de la Plata, Senegal, Niger.

with political systems unable to deliver jobs, and repeated failure to do so could cause political instability and possibly violent conflict. Finally, it may be that climate change stimulates more migration which can be a factor in violent conflict" (Barnett, 2003).

More important is the emergence of conflicts at local and sub-national level, where a variety of factors are at play. Therefore, the case studies below are illustrations of (a) conflicts between three groups (at community level) which are triggered by poor governance; (b) conflict between industry and communities over environmental pollution at sub-regional level; and (c) conflict between India and Bangladesh over the water flow of the Ganges. The last case study will be a slightly less detailed case study, as additional information can be found in the references.

CASE STUDY 1: ACCESS TO FISHING GROUNDS IN PATKELBARI, CHANDA BEEL[22]: COMMUNITY LEVEL

Chanda Beel[23] is one of the largest non-saline tidal floodplains of Bangladesh, located in Gopalganj District in the south-central region of Bangladesh. Chanda Beel is rich in biodiversity and encompasses "a large variety of wild fish, aquatic animals, aquatic weeds and plants and a large quantity of peat" (Clemett et al., 2000). In the wet season, Chanda Beel harbors frogs, turtles, crabs, mollusks and aquatic snakes. It is the breeding ground for several resident water-birds, and in the winter period, a large variety of (migratory) birds can be spotted here. Between June and October, the monsoon period, the *beel* is usually inundated. From October to December the water level decreases and the major part of the *beel* becomes dryland by December, with an exception for some *kuas* and ponds (estimated number to be 5,749).

The livelihood strategies of the population in the *beel* are concentrated around these resources which provides them with food and income from agriculture, fisheries, aquaculture, selling peat. Huq and Alam (2000) indicate that according to the 1991 Census among the local population, approximately 53% were involved in the cultivation of crops; 24% were involved in agricultural day labor; 8% were small business owners, 6% were civil servants (including teachers); 3% were fulltime fishermen, and 6% belonged to other occupational groups. Many people would, however, be involved in multiple activities, some of which are concentrated in

---

[22] Information based on focus group discussion with fishermen in Patkelbari – 18 July 2005. Assisted by Mr. Saifuddin and Mr. Bidhan, BCAS.

[23] A 'beel' is a natural depression in the floodplain, perennially/seasonally flooded (UNDP/ IUCN, 2004, p. 4).

particular seasons of the year. For example, although the percentage of fulltime fishermen is rather limited, in total 81% of the households catch fish in the *beel* and canals, either as fulltime activity, as part-time activity or on occasional basis (Huq and Alam, in: Clemett et al., 2000). Unfortunately, fish and aquatic resources are said to be declining due to indiscriminate catching and loss of over-wintering and breeding grounds (UNDP/IUCN project brief, undated; field notes).

The total population of *Patkelbari* is 3000. Most villagers are involved in both agriculture and fisheries[24]. Sharecropping is a common feature in this area even though 90% to 95% of the village households own some land: 50% of the households own a plot of land below one acre; 25% owns land between one and two acres; and another 25% is owner of a plot of land of more than two acres. Out of these, 30-35% are owners of a ditch from which the water can be withdrawn during the dry season, thereby enabling the easy catch of all fish captured within the earthen bunds. Such plots are referred to as 'kuas'. In the rainy season these plots are completely flooded and part of the larger floodplains. The extra annual income from fish catching in *kuas* fluctuates between 30,000 and 60,000 Taka/acre. Fish is either used for consumption and sold at local markets or sold to wholesalers who transport the fish to bigger cities such as Dhaka, Khulna, Rashayi and Burshel, or to Rangpur, Sylhet and Kurigram. The fishing season is from June – December; paddy cultivation (4 months High Breed Variety) from December to the end of March or mid-April.

Approximately 5% of the villagers also work as carpenter in addition to farming and fishing, and only 1% work as civil servant (school teacher, NGO worker). Carpenting is done throughout the year except during the period of paddy cultivation.

Two typical conflicts related to fishing can be identified in Chanda Beel. First of all, there is the conflict between interests of fishermen and *kua*-owners in the period December–February. *Kua* owners start to protect their *kuas* when the water level is only 2-3 feet (knee deep). The fishermen are not allowed to fish there during that period of time[25]. This conflict of interest usually does not result in any violent conflict, as many fishermen are also involved in agricultural activities and day labor.

Secondly, individuals from the local elite (Village level, *Thana*-level and Union level politicians) give out illegal leases for part of the government canals in the period November–March. At some places in the Boushar

---

[24] Fisheries is an activity strongly dominated by men in this area. In exceptional cases women also catch fish, but only for their subsistence. Women are involved in fisheries related activities.

[25] There is one exception: poor families may search the last fish inside *kuas* after the landowner has withdrawn the water form his kua, and completed fishing himself. This practice is called 'mas dharra' [fish harvesting].

Canal (close to Patikel Bari), a ditch is created in the canal bed by digging soil. As the hollow is created, the other portion of the canal receives less water and as soon as the hollow is filled with water, many fish enter this part. Without a lease over part of the canal, fishermen are not allowed to fish in the canal, unless it is for subsistence purposes. The politicians involved in this get the support from the police at *Thana* level for 'enforcement' of this illegal lease system (to the benefit of both).

In a recent conflict (2005), two groups of this village leased the same part of the canal. To get (informal) permission to catch fish in the Boushar Canal, fishermen were required to pay a lease of 10,000 Taka[26] on behalf of one of the clubs of western Partical Bari through the *Tehsil* Officer, who is in charge of collection of revenue of Mousudpur Union. According to the fishermen, he 'sold' a lease for part of the canal for 37,000 Thaka:

- 10,000 Taka to be paid by a group of 200 people from the east side of the village;
- 20,000 Taka to be paid by a group of 300 people from the west side of the village; and
- 7,000 Taka to be paid by the *Thana* police and other parties.

Although it was already finalized that after *Eid* vacation the members of western Partikel Bari would take lease at 10,000 Taka, the fishermen of the Eastern Partikel Bari took the lease instead through the same *Tehsil* Officer of administration of Mousudpur Union, who tried to keep the matter secret. As soon as the people of western Partikel Bari heard about this, they asked some political leaders of that district that the lease should be granted to them. The competition over this part of the canal created a fight between the two groups, who sought redress at the *Thana* (police station). The case is now in the court. While the two groups were fighting over the lease of this canal, the *Upazilla Tehsil* Officer and the police caught the fish in the deepest area of the canal. Although both sides got some amount of money returned, after intervention of politicians, the case has not been withdrawn. This conflict was not the first, as there has always been a division between the eastern and western part of the village created by political differences: one group (west) supports Awami League; the other one supports Bangaldesh Nationalist Party (BNP).

Identification of the Issues: what is the conflict about?

In Partikel Bari and the Boushar Canal, one can observe competition over access to fishing grounds for productive purposes. This competition is

---

[26] During a focus group discussion near Patkelbari, farmers/fishermen indicated that a standard unit of river/canal (75* 30 ft.) is leased out for 200–500 Taka (per person for one season). The net income of leasing a standard size is estimated by local fishermen in Chanda Beel to be 1000-2000 Taka per person.

seasonal (November–March); and spatially based (pre-determined part of the canal), and access to fishing grounds in Boushar Canal (using this particular method) depends on purpose: access is, during these months, exclusive for fishermen with lease.

It is interesting to see that public property is privately managed, through illegal temporary lease of the canal; enforced with private assistance of police. Illegal lease systems can be seen as competing mechanisms of ownership. Although the lease is related to a particular method of catching fish in the canal, this is not, in itself issue of conflict; it is the mechanisms of allocation of lease and political influence which underlies the conflict. Several members of the political elite have invested interest and control the size of the production group with the support of the police, whereas these mechanisms are supposed to be in the hands of the fishermen themselves. Access to political support seems to be equally relevant as access to fishing grounds: one can observe competition over access to political support and political resources between the eastern and western side of the village.

Identification of the Actors: who are involved and what are their features

For an overview of all the actors involved, see Table 3. The relationships between the actors are dynamic; these relationships change over time, which can be seen most clearly if one ruling party is being replaced by another one. The relationship between the eastern and western group can be described as problematic but not broken; the relationship between both groups and the political leaders as dependent; the relationship between political leaders and police as dependency and reciprocal (mutual benefit); between administration and fishermen problematic and dependent.

History and Distribution of the Conflict

The East and West bank village had an inner-clash, for a long time. During the period of the last Government, both parties were the supporter of Awami League (ruling party during previous period). However, since their location was at two banks of the Boushar canal, each group tried their best to gather more influence and was lobbying at the Police Station and at District Level. Since BNP is in power, both groups joined the BNP. Since the clashes or conflicts have never been resolved they still persist.

Grievances
Villagers have expressed grievances of the poor against the rent-seeking nature of politicians. Despite these grievances, both groups expect to gain significant benefit by exclusive access to political resources.

Table 3. Actor analysis of conflict in Patkelbari village; Chanda Beel, Bangladesh

Actors	Role in conflict	Interests, needs, fears	Access to, ownership over, use of natural resources	Access to and control over political resources	Access to and control over economic resources	Economic benefits from political position	Access to decision-making / implementation
Fishermen west side	Actor in and affected by conflict	*Interests*: Financial interests, gaining access to political resources and benefits from political resources. *Needs*: securing livelihoods. *Fears*: loosing means of income, poverty	Subject to negotiation with political leaders during 'lease' season (6 months)	Yes, but strong competition between two groups within the village	Limited	No, but opportunity for financial benefits from supporting ruling party	No
Fishermen east side	Actor in and affected by conflict	*Interests*: Financial interests, gaining access to political resources and benefits. *Needs*: securing livelihoods. *Fears*: loosing means of income, poverty	Subject to negotiation with political leaders during 'lease' season (6 months)	Yes, but strong competition between two groups within the village	Limited	No, but opportunity for financial benefits from supporting ruling party	No
Political leader / Upazilla Tehsil officer	Cause and initial benefit by catch and sale of fish	*Interest*: Financial interests, profit-maximization; securing political support. *Fears*: lack of political support; limited benefits	Encroached management of public property (canal)	Direct control	Yes, limited control through allocation of means	Yes, 37,000 Taka (initially)	Yes
Police	Financial benefit and cooperation with political leaders	*Interest*: Financial interests, profit-maximization; support from political leaders. *Fears*: deterioration of cooperation with political leaders; limited benefits	Enforcement of illegal lease system for own benefit	Access through cooperation with political leaders	Limited	No, but opportunity for finan-cial and other benefits from supporting political leaders	Yes, but only in enforcement
BNP	Indirect role as ruling party			BNP allows continuation of illegal lease	Control as long as ruling party	Yes, control through allocation of means	Yes	Direct
AL	Indirect role, opposition party		AL banned lease system	Previous control	Not at present	Not at present	Indirect
Court	Conflict resolution	Resolution of the conflict	n.a.	Indirect	No	No	Resolution

Geographic, demographic and temporal distribution
The conflict occurs in Patkelbari, a small village divided by the Boushar canal in Chanda Beel. Chanda Beel spreads in three medium-sized administrative units within Gopalganj District, (a) Gopalganj Sadar *Thana*; (b) Muksudpur *Thana* and (c) Kashiani *Thana* (Huq and Alam, 2000). Chanda Beel encompasses 45 villages with a total population of approximately 58,000 people, of which 76% are Hindus, 17% are Muslims and 7% are Christians (Asia Foundation/BCAS, 1997)[27].

Temporal distribution
The lease for catching fish in the Canal is required between December and February. Apart from the seasonality in terms of access to fishing grounds, one can also observe political opportunism: support to the party in power.

Level and Intensity of Conflict

This conflict can be characterized as low level, only temporary violent. Although there was a clash in which several were wounded, this was reported as incidental. Although the conflict plays directly at *Thana* level and community level, the (abandoning of the) lease system was decided at national level.

Underlying Social, Political, Economic and Institutional Structures

Officially, the lease-system has been abandoned and the water bodies are supposed to have open access. In the recent past, members of the co-operative society of fisherman were able to lease the canal through secret tenders. In order to defend the rights of the poor fisherman, the former Awami League Government started a policy called *"Jal Jar Jola Tar"* to ban the canal lease tradition during their rule. Although officially banned, in reality, the lease system has not yet been abandoned, as it is profitable for both the local elite (businessmen and local politicians) and the police who seem to have established a 'joint venture'.

The case study on Chanda Beel shows that local elite (local political leaders) have invested in exclusion and control mechanisms to control the group, thereby contributing to their own profit. In this case they have been effective through cooperation with the police, who is supposedly the law-enforcing authority in the area. This case confirms Barnett's observation who stated that "the issue is not competition over scarce resources, but rather competition to gain [dominant] control over substantial income generating resources, or more equitable access to the spoils of resource extraction" (Barnett, 2003).

---

[27] Asia Foundation/BCAS. 1997. *Promoting Grassroots Participation through Advocacy on Environment and Natural Resources Management of Chanda Beel, Gopalganj.*

One can observe a failure of existing dispute resolution mechanisms which have been bypassed by the villagers by seeking intervention through politicians and by going to court. Conflict is clearly affected by institutional failure (in particular in relation to rule of law and enforcement of law), and unequal relations of resource control and distribution (benefit to supporters of the ruling party). The visit has been too brief to identify social structures which may have contributed to the conflict.

Impact of the conflict on Human Security

None of the two groups has been able to benefit from the lease they had to pay. The costs of conflict resolution by the court are relatively high. Both groups have gone to court, for which each groups had to pay 50,000 Taka, which equals 167 to 250 Taka per household. One of the groups raised this money from a village club deposit; the other group paid this by collecting money among the public.

CASE STUDY 2: INDUSTRIAL POLLUTION IN MOKESH BEEL,
SUB-REGIONAL LEVEL[28]

Kaliakoir *Thana* in Gazipur District is situated approximately 25 km north-east of Dhaka and located in the surroundings of Mokesh Beel, one of the largest wetlands in this area. Over the last ten years, several industries were established in the area and the number of industries has increased rapidly to more than 82. The establishment of industries caused serious environmental problems for the local communities, whose livelihoods and food supply are based on the availability and quality of natural resources for farming and fishing. Chowdhury and Huq (2005) indicate:

> "These [industries] are established in clusters for availability of infrastructure and utilities, market accessibility, easy disposal of effluent but there has been a lack of Government of Bangladesh policy on zoning of industries. These industries are mostly dominated by fully export oriented textile and dyeing factories. It has been observed that these industries are using the beel as a disposal ground for their industrial waste. As a result, water quality has gradually deteriorated to a level that is unsuitable for certain types of aquatic life. This is particularly true for Ratanpur khal, which is the conduit for waste from many of the factories to the beel and which in the dry season is fed only by industrial effluent" (Chowdhury and Huq, 2005).

---

[28] Information provided by Chowdhury and Huq (2005) from the Bangladesh Centre for Advanced Studies.

Five of the textile factories studied were found to contribute already, on yearly basis, several hundred metric tones of acetic acid, soda ash, glauber salt, common salt, and dyes to the system. Chowdhury and Huq indicate that "given the huge increase in the number of textile dyeing units in the area over the past three years, this figure is likely to have increased several fold" (Chowdhury and Huq, 2005). They expressed their particular concern about the high chemical oxygen demand (COD); the high biological oxygen demand (BOD), the level of pH and sulfide, and low dissolved oxygen levels (DO) (Chowdhury and Huq, 2005)[29]. Fishermen have indicated that the environmental pollution had a negative impact on fish stocks in the areas, that customers do not want to buy tainted fish caught by them, as a result they have to search for alternative ways of income generation. Farmers have expressed similar concerns about the crops they cultivate, both in terms of yields and in terms of quality (taste). The increased industrial activity in the area encouraged people to migrate to the area, which has resulted in increased domestic garbage and sewage contamination in surrounding water body. Not surprisingly, the combination of industrial pollution, poultry farm waste, lack of proper sanitation systems, and increased garbage and sewage contamination has intensified a number of health problems between 1994 and 2004, in particular skin diseases, dysentery, and diarrhea. The impacts of the industrial pollution on peoples' health and their livelihoods has not only created conflict between the community and local industries, but also between the community and relevant government departments who are not capable or willing to enforce existing legislation and policies. Furthermore, local communities hold banks responsible for supporting polluting industries through loans, without ensuring environmental standards (Chowdhury and Huq, 2005). Finally, the situation has also resulted in intra-household conflicts, in households where the sons or daughters of farmers or fishermen are employed by one of the polluting industries. The conflict between community and industries was highlighted during a workshop in September 1999, arranged by the Managing Aquatic Ecosystems through Community Husbandry (MACH) project, and the community has become active since then, by inviting environmental activists, and others, who were perceived as capable of giving more weight to their concerns or who were perceived as potential actors in contributing to resolution of the problems.

Identification of issues

The conflict is primarily over environmental degradation, caused by the use of natural resources as industrial waste dump. Additionally, there is

---

[29] See Chowdhury and Huq (2005) for the results on each of the parameters.

also some conflict over management and implementation of policies. The study by Nishat and Huq indicates that overlapping policy jurisdiction between the Water Resources Planning Organization (WARPO), the Department of Environment (DoE), and the Bangladesh Water Development Board (BWDB), as well as lack of endorsement of the National Water Policy by the Parliament, result in failure to implement existing policies and standards.

Identification of actors

The relationships between the actors are dynamic; these relationships change over time, as in the previous case study. The relationships between projects and several industries improved significantly during recent years; the interaction between communities and environmental activists has increased, and the relationships between the project staff and communities enable these communities to communicate their concerns to the industries through project staff. The relation between industries and communities is a complex one. On the one hand the industries employ community members for the production at their factories, and on the other hand, one can observe a relation between communities opposing the pollution caused by the industries. The relationship varies among industries willing to change their production processes and those not interested in changing anything.

This means that the scope for conflict resolution and prevention of further pollution has gradually improved over the years. However, part of this is nullified by the rapid increase in number of industries in the area (including those who are not willing to change their production processes or waste disposal).

Table 4 gives an overview of actors, their role, interests, fears, and access to various resources.

History and distribution of the Conflict

The conflict between communities and industries increased gradually since 1994, and was highlighted in September 1999. One can observe a seasonal variability in the level of water pollution (highest in the dry season, lowest in the wet season). The concentration of industries is provided in Figure 3.

Level and Intensity of the Conflict

This conflict can be characterized as low level, non-violent conflict. Relevant policies, such as the National Water Policy, the National Environmental

Table 4. Actor analysis of conflict in Mokesh Beel, Bangladesh

Actors	Role in conflict	Interests, fears	Access to, ownership over, use of natural resources	Access to and control over political resources	Access to and control over economic resources	Economic benefits from political position	Access to decision-making / implementation
Community	Access to clean and safe water for various purposes constrained	*Interests:* use clean water for domestic use; employment in industries; clean environment for new generations; health; *Fears:* increasing health problems; lack of drinking water; lack of income-generation; food insecurity	Distance to clean water for domestic purposes increased; use of polluted water likely to contribute to health problems	Yes, access to local political leaders, but no control, no guarantees	No other than their means of income	Limited	Slightly increased since 1999 through presence of projects, but relatively low
Farmers	Affected by reduction of yield	*Interests:* safe water for cultivation; same yield and quality; *Fears:* further reduction of yield and deterioration of quality	Farmers are sometimes forced to use polluted *beel* water due to low groundwater	Yes, access to local political leaders, but no control, no guarantees	No other than means of income	Limited	Low
Fishermen	Affected by reduction of fish stock and tainted fish	*Interests:* being able to fish in all water bodies; to continue as fishermen *Fears:* further reduction of fish stock, polluted fish; need to search alternative means of income	Access to 'clean' fishing grounds limited. Use of other fishing grounds likely to result in clashes	Yes, access to local political leaders, but no control, no guarantees	No other than means of income	Limited	Low
Environmental Activists	Highlighting environmental problems	*Interests:* saving the environment; cooperation with local communities	n.a.	Unclear	n.a.	Limited	Low
Sons / daughters employed by industry	Benefit from employment in industries	*Interests:* continued employment opportunities; clean environment; *Fears:* closing of industries and lack of employment	Similar to parents	Yes, access to local political leaders, but no control, no	No other than means of income	Limited	Low

Contd.

Table 4. *Contd.*

Actors	Role in conflict	Interests, fears	Access to, ownership over, use of natural resources	Access to and control over political resources	Access to and control over economic resources	Economic benefits from political position	Access to decision-making / implementation
				guarantees			
Industries	Contribute heavily to environmental pollution of water bodies	*Interests*: stay in business, earn foreign currencies, comply with buyers, criteria / *Fears*: loosing profit, closure	Uses water bodies as disposal ground for their industrial waste	Yes, access to local, regional and national political leaders	High	Good relations with political leaders may pay off very well	High
International buyers	Formulated inadequate standards in codes of conduct	*Interests*: improve image among customers through corp.resp. / *Fears*: loosing markets and customers due to scandals	n.a.	Unclear to what extent international buyers try to lobby with political leaders	High	Indirect economic benefits from social and environmental responsibility	Indirect
Department of Environment	Are not able/ willing to protect rights of citizens to sufficient supply of water of good quality	*Interests*: maintain water quality; enforcement environmental standards / *Fears*: loosing discretionary power vis-à-vis other departments	n.a.	Yes, access to political leaders; but not sufficient for endorsement of relevant policies	Low	Undefined	High
Projects	Conflict resolution and pollution prevention through cooperation with all actors	*Interests*: identification of efficient production process; less waste; / *Fears*: not being successful in convincing industries to change their production processes	n.a.	Yes, but not sufficient to get National Water Policy endorsed yet	Low	n.a.	Indirect

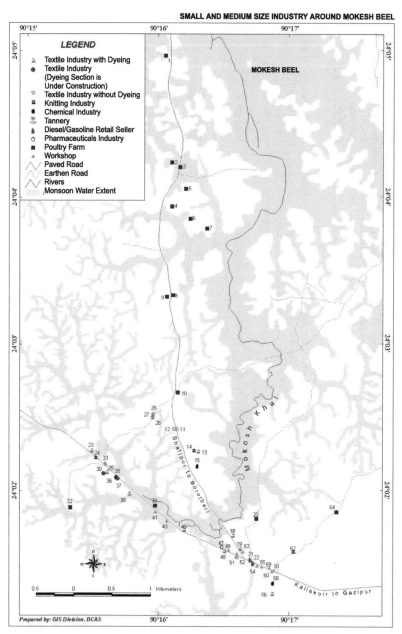

Fig. 3. Industries around Mokesh Beel (*Source:* Chowdhury and Huq, 2005)

Policy and the Environmental Conservation Act are decided (and should be endorsed) at national level. Specific provisions are made under the National Water Policy with regard to agro-chemicals, industrial effluents, and effluent disposal in common watercourses, set by WARPO together with DoE (Chowdhury and Huq, 2005).

Underlying Social, Political, Economic and Institutional Structures

The most relevant structures in this case are the economic priorities of the industries and Government of Bangladesh; and the institutional failure to implement existing legislation and procedures (described in more detail in Chowdhury and Huq, 2005). Communities blame government departments for lack of enforcement; and they blame banks for financing polluting industries.

Impact of the Conflict on Human Security

Human security is directly threatened by pollution (rather than by the conflict) in terms of food security and income security of fishermen, who are in need of finding alternative means of income generation. Human security is further threatened in terms of human health, even though the increase in health problems can not be attributed to industrial pollution only. In the unlikely event that the conflict would further escalate in a way that these industries have to close down, this would result in unemployment of people from the same communities affected by the pollution.

INTERNATIONAL LEVEL: FARAKKA DAM AND SHARED GANGES AGREEMENT

(i) Identification of issues and (ii) identification of key actors

Since the mid-seventies, the governments of India and Bangladesh[30] have played a key role in the dispute over the shared use of the Ganges River. The Ganges flows largely through Indian territory before it reaches Bangladesh and enters the Brahmaputra and Meghna rivers in Bangladesh (McGregor, 2000). In 1961, the Indian government announced its intentions to construct a barrage, the Farakka barrage. The perceived need for such an ambitious plan emerged from: (i) the large and increasing population

---

[30] The Joint Rivers Commission (JRC) was formed in 1972 by both governments to facilitate water related issues. Nepal remained on the sideline. Between 1982 and 1988, "Nepal showed keen interest in a tri-lateral initiative for resolving the water crisis … [but] India refused Nepal's involvement as a third party, emphasizing the bilateral nature of the negotiations between India and Bangladesh" (Tanzeema and Faisal, 2001, p. 18)

pressure of Uttar Pradesh, Bihar and West Bengal in India, (ii) the intensification of dry season agriculture, and (iii) the diversion of the Ganges flow into the Bhagirathi-Hooghly River to improve navigability of the Kolkata Port (Tanzeema and Faisal, 2001). In other words, growing populations, increasing demands (demand-induced scarcity) and siltation of the Kolkata Port (supply-induced vulnerability) resulted in the announcement and realization of adaptation measures. These adaptation measures created much concern for the Government of Bangladesh, as they realized that the diversion of water at Farakka, would have great environmental and economic impacts for Bangladesh (adaptation-induced vulnerability).

From the start, the Farakka Barrage was perceived to be a threat to the food security and livelihoods of approximately 30 million people in Bangladesh depending on water from the Ganges River during the dry season (McGregor, 2000; Tanzeema and Faisal, 2001). Since the barrage was not affecting the flow of the river during the rainy season, the debate was mostly about the dry season flow and the diversion of 40,000 cusecs from the Ganges River into the Bhagirathi-Hooghly River. The concern of Bangladesh (previously Eastern Pakistan) was, and still is, not surprising, given the strong dependence of Bangladesh on a regular and sufficient flow from the Ganges River, for its agricultural sector, its fisheries, navigation and salinity control (Mirza, 2002).

History and Temporal distribution of the Conflict

The conflict over the Ganges River goes back to the partition of India and Pakistan in 1947. Subedi (1999) refers to many quarrels, which emerged shortly after partition. The distribution of territory and water, which previously belonged to one territory, suddenly had to be adjusted to a different political reality. Shortly after the war between West and East Pakistan in 1971, India and Bangladesh signed a Treaty of Peace and Friendship, which facilitated the establishment a Joint Rivers Commission in 1972 (McGregor, 2000). In 1975, a temporary agreement (41 days) was concluded between India and Bangladesh. According to this agreement, Bangladesh would receive 75% of the flow for the period between 21 April and 31 May, and India was "allowed to divert 11,000 cusecs of flow into the feeder canal in April, and 16,000 cusecs in May 1975" (Subedi, 1999). In 1976 and 1977, unilateral withdrawal of water from the Ganges caused a major water crisis in the Southwest region of Bangladesh (Tanzeema and Faisal, 2001).

Another agreement (from 1977 to 1982) was concluded in 1977, according to which Bangladesh would receive 60% of the Ganges flow at Farakka.

This five-year treaty incorporated a guarantee clause[31] to Bangladesh, which was neither repeated in the Memorandum of Understanding of 1982 (for a period of 18 months) and the Memorandum of Understanding of November1985 (for 3 years), nor in the 1996 Agreement between India and Bangladesh. In 1988, negotiations came to a standstill. Between 1988 and 1996, the Government of India continued to withdraw water from the Ganges. Finally, in 1996, another Agreement was signed, this time for a time-span of 30 years[32].

Despite the agreements and treaties (1975, 1977-1982, and the 30-year Treaty of 1996), the average dry season flow at Harding Bridge in Bangladesh has dropped significantly in the post-Farakka period[33] (after 1975) when compared to the pre-Farakka period (Swain, 2001). Several authors (Swain, 2001; McGregor, 2000; Mirza, 2002) indicate that the 1996 agreement did not take the longer-term reduction of dry season discharge of the Ganges River into account, in other words, the simulated flows which were used for the 1996 agreement were not in correspondence with the real hydrological situation after 1996 (Swain, 2001; Mirza, 2002; Tanzeema and Faisal, 2001).

According to McGregor, four problems are still remaining: (a) there is little provision for augmenting the river, even though both India and Bangladesh see this as essential element; (b) there are severe data flaws; (c) topics such as flooding, environmental impacts and social mobilization are not mentioned; and (d) all treaties seem to be temporary solutions, without a permanent plan for augmentation, distribution and enforcement of the agreement (summarized from: McGregor, 2000).

Level and Intensity of the Conflict/Cooperation

This conflict can be characterized as low level, non-violent conflict at international level, involving primarily India and Bangladesh (former East

---

[31] "Under that Agreement, Bangladesh received two types of guarantees from India: (1) if the actual availability at Farakka of Ganges waters during the ten-day period should be lower or higher than the average flows given in the schedule annexed to the agreement, they would 'be shared in the proportion applicable to the period' and (2) if during a particular ten-day period the flows at Farakka diminished to such a level that Bangladesh's share was lower than 80% of the value shown in the schedule annexed to the agreement, 'the release of waters to Bangladesh, during that 10-day period shall not fall below 80 percent of the value shown'" (Subedi, 1999, p. 960).

[32] For an overview of historical phases of Indo-Bangladeshi Negotiation, see Tanzeema and Faisal (2001, pp. 17-18).

[33] The decline in the average dry season flow is estimated to be approximately 46,433 cusecs, which is a nearly 51% reduction in flow (Tanzeema and Faisal, 2001, p. 23).

Pakistan)[34]. The intensity of cooperation seemed to be strongly influenced by the political climate in both countries. Subedi (1999) indicates how the political climate affected the willingness of both governments for negotiations:

> "During most of the intervening years, Bangladesh was suffering from political instability and India was going through a period of reckoning and reshaping its relations with its immediate neighbors. As a result, by 1996, there was a marked change in New Delhi's attitude toward these states. The political climate in Bangladesh as well had become conducive to cooperation with India. Sheikh Hasina Wajed, known for her friendly attitude toward India, had been elected as Prime Minister" (Subedi, 1999).

Despite the positive changes in the political climate the negotiation process was prolonged due to the 'non-compromising attitude of the technical members of the Joint Rivers Commission" who were debating minor technical details without responding to the need for a settlement (Tanzeema and Faisal, 2001). Only after technical members were temporary withdrawn from the immediate negotiation process, both Governments were able to reach agreement in 1996 (personal notes, Dhaka, 2005).

Underlying Political Structures

The willingness of Bangladesh and India to negotiate was not only influenced by the national political climate, but also by their position as lower- and upper-riparian state. As stated by McGregor: "India understandably assumes that it is in a better position to get what it wants if it works bilaterally. It is the bigger nation, it is the upper-riparian, and therefore it can push to get the domestic aims that it sees as vitally important. Bangladesh is also trapped in this way of thinking. It has little option but to attempt to pursue bilateral gains, as it has neither the strength nor the riparian advantage to unilaterally gain what it wants" (McGregor, 2000).

Impact of the Conflict on Human Security

An estimation of the impact of the construction of the Farakka barrage on water availability has been given by Tanzeema and Faisal (2001) and Mirza (2002). The impact of the conflict itself can be estimated, first of all, by looking at the periods not covered by treaties and agreements, such as the

---

[34] According to Tanzeema and Faisal, between 1982 and 1988, "Nepal showed keen interest in a tri-lateral initiative for resolving the water crisis ... [but] India refused Nepal's involvement as a third party, emphasizing the bilateral nature of the negotiations between India and Bangladesh" (Tanzeema and Faisal, 2001, p. 18)

period of 1976 and 1977, when the southwest region of Bangladesh faced a major water crisis. Secondly, one would have to look at the actual implementation and enforcement of the 1996 agreements, and the impacts of water shortage on the economy and environment of Bangladesh and India.

CONCLUSIONS

The environment-conflict debate evolves around the question to what extent natural resource scarcity and environmental degradation play a role in the escalation of armed conflict. Given the impact of climate change on Bangladesh, the question was adjusted to the question whether climate change is likely to contribute to future violent conflicts [between countries]. We concluded that this remains highly uncertain, as future conflict is likely to operate through complex environmental, social and political processes. Poor implementation and enforcement may aggravate existing tensions.

When looking at the conflict from the first case study, one can see elements of both the grievance and greed theory, although it provides only limited explanation. Whereas the vivid environment-conflict debate has concentrated on the potential for armed conflict, the three case studies used in this chapter illustrated how *non-violent* conflict evolves, in situations when the observed problems directly affects people's health, their livelihoods, and their security. This resembles the observation of Barnett, "The issues that should be of more concern ... are the day-to-day insecurities associated with the erosion of individual and group welfare and resilience" (Barnett, 2000).

The first case showed that the conflict was not only about access to fishing grounds, but also to feelings of unfair treatment and relative deprivation (through encroachment, manipulation and thuggery). When looking at access to, and ownership of natural resources, it is therefore important to realize how competing mechanisms interfere with formal regulations. The villagers expressed their grievance about the rent-seeking nature of political leaders and the corruption from the police. At the same time, however, the two communities use the same underlying political and institutional structures to maximize their benefits, and use both legal instruments and political mechanisms to get redress.

The conflict in Chanda Beel was taken higher up by the communities themselves, whereas the communities in the second case study seemed to play a less active role in the conflict. From a rational choice perspective, the relative resigning attitude of communities in the second case study can be explained by their limited opportunity structure. Given the poor record of enforcement of environmental standards, they cannot expect support from relevant government departments. They are in a poor social, economic and political position to fight the industries causing environmental degradation. This is further complicated as part of the community (those employed by the industries) have an interest in continuation of industrial activity in the region. The third case study of the dispute over the water flow from the Ganges, between India and Bangladesh, is an example of the Homer-Dixon theory which concentrates on the potential contribution of scarcity to conflict. The distinction between demand-induced, supply-induced scarcity and conflict related to adaptation measures can be successfully applied in this case.

ACKNOWLEDGEMENTS

This chapter could not have been written without the support from the Bangladesh Centre for Advanced Studies (BCAS), in particular Dr. Atiq Rahman; Dr. Saheemul Huq, Dr. Sharif, Mrs. Nishat Chowdhury, Mr. Belayat Hossain, Mrs. Olena Reza, Mr. Bhidan, and Mr. Saifuddin and information from Mrs. Engr. Tanzeba Huq.

REFERENCES

Agrawal, A. 2001. 'Common property institutions and sustainable governance of resources' in: World Development, 29-10, pp 1649-1672.
Austin, A., Fischer, M. and Ropers, N (eds.). 2004. Transforming Ethnopolitical Conflict: *In:* The Berghof Handbook. Berghof Research Center for Constructive Conflict Management. Wiesbaden.
Barnett, J., 2000. Destabilising the environment-conflict thesis. Review of International Studies 26 (2), 271–288.
Barnett. J. 2003. 'Security and Climate Change' *In:* Global Environmental Change, Vol. 13, 7-17.
Bennett, E.; Neilanda, A; Anang, E.; Bannerman, P.; Rahman, A.A.; Huq, S.; Bhuiyac, S.; Day, M.; Fulford-Gardiner, M.; Clerveaux, W. 2001. Towards a better understanding of conflict management in tropical fisheries: evidence from Ghana, Bangladesh and the Caribbean. Marine Policy, Vol. 25(2001), 365-376.
Buchanan, J. 1965. An economic theory of clubs. Economica, 32 (125), 1-14.
Chadwick, M. T. 2000. Impacts of the 1998 Flood on Floodplain Livelihoods. *In:* Clemett, M.T. et al. 2000. People's Livelihoods at the Land-Water Interface: Emerging Perspectives on Interactions between People and the Floodplain Environment; Symposium Proceedings, Dhaka, January 2000, 21-31.
Chowdhury, N. and Saleemul Huq, 2005. Conflict in areas of industrial pollution. Paper prepared for the Training of Trainers Workshop on Conflict Resolution and Negotiation Skills for IWRM, BCAS, Dhaka; 1-5 August 2005.
Clemett, M.T. et al. 2000. People's Livelihoods at the Land-Water Interface: Emerging Perspectives on Interactions between People and the Floodplain Environment; Symposium Proceedings, Dhaka, January 2000.
Collier, P. 2000. Economic Causes of Civil Conflict and Their Implications for Policy. The World Bank, Washington.

Davies, S. 1993. Are Coping Strategies a Cop Out? IDS Bulletin, 24 (4), 60-72.

Department for International Development (DfID), undated. Sustainable Livelihoods Guidance Sheets (1.1), www.livelihoods.org, accessed on 22 August 2005.

Dessler, D. 1994. How to sort causes in the study of environmental change and violent conflict. *In* Environment, Poverty, Conflict. Graeger, N. and D. Smith (eds), International Peace Research Institute. Oslo.

Fumerton, M. 2005. Notes for lectures on conflict management. Unpublished.

Gleditsch, N. 1998. Armed conflict and the environment: a critique of the literature. Journal of Peace Research 35 (3), 381–400.

Gleick, P.H. 1993. Water and Conflict: Fresh Water Resources and International Security. International Security, 18(1), 79-112.

Goodhand, J. and Hulme, D. 1999. From Wars to Complex Political Emergencies: Understanding Conflict and Peace-building in the New World Disorder. Third World Quarterly, 20(1), 24.

Homer-Dixon, T. 1991. On the threshold: environmental changes as causes of acute conflict. International Security 16 (2), 76–116.

Homer-Dixon, T. and Blitt, J. (eds) 1998. Ecoviolence: links among environment, population, and security. Lanham, MD: Rowman and Littlefield.

Homer-Dixon, T. 1999. Environmental Scarcity and Violence. Princeton University Press, Princeton. USA.

Huntington, S.P. 1998. The Clash of Civilizations and the Remaking of the World Order. Simon and Schuster, New York. USA.

Kaplan, R.D. 1993. Balkan Ghosts: A Journey through History. MacMillan, London.

Kaplan, R.D. 1994. The coming anarchy: how scarcity, crime, overpopulation and disease are rapidly destroying the social fabric of our planet. Atlantic Monthly, February 2004, pp. 44-76.

Kliot, N. et al. (2001) Institutions for management of transboundary water resources: their nature, characteristics and shortcomings. Water Policy, 3, 229-255.

Klugman, J. 1999. Social and Economic Policies to Prevent Complex Humanitarian Emergencies: Lessons from Experience. Helsinki: United Nations University World Institute for Development Economics Research.

Klugman, J.; Neyapti, B; Steward, F. 1999. Conflict and Growth in Africa. Vol. 2: Kenya, Tanzania and Uganda. OECD, Paris.

Le Billon, P. 2001. The political ecology of war: natural resources and armed conflicts. Political Geography. 20, 561-584.

Mbonile, J.J. 2005. Migration and intensification of water conflicts in the Pangani Basin, Tanzania. Habitat International, 29, 41-67.

McGregor, J.M. 2000. The Internationalization of Disputes over Water: The Case of Bangladesh and India. Paper presented at the Australasian Political Studies Association Conference, ANU, Cranberra, 3rd-6th October 2000.

Mirza, M.M.Q. 2002. The Ganges water-sharing treaty: risk analysis of the negotiated discharge'. Int. J. Water, 2(1), 57-74.

Mirza, M.M.Q. 2002a. Global warming and changes in the probability of occurence of floods in Bangladesh and implications. Global Environmental Change. 12 (2002) 127-138.

Molen, I. van der and Hildering, A. 2005. Water: cause for conflict or co-operation? ISYP Journal on Science and World Affairs 1 (2): 133-143.

Muscat, R.J. 2002. Investing in Peace; How Development Aid Can Prevent or Promote Conflict, Sharpe Inc. Armonk, New York; London, England.

Neefjes, K. Environments and Livelihoods: Strategies for Sustainability; Development Guidelines, Oxfam.

Ohlsson, L. 2000. Water Conflicts and Social Resource Scarcity. Physics and Chemistry of the Earth, Part B: Hydrology, Oceans and Atmosphere. 25 (3), 213-220.

Olson, M. 1965. The Logic of Collective Action, Harvard University Press.

Ostrom, E. 1999. Coping with the tragedy of the Commons, workshop in Political Theory and Policy Analysis, CSIPEC, Indiana University, Bloomington.

Ragab, R. and Prudhomme, C. 2002. Climate Change and Water Resources Management in Arid and Semi-arid Regions: Prospective and Challenges for the 21st Century. Keynote Paper. Biosystems Engineering. 81(1), 3-34. Silsoe Research Institute.

Requier-desjardins D. undated. Produced Common Pool Resources, Collective Action and sustainable local development: the case of food-processing clusters. (C3ED UMR IRD/ UVSQ), est. 2004.

Smith, D. 2004. Trends and causes of armed conflict, *In*: Transforming Ethnopolitical Conflict: the Berghof Handbook. Austin, A; Fischer, M.; Ropers, N (eds.). Berghof Research Center for Constructive Conflict Management. Wiesbaden.

Soussan, J.G. 2000. Leeds/BCAS Research into Sustainable Local Water Resources Management – Meeting Needs and Resolving Conflicts *In:* Clemett, M.T. et al. 2000. People's Livelihoods at the Land-Water Interface: Emerging Perspectives on Interactions Between People and the Floodplain Environment; Symposium Proceedings, Dhaka, January 2000, pp. 1-8.

Subedi, S.P. 1999. Hydro-diplomacy in South Asia: the Conclusion of the Mahakali and Ganges River Treaties. The American Journal of International Law, 93 (4), 953-962.

Swain, A. 1993. Conflicts over water: the Ganges water dispute. Security Dialogue 24 (4), 429-439.

Swain, A. 2001. Water wars: fact or fiction? Futures, 33, 769-781.

Tanzeema, S. and Faisal, I.M. 2001. Sharing the Ganges: a critical analysis of the water sharing treaties. Water Policy, 3, 13-28.

Toset, H.P.W. et al. 2000. Shared rivers and interstate conflict. Political Geography, 19, 971-996.

Warner M. 2000. Conflict management in community-based natural resource projects: experiences from Fiji and Papua New Guinea. Working Paper, 135 edn. London: ODI. *Cited in* Bennett et al. 2001, p. 367.

Wolf, A.T. 1998. Conflict and cooperation along international waterways. Water Policy, 1, 251-265.

Wolf, A., 1999. 'Water wars' and water reality: conflict and cooperation along international waterways' *In*: Environmental Change, Adaptation, and Security. Lonergan, S. (Ed.), Kluwer Academic Publishers, Dordrecht, pp. 251–265.

Wolf, A.T., Yoffe, S.B. and Giordano, M. (2003) International waters; identifying basins at risk. Water Policy, 5(1), 29-60.

Yoffe, S.B., Wolf, A.T. and Giordano, M. 2000. Conflict and cooperation over international freshwater resources: indicators and findings of the basins at risk project. *In*: Basins At Risk: Conflict and Cooperation Over International Freshwater Resources. S.B.Yoffe et al. pp. 64-12.

Africa

6

The Hydropolitics of Cooperation: South Africa during the Cold War

Anthony Turton

Gibb SERA Chair in IWRM
Environmentek
Council for Scientific and Industrial Research (CSIR)
African Water Issues Research Unit (AWIRU)

INTRODUCTION

The hydropolitics literature is strongly biased towards river basins in conflict. This is informed by the logic that is inherent to the Environmental Scarcity School of Thought, which posits a causal linkage between the scarcity of a strategic natural resource and the propensity towards interstate conflict, usually of a violent nature (Bulloch and Darwish, 1993; de Villiers, 1999; Gleick, 1991; 1993; Homer-Dixon, 1991; 1994; 1996; Irani, 1991; Klare, 2001a; 2001b; Mathews, 1989; Myers, 1993). Admittedly this logic is seductively simple, but it is also flawed, at least in the case of Southern Africa and certainly in the case of water. While the Southern African region was a hotbed of violent conflict during the second half of the twentieth century, this was closely linked to the political dynamics of the Cold War, and nowhere is there evidence that water was ever causally linked as a driver of that conflict (Turton, 2005a). While it is true that in some cases water infrastructure became a target of war, this was for clearly defined tactical reasons, relevant only to the logic of the commander of a combat unit as he (they are always men) seeks to shape the battlefield in a way that favors his strategic objectives. Southern Africa as a region therefore makes an interesting case study of water and conflict, for four good reasons. Firstly, there were many armed conflicts in the region. Secondly, some of the international river basins in the region were the theatres of substantial military confrontations (Turton, 2005a), with at least one being the location of the largest single military defeat since World War II (WW II) (Turner, 1998). Thirdly, South Africa was the regional hegemon (Turton, 2005b), with a history of projecting its power beyond its own borders (Bernstein and Strasburg, 1988). Finally, four of the most developed countries in the Southern African Development Community (SADC) Region have reached a point where endemic water scarcity is starting to pose constraints on their future economic development potential (Ashton and Turton, in press; Turton, 2003). One of these countries is South Africa with an alleged history of resource-driven conflict (Percival and Homer-Dixon, 1998; 2001). We therefore have a classic case in which all the essential elements of environmental conflict are in place. The question therefore arises, to what extent the Environmental Security literature is capable of explaining and predicting the political dynamics that have actually occurred over the last quarter century? This chapter will argue that the Environmental Security pundits have got it wrong, at least in the case of transboundary water resources in the context of South Africa.

BRIEF HYDROPOLITICAL HISTORY OF SOUTH AFRICA

South Africa is a highly water-stressed country, so it has a long and complex hydropolitical history (Turton et al., 2004). It is useful to study this history by using a concept that is known as the hydraulic mission, which is the official policy that seeks to mobilize water and improve the security of supply as a foundation for social and economic development (adapted from Reisner, 1993; Turton and Meissner, 2002).

FIRST PHASE OF THE SOUTH AFRICAN HYDRAULIC MISSION

The earliest record of water as a potential constraint to human development dates back to two books that were written by a botanist in the 1870s, both describing the conditions of aridity and suggesting dams as a remedy (Brown, 1875; 1877). This can be considered the birth of the hydraulic mission, with the first known technical drawing of a dam on the Orange River coming from the pen of Thomas Baines in 1885 (see Figure 1) (Turton et al., 2004). Baines went on to become a famous explorer and naturalist, with many species and geographic places named after him.

The hydraulic mission could not become manifest as a policy at this stage because the frontier had yet to be closed in the new territories (Turton et al., 2004). A series of border wars ensued, each expanding the frontier over time. Non-renewable natural resources played a major role in this process, notably copper, diamonds and gold (Turton et al., 2006).

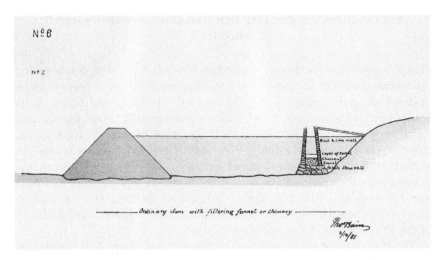

Fig. 1. The earliest known technical drawing of a dam by Thomas Baines dated 1885 (Turton et al., 2004).

The resource war that was the most closely linked with South Africa's hydraulic mission was driven by gold, however, with the discoveries around the Witwatersrand areas in 1886 being a major turning point (Cunningham, 1987). The British Government coveted this gold, and immediately set about planning how to gain access to it (Evans, 1999; Farwell, 1999; Nasson, 1999; Porch, 2000). At first they tried to overthrow the legitimate government of the Zuid-Afrikaanse Republiek (ZAR) – one of the first sovereign states in Africa that had been created as the result of the Westphalian State system (Turton et al., 2004) – by means of a band of mercenaries under the leadership of Dr. Jamieson (Longford, 1982; Seymour Fort, 1908). When the Jamieson Raid failed, an alternative plan was developed. This evolved into the Anglo-Boer War (Evans, 1999; Farwell, 1999), also known as the South African War (Nasson, 1999), in which the ZAR was overthrown militarily using a brutal system of concentration camps that was supported by a scorched earth policy that virtually decimated the entire Afrikaner nation (Lee, 2002; Raath, 1999; Spies, 1977; Van Rensburg, 1980). It is the contention of the author that it was this brutality that fed into the Afrikaner psyche that ultimately led to the inhumanity that was perpetrated during the years of Apartheid (Turton et al., 2006). Barber (1999) has concluded that the Anglo-Boer War was a resource war triggered by the desire of the British to gain control of South African gold.

SECOND PHASE OF THE SOUTH AFRICAN HYDRAULIC MISSION

Once the goldfields had been captured and the Afrikaner nation subjugated, the British set about mobilizing water resources with the express purpose of developing the mines. This was the birth of the second phase of the South African hydraulic mission, which was essentially internally-focused, based on the desire to develop a high assurance of supply. Significantly, this could only occur after the frontier had been closed, because it was only at that stage that sufficient national security existed to support economic development in a sustainable way. The chosen vehicle for this hydraulic mission was the Rand Water Board, which immediately set about developing the necessary hydraulic infrastructure to sustain the goldfields of the Witwatersrand (Tempelhoff, 2003).

The British governed the Union of South Africa from 1910 until 1961. This consisted of the two former British colonies of the Cape and Natal, which had been forcefully united with the two Boer Republics of the ZAR and the Orange Free State (OFS). During this time water resource management was seriously considered to be a fundamental aspect of the future successful development of the natural resource wealth of the country. During the First World War, what was known as German South West Africa was captured by allied forces under South African command. This was given to South Africa to administer as a League of Nations Mandated Territory. Immediately this expanded the territorial area under South African control, albeit *via* British tutelage in the form of the Union of South Africa.

The earliest record of South African interest in transboundary rivers goes back to a plan in 1906 when the Victoria Falls and Transvaal Power Company considered the importation of water from the Zambezi River into South Africa (Scudder et al., 1993). Actual transboundary river basin management dates back to 1926 when the so-called *First Use Agreement* was reached between South Africa and Portugal on the use of the Cunene River (Treaty, 1926a). This took place immediately after the border had been demarcated between what is now Namibia and Angola (Treaty, 1926b). Similar agreements were reached on the border delimitation between what is now Mozambique and South Africa (Treaty, 1926c). Some uncertainty still existed regarding the border between Namibia and Angola as it pertained to the Okavango River basin, so a second border treaty was reached in 1928 (Treaty, 1928). The border agreements were strengthened further during the 1930s and 1940s with various treaties between Portugal, Britain and South Africa (Treaty, 1931a; 1931b; 1933; 1943). This suite of agreements set the legal foundation for future cooperation in the management of transboundary river basins in South Africa.

In 1948 the National Party (NP) won an election victory, which many Afrikaners saw as being some vindication for their humiliation at the hands of the British during the Anglo-Boer War (Turton, 2005b). Faced with the massive unemployment of the Afrikaner people, a fate that had arisen through the scorched earth policy and the decimation of families in the concentration camps (in which more civilian women and children died than combat soldiers on both sides), the NP decided to launch an ambitious job-creation program. This was designed to favor Afrikaners in an aggressive affirmative action campaign, through a system that evolved over time into what the world now knows as Apartheid. There were two elements to this job-creation strategy. The first was focused on agriculture, with large public works being launched to construct hydraulic infrastructure like dams, canals and irrigation systems. Over time this led to the forced removal of many Black people from the land, as Afrikaners claimed it for themselves, with the support of the NP government. The second was focused on industrialization, which was centered on mining, but later embracing heavy industry like steel mills and the energy sector. Both of these needed water, and it was soon discovered that South Africa simply did not have enough to meet the future aspirations of the NP government. This spurred the early tentative reconnaissance studies that were designed to determine where in Africa future water resources could be secured. The work by Ninham Shand (1956) arose from this initiative, which became the foundation for the future Lesotho Highlands Water Project (LHWP). Similar studies were done on the Okavango and Zambezi River basins (Borchert, 1987; Borchert and Kemp, 1985; Davies et al., 1993; Midgley, 1987; Scudder, 1989; Scudder et al., 1993; Trolldalen, 1992).

THIRD PHASE OF THE SOUTH AFRICAN HYDRAULIC MISSION

The 1961 Commonwealth Conference in London saw South Africa come under heavy criticism for its Apartheid policies (Geldenhuys, 1984). This led ultimately to the expulsion of South Africa, which the Prime Minister at the time (H.F. Verwoerd) spun as a victory for the Afrikaner nation and final vindication of their defeat during the Anglo-Boer War. Associated with the de-colonization of South Africa in 1961, was an incident that came to be known as the Sharpeville Massacre, in which sixty-nine people were killed and one hundred and eighty wounded. The significance of these two events – the expulsion of South Africa from the Commonwealth and the Sharpeville Massacre – was encapsulated in the Afrikaner spirit of independence, which took the form of a virulent nationalism that had at its heart the desire to bring prosperity by making the desert bloom. This potent form of ideology grew from the fusion of grand socio-economic aspirations with the perceptions of threat that later came to be known officially as the *Total Onslaught*, resulting in a *puissance*-styled political culture that rose to prominence during the Cold War. Closely associated with this was the birth of the armed struggle and the repression of the various liberation movements, including the incarceration of political leaders like Nelson Mandela, Walter Sisulu and Govan Mbeki.

Immediately after the Sharpeville Massacre, there was massive loss in investor confidence with the flight of foreign capital from South Africa. This prompted the NP government to re-visit the hydraulic mission, with the result that the Orange River Project (ORP) was born. This project was extremely ambitious, taking water from the Orange River and diverting it through a major watershed *via* a 5.35-m diameter, 82.8-km long delivery tunnel, into the Fish River, and then cascading it across a smaller watershed into the Sundays River (Conley and van Niekerk, 1998; Turton et al., 2004). The whole project was profoundly political in nature, because at its heart was the need to develop an area that was the political home of the African National Congress (ANC), in the hope that economic employment would become sufficient incentive to young men not to join the armed struggle. When speaking of the ORP, Prime Minister Verwoerd hailed it as a triumph of Afrikaner independence and technical ingenuity, claiming that it was a symbol of the determination of the white civilization in Southern Africa to stay on the African continent (Turton et al., 2004).

The ORP was the first manifestation of an aggressive hydraulic mission, with rhetoric closely associating economic development with the construction of large dams. So, for example, mention was made in speeches of the value of the Boulder Dam on the Colorado River; the Volta River Project in Ghana; and the Kariba Dam on the Zambezi River, suggesting that the ORP would "transform the desert into a paradise" and eventually become much larger than the Tennessee Valley Authority (TVA) in the USA (Turton et al., 2004). Closely associated with this process was a spate of new international agreements on the management of transboundary rivers (Ashton et al., 2005; Treaty, 1964; 1969a; 1969b). These treaties all resulted in major hydraulic infrastructure being built on international rivers in which Portugal had an interest – the Cunene and the Zambezi.

On the political front, the armed struggle was gaining momentum since its birth after the Sharpeville Massacre, with the first guerrilla incursions taking place around the Caprivi Strip in present day Namibia (Frankel, 1984). The South African government responded in a way that came to define the rest of the Third Phase of the Hydraulic Mission period. On the one had military forces were committed to what became known as "the border" in a *puissance*-styled approach to the problem (see Figure 2).

On the other hand, a concerted effort was made by the South African government to engage African leaders in a process of dialogue and détente.

Fig. 2. South African armed forces were committed to a counter-insurgency role when guerrilla forces infiltrated into South-West Africa in 1966. This shows a Special Operations unit known as Koevoet being supported by a counter insurgency unit of the South African Defence Force (SADF). (Turton, 2006)

Spearheading this was General Hendrik van der Bergh, who was tasked in 1968 with the establishment of the first South African intelligence structure that came to be known as the Bureau of State Security (BOSS) (NIS, 1994). Lang Hendrik (tall Hendrik) as he was known, was fascinated by the concept of power as control, which is captured in the French word of *pouvoir*. Rather than confronting an opponent with military force (power as might or force and translated as *puissance* in French), General van der Bergh was of the opinion that more formidable power could be wielded through control and engagement (Turton, 2006). This *pouvoir*-styled approach therefore became the alternative form of diplomacy, mostly conducted in secret by BOSS, because of the absence of formal embassy representation. An early success took the form of engagement with Chief Leabua Jonathan in his capacity as Head of State in the newly-independent Lesotho. This period therefore saw South African foreign policy crystallize around what is easiest to understand as the carrot of inducement to cooperate peacefully (*pouvoir*-styled politics) and the stick of sanction for non-cooperation (*puissance*-styled politics). Significantly, the focus of this new approach became water resource management for two basic reasons. Firstly, the first guerrilla incursions took place in areas where hydraulic infrastructure was being built on shared rivers in the Cunene and Zambezi

River basins. Secondly, the Hoover-styled New Deal that navigated the American economy through the Great Depression was seen as a model for industrial development with large hydraulic infrastructure as key components.

During 1971 an agreement was reached with Mozambique on the construction of Massingir Dam in the Limpopo Basin (Treaty, 1871). In 1979 serious consideration was given by South Africa to the importation of water *via* the Thamalakane and Boteti Rivers in the Okavango/ Makgadikgadi Basin, to the industrial hub of the Transvaal (Borchert, 1987; Midgley, 1987; Scudder et al., 1993). These plans were frustrated, however, so alternatives were considered and the earlier Ninham Shand (1956) proposal was revisited with greater interest. During 1980 an agreement was reached between South Africa and Swaziland on the construction of Pongolapoort Dam in the Maputo River Basin (Treaty, 1980). Negotiations around this event led to the realization that a river basin commission was needed, so the Tripartite Permanent Technical Committee (TPTC) was established in 1983 between South Africa, Swaziland and Mozambique (Treaty, 1983).

There was great political tension in the region, however, with the Rhodesian Bush War, also known as the Second War of Chimurenga (Frederikse, 1982), coming to an end. The Lancaster House Conference in 1979 laid the foundation for the cessation of hostilities in Rhodesia. South African security force officials were expecting Bishop Abel Muzorewa to win the elections in Zimbabwe, so it came as a great surprise when Robert Mugabe emerged as the first Zimbabwean Head of Government. To understand the significance of this, one needs to travel back to 1974 when Prime Minister John Vorster made a speech in South Africa that made reference to a power block of states in Southern Africa (Geldenhuys, 1984). This idea was developed further when he spoke of a constellation of politically independent states maintaining close economic ties. When P.W. Botha came to power in South Africa, he morphed this idea into what he called the Constellation of Southern African States (CONSAS), which became the basis of his foreign policy (Geldenhuys, 1984). Foreign Minister Pik Botha (no relation to P.W. Botha) subsequently announced that this CONSAS vision embraced some forty million people south of the Cunene and Zambezi Rivers, all joining forces in a common security, economic and political approach. In short, CONSAS was a non-aggression pact using the carrot of development as a form of inducement, linking water resource management to national security. Central to this was the mobilization of water from transboundary river basins, specifically the Zambezi and Okavango (Midgley, 1987). Great was the shock in South Africa when Robert Mugabe came to power, because he immediately announced his rejection of the CONSAS proposal. Instead, Prime Minister Mugabe proposed that the so-called Front Line States (FLS) (Bernstein and Strasburg, 1988) join forces to fight capitalism, colonialism and racism, using the Southern African Development Coordination Conference (SADCC) as the vehicle. This grouping was quickly dubbed the counter-constellation (Baynham, 1989; Conley and van Niekerk, 1998; Geldenhuys, 1984). The establishment of SADCC was thus a response to the perceived threat of South African destabilization in the region. It was against this background that subsequent developments must be interpreted. This series of events gave rise to the *Total National Strategy* approach that saw South African foreign policy become captive to the State Security Council (SSC), which had an all-consuming security focus to it (Frankel, 1984). This securitized water resource management, given the strategic importance of rivers to the future economic security of South Africa.

With Zimbabwe in the lead of the FLS, Mozambique quickly threw in its support for the struggle against capitalism, colonialism and racism, offering safe havens to guerrilla forces of the various Liberation Movements, many of whom went into exile after the Soweto Riots in June 1976. In 1980 the armed struggle intensified when the ANC announced an increase in military engagement (Gutteridge, 1990). A series of guerrilla raids took place inside South Africa as a result of this (Gutteridge, 1981), which the SSC took as evidence of a *Total Onslaught* (Geldenhuys, 1984; Simon, 1991; Turner, 1998). Arising from this was the first cross-border military action by South African Special Forces, with a raid on ANC bases at Matola near Maputo in Mozambique (Geldenhuys, 1984; Gutteridge, 1981). This was followed in 1982 with raids into Lesotho (Gutteridge, 1983). A car bomb was detonated in Pretoria in 1983 with significant casualties, as part of the intensification of the ANC-led campaign to liberate South Africa, which served to polarize society further. This locked South Africa into a bitter relationship with the FLS, sowing the seeds of fear and mistrust because of the military activities of the liberation movements and security force reprisal. Right in the heart of this struggle was Mozambique, with military attacks on the electricity power lines from the Cahora Bassa Dam rendering the project useless. As a response to this, South Africa started giving support to the small rebel group operating inside Mozambique called the Resistencia Nacional de Moçambique (RENAMO) (Turner, 1998). This became a feature of the Mozambique Civil War, to the extent that Zimbabwe became embroiled in the military defence of the Beira Corridor from 1882 onwards (Turner, 1998). This placed such stress on the Mozambique Government that they entered into negotiations leading to the non-aggression pact known as the *Nkomati Peace Accords*, being signed in 1984 (Treaty, 1984a). This was immediately followed by a treaty on the rejuvenation of the Cahora Bassa Project, which brought much-needed revenue into the coffers of the Mozambique Government (Treaty, 1984b).

This period of heightened tension created a major dilemma for South Africa. On the one hand, the government wanted to cooperate peacefully with its neighboring states using water resource infrastructure as a vehicle. On the other hand, Zimbabwe was creating a major stumbling block with its belligerent rhetoric and confrontational-style politics. This saw what can only be described as a hybrid-type solution being generated. In February 1983, the Tripartite Permanent Technical Committee (TPTC) was established between South Africa, Mozambique and Swaziland (Treaty, 1983). The stated purpose of the TPTC was to make recommendations on the management of water shortages that were being experienced in the Limpopo, Incomati and Maputo River basins (Turton, 2004). This meant that the TPTC was more than an RBO because it was designed to manage three different river basins, which is something that is unusual in the management of transboundary rivers. Significantly however, Zimbabwe was excluded from this arrangement even though it is a riparian on the Limpopo. The strategic objective was to break the linkage between Zimbabwe and Mozambique and therefore try to woo the latter into a cooperative relationship with South Africa. Commentators say that the TPTC was a failure (Heyns, 1995; Ohlsson, 1995; Vas and Pereira, 1998) – which it was in the narrow sense of a water resource management institution – but they do not recognize the strategic context in which it was operating. While it failed as an RBO designed to manage three transboundary river basins – something way beyond the scope of our contemporary understanding of the management of transboundary rivers – it played a significant role in the diplomatic re-engagement between South Africa and Mozambique at a time when tensions were at their very highest, immediately prior to the signing of the *Nkomati Peace Accords*. This subtle fact indicates the significance of the hydropolitical dimension to international politics in Southern Africa and shows the role that was played by water resource management during a time of Cold War tension and open military conflict.

After the signing of the *Nkomati Peace Accords*, a spate of treaties relating to water resource management was signed. Agreement was reached between South Africa and Lesotho on the Lesotho Highlands Water Project (LHWP) (Treaty, 1986a; 1986b). Almost concurrently, an agreement was reached between all of the riparian states on the establishment of the Limpopo Basin Permanent Technical Committee (LBPTC) (Treaty, 1986c). A year later an agreement was reached between South Africa and Namibia on the co-development of the lower Orange River (Treaty, 1987). In 1988 an agreement was reached between the Bantustan Government of Bophuthatswana and the Water Utilities Corporation (WUC) of Botswana on the supply of water from the Molatedi Dam in South Africa to Gaborone (Treaty, 1988). The politics of the latter agreement is very interesting, because Botswana did not want to recognize Bophuthatswana as a sovereign entity, so the agreement was effectively reached between the WUC and the South African Government.

Lessons learned from the LHWP informed the decision-making process in other river basins. In the Incomati for example, the Komati Basin Water Authority (KOBWA) was created by agreement reached in 1992 (Treaty, 1992a). This led to the creation of the Joint Water Commission (JWC) between South Africa and Swaziland (Treaty, 1992b). The significance of this set of agreements is derived from a non-aggression pact between South Africa and Swaziland that was never made public knowledge, but which functioned so well that Swaziland was seldom used by guerrilla forces of the various liberation movements during the armed struggle period. This is of great significance because Swaziland borders on Mozambique and offers an ideal infiltration route into South Africa. Part of the reward to Swaziland arising from the carrot and stick approach of the Total National Strategy, is the railway line that was negotiated at the time of the *Nkomati Peace Accords* (Treaty, 1983c).

FINAL PHASE OF THE HYDRAULIC MISSION

The final phase of the South African hydraulic mission is associated with the demise of the Cold War and the normalization of the international politics of the Southern African region. This period was closely associated with ending of military hostilities and the outbreak of peace in most of the SADCC area. Part of this process was the independence of Namibia, which resulted in a plethora of agreements on water resource management being signed by that state and its various neighbors. This included the establishment of the Permanent Water Commission (PWC) (Treaty, 1992a) to manage the Vioolsdrift and Noordoewer Joint Irrigation Scheme (VNJIS) (Treaty, 1994d). Of even greater importance was the change to the strategic political landscape that the ending of the Cold War had in the context of SADCC. Almost overnight, the *raison d'etre* of SADCC was no longer relevant. This saw the transformation of SADCC into the Southern African Development Community (SADC) (Treaty, 1992e), based on the European Union (EU) model of regional integration. The very first Protocol to be signed within the context of the SADC Treaty occurred when South Africa became a member – the *SADC Water Protocol* – showing the significance that South Africa placed on the management of transboundary rivers as a vehicle of peaceful cooperation, regional integration and post-conflict reconstruction (Treaty, 1995).

Great effort was made by South Africa to normalize its relations with all of its neighboring states, resulting in the negotiation of formal agreements for the management of all transboundary river basins to which it is a riparian. The significance of this becomes apparent when viewed against the recent finding by Conca (2006) that two-thirds of the world's international river basins have more than two riparian states within their hydrological configuration, with more than three quarters of the international agreements that exist for the joint management of transboundary rivers coming from these basins. Yet, within these so-called multilateral basins (basins with more than two riparian states), the most common type of agreement by a ratio of 2:1, is a bilateral arrangement. Stated differently, the global norm consists of bilateral agreements rather than basin-wide agreements for the management of international river basins. In the case of contemporary South Africa, every international river to which it is a riparian – Orange, Limpopo, Incomati and Maputo – is managed by a basin-wide agreement, many of which have a history of evolution that dates back to the period of time before the Cold War. The Orange is governed by ORASECOM (Treaty, 2000). The Limpopo is governed by the Limpopo Watercourse Commission (LWC) (Treaty, 2003) and the Incomati and Maputo Basins are governed by the so-called *Incomaputo Agreement* (Treaty, 2002). All of these basin-wide agreements have evolved over time from the experiences gained in the *First Use Agreement* (Treaty, 1926a), going through different iterations as bilateral arrangements, and surviving the test of fire caused by the Cold War experience.

In order to truly grasp the relevance of these achievements, one needs to evaluate the cost of conflict versus the benefit of cooperation. The inflation trends in South Africa are shown in Figure 3. The log inflation rate shows the general trend over time. Unique trends are evident over specific periods of time. For example, the conflict period associated with WW II resulted in a level of inflation above the average rate. The period of relative peace between the end of WW II and the outbreak of political violence in 1976 (associated with the Portuguese *coup d'etat* in 1975 and the resultant outbreak of civil war in Mozambique and Angola), and the associated internal political unrest that occurred after the Soweto Riots of 16 June 1976, coincided with a period of below average inflation. The period of heightened political tension between 1976 and 1994 is associated with an above average inflation rate. This period was dominated by South African *puissance*-styled military engagement in all of the Civil Wars on the whole Southern African sub-continent, earning it the reputation of a regional destabilizer. The transition to peace arising from the ending of the Cold War, and in particular the transition to democracy in 1994, is again associated with a discernable trend to a level of inflation substantially below the average.

This data provides graphic proof of some of the invisible costs of conflict. It also bears testimony to the recent finding by Gleditsch et al.,

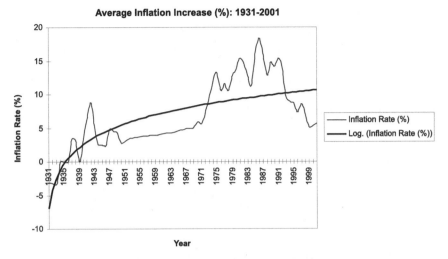

Fig. 3. Inflation trends in South Africa since 1931 (Turton et al., 2004).
Source: Statistics South Africa, 2002

(2005) when they suggest that where endemic water scarcity occurs in a shared river basin, there are substantial long-term incentives for the investment in water management measures that avoid conflictual outcomes. Similarly, when Wolf et al. (2003) did their global study of transboundary river basins in 1999, they found a number of so-called Basins at Risk, six of which were in Southern Africa, three of which had South Africa as a riparian state – the Orange, Incomati and Limpopo. A more recent study (Turton, 2005a) showed the extent to which that situation had changed, with four of the Southern African Basins at Risk having basin-wide management arrangements that were negotiated after the study by Wolf et al. (2003). This shows a strong desire to cooperate peacefully using *pouvoir*-styled hydropolitical diplomacy.

CONCLUSION

The conflict in Southern Africa was about issues of a high politics nature, mostly associated with national liberation from the oppression of colonialism, but this was given additional energy from the protagonists of the Cold War because, each of the local parties drew in financial and military support from the USA, the USSR or China. At no time was the conflict ever driven by the desire to gain access to water, but water resource management projects were used as inducements to entice cooperative inter-state behavior, in particular being linked to the non-harboring of guerrilla fighters from the various liberation movements. This resulted in the securitization of water resource management, with all negotiations about shared river systems falling under the banner of foreign policy, which in turn was dominated by the State Security Council and therefore the security force community. This is called *puissance*-styled hydropolitics.

But this is not the whole story. The preferred option of South Africa as the regional hegemon was always a cooperative approach using international agreements to structure inter-state relations. This is called *pouvoir*-styled hydropolitics and was initially envisaged through CONSAS. Only when this was opposed by Zimbabwe with the formation of SADCC, alternative bilateral agreements were reached. There has been substantial River Basin Organization (RBO) development as a result of this. These RBOs were forged of fire and tended to float like islands of peaceful cooperation on a seething sea of high political violence. The foundation of regional cooperation within SADC is evidenced by the *SADC Water Protocol*. The management of transboundary rivers thus plays a major role in regional integration and post-conflict reconstruction, to the extent that a Hydropolitical Complex is said to exist in Southern Africa that structures international politics in a way that is likely to see the desire for hydrological security becoming a significant driver of future regional integration (Turton, 2005a; Ashton and Turton, in press).

REFERENCES

P.J. Ashton and A.R. Turton (in press). Water and Security in Sub-Saharan Africa: Emerging Concepts and their Implications for Effective Water Resource Management in the Southern African Region, In: Globalisation and Environmental Challenges. H.G., Brauch, J., Grin, C., Mesjasz, N.C., Behera, B., Chourou, U.O., Spring, P.H. Liotta, and P. Kameira-Mbote, (Eds.) Berlin: Springer Verlag.

Ashton, P.J., Earle, A., Malzbender, D., Moloi, B., Patrick, M.J. and Turton, A.R. 2005. Compilation of all the International Freshwater Agreements entered into by South Africa with other States. Final Water Research Commission Report for Project No. K5/1515. Pretoria: Water Research Commission (WRC).

Barber, J. 1999. South Africa in the Twentieth Century. Oxford: Blackwell Publishers.

Baynham, S. 1989. SADCC Security Issues. Africa Insight, 19(2), 88-95.

Bernstein, K. and Strasburg, T. 1988. Frontline Southern Africa. London: Christopher Helm.

Borchert, G. 1987. Zambezi-Aqueduct. Institute of Geography and Economic Geography, University of Hamburg, Hamburg.

Borchert, G. and Kemp, S. 1985. A Zambezi Aqueduct. SCOPE/UNEP Sonderband Heft. No. 58; 443-457.

Brown, J.C. 1875. Hydrology of South Africa; or Details of the Former Hydrographic Conditions of the Cape of Good Hope, and causes of its Present Aridity, with Suggestions of Appropriate Remedies for this Aridity. London: Kirkaldy.

Brown, J.C. 1877. Water Supply of South Africa and the Facilitation for the Storage of It. Edinburgh: Oliver Boyd, Tweedale Court.

Bulloch, J. and Darwish, A. (Eds.) 1993. Water Wars. Coming Conflicts in the Middle East. London: Victor Gollancz.

Conca, K. 2006. Governing Water: Contentious Transnational Politics and Global Institution Building. Cambridge, MA.: MIT Press.

Conley, A. and van Niekerk, P. 1998. Sustainable Management of International Waters: The Orange River Case. *In:* The Management of Shared River Basins: Experiences from SADC and EU. H. Savenije and P. van der Zaag (Eds.) pp 142-159. The Hague: Netherlands Ministry of Foreign Affairs.

Cunningham, A. 1987. The Strubens and Gold. Johannesburg: Ad. Donker Ltd.

Davies, B.R., O'Keefe, J.H. and Snaddon, C.D. 1993. A Synthesis of the Ecological Functioning, Conservation and Management of South African River Ecosystems. Water Research Commission Report No. TT 62/93. Pretoria: Water Research Commission.

de Villiers, M. 1999. Water Wars: Is the World's Water Running Out? London: Weidenfeld and Nicolson.

Evans, M.M. 1999. The Boer War: South Africa 1899 – 1902. Osprey Military. Mechanicsburg: Stackpole Books.

Farwell, B. 1999. The Great Boer War. London: Wordsworth Editions.

Frankel, P.H. 1984. Pretoria's Praetorians: Civil-Military Relations in South Africa. London: Cambridge University Press.

Frederikse, J. 1982. None but Ourselves: Masses vs. Media in the Making of Zimbabwe. Johannesburg: Ravan Press.

Geldenhuys, D. 1984. The Diplomacy of Isolation: South African Foreign Policy Making. Johannesburg: Macmillan, South Africa.

Gleditsch, N.P., Furlong, K., Hegre, H., Lacina, B. and Owen, T. 2005. Conflicts over Shared Rivers: Resource Scarcity or Fuzzy Boundaries? Oslo: International Peace Research Institute (PRIO).

Gleick, P.H. 1991. Environment and Security: The Clear Connections. Bulletin of the Atomic Scientists, 47, 17-21.

Gleick, P.H. 1993. Water and Conflict. Fresh Water Resources and International Security. International Security, 18(1), Summer 1993; 79-112.

Gutteridge, W. 1981. South Africa: Strategy for Survival? In Conflict Studies, No. 131; 1-33. *Reprinted in* Gutteridge, W. (Ed.) 1995. South Africa: From Apartheid to National Unity, 1981-1994. Page 1-32. Aldershot, Hants and Brookfield, VT: Dartmouth Publishing.

Gutteridge, W. 1983. South Africa's National Strategy: Implications for Regional Security. Conflict Studies, No. 148; 3-9. *Reprinted in* Gutteridge, W. (Ed.) 1995. South Africa: From Apartheid to National Unity, 1981-1994. Page 35-41. Aldershot, Hants and Brookfield, VT: Dartmouth Publishing.

Gutteridge, W. 1990. South Africa: Apartheid's Endgame, in Conflict Studies, No. 228; 1-37. *Reprinted in* Gutteridge, W. (Ed.) 1995. South Africa: From Apartheid to National Unity, 1981-1994. Page 147-182. Aldershot, Hants and Brookfield, VT: Dartmouth Publishing.

Heyns, P.S. 1995. Existing and Planned Development Projects on International Rivers within the SADC Region. *In:* Proceedings of the Conference of SADC Ministers Responsible for Water Resources Management. Pretoria, 23-24 November 1995.

Homer-Dixon, T.F. 1991. On the Threshold: Environmental Changes as Causes of Acute Conflict. International Security, 16(2), Fall; 76-116.

Homer-Dixon, T.F. 1994. Environmental Scarcities and Violent Conflict: Evidence from Cases. International Security, (19)1; 5-40.

Homer-Dixon, T.F. 1996. Environmental Scarcity, Mass Violence and the Limits to Ingenuity. Current History. 95, 359-365.

Irani, R. 1991. Water Wars. New Statesman and Society, 4(149): 24-25.

Klare, M.T. 2001(a). The New Geography of Conflict. Foreign Affairs, 80(3): 49-61.

Klare, M.T. 2001(b). Resource Wars: The New Landscape of Global Conflict. New York: Metropolitan.

Lee, E. 2002. To the Bitter End: A Photographic History of the Boer War 1899–1902. Pretoria: Protea Book House.

Longford, E. 1982. Jamieson's Raid: the Prelude to the Boer War. London: Weidenfeld and Nicholson.

Mathews, J.T. 1989. Redefining Security. Foreign Affairs, 68(2); 162-177.

Midgley, D.C. 1987. Inter-State Water Links for the Future. Die Suid-Afrikaanse Akademie vir Wetenskap en Kuns Symposium: Water for Survival. August, 1987.

Myers, N. 1993. Ultimate Security: the Environmental Basis of Political Stability. New York: Norton and Co.

Nasson, B. 1999. The South African War 1899–1902. London: Arnold.

Ninham Shand. 1956. Report on the Regional Development of the Water Resources of Basutoland. Report Commissioned for the Government of Basutoland by the Director of Public Works under Terms of Reference 1290/W30 dated 11 October 1955. Cape Town: Ninham Shand.

NIS. 1994. National Intelligence Service: 1969-1994. Pretoria: National Intelligence Service.

Ohlsson, L. 1995. Water and Security in Southern Africa. Publications on Water Resources: No. 1. SIDA: Department for Natural Resources and the Environment.

Percival, V. and Homer-Dixon, T. 1998. Environmental Scarcity and Violent Conflict: The Case of South Africa. Journal of Peace Research. 35 (3); 279-298.

Percival, V. and Homer-Dixon, T. 2001. The Case of South Africa, In: Environmental Conflict. Diehl, P.F. and Gleditsch, N.P. (Eds.) Boulder: Westview Press. pp 13–35.

Porch, D. 2000. Wars of Empire. London: Cassell & Co.

Raath, A.W.G. 1999. The British Concentration Camps of the Anglo Boer War 1899-1902: Reports on the Camps. Bloemfontein: The War Museum. (Available from Thorrold's Africana Books.)

Reisner, M. 1993. Cadillac Desert: The American West and its Disappearing Water. Revised Edition. New York: Penguin.

Scudder, T. 1989. River Basin Projects in Africa. Environment, 31(2), 4-32.

Scudder, T., Manley, R.E., Coley, R.W., Davis, R.K., Green, J., Howard, G.W., Lawry, S.W., Martz, P.P., Rogers, P.P., Taylor, A.R.D., Turner, S.D., White, G.F. and Wright, E.P. 1993. The IUCN Review of the Southern Okavango Integrated Water Development Project. Gland: IUCN Communications Division.

Seymour Fort, G. 1908. Dr. Jamieson. London: Hurst & Blackett Ltd.

Simon, D. 1991. Independent Namibia One Year On, in Conflict Studies, No. 239; 1-27. *Reprinted in* Gutteridge, W. (Ed.) 1995. South Africa: From Apartheid to National Unity, 1981-1994. Page 185-211. Aldershot, Hants & Brookfield, VT: Dartmouth Publishing.

Spies, S.B. 1977. Methods of Barbarism: Roberts, Kitchener and Civilians in the Boer Republics January 1900–May 1902. Cape Town: Human & Rousseau.

Tempelhoff, J.W.N. 2003. The Substance of Ubiquity: Rand Water 1903–2003. Vanderbijlpark: Kleio Publishers.

Treaty. 1926(a). Agreement Between South Africa and Portugal Regulating the Use of the Kunene [sic] River for the Purposes of Generating Hydraulic Power and of Inundation and Irrigation of the Mandated Territory of South-West Africa. League of Nations Treaty Series, Vol. LXX, No. 1643; 315.

Treaty. 1926(b). Agreement Between the Government of the Union of South Africa and the Government of the Republic of Portugal in Relation to the Boundary Between the Mandated Territory of South-West Africa and Angola. League of Nations Treaty Series, 70(1642); 306.

Treaty. 1926(c). Agreement Between the Union of South Africa and Portugal on the Settlement of the Boundary Between the Union of South Africa and the Province of Mozambique. See Ashton et al., 2005.

Treaty. 1928. South-West Africa–Angola Boundary Delimitation Commission: Detailed Definition of Boundary Position Between Angola and South West Africa. See Ashton et al., 2005.

Treaty. 1931(a). Exchange of Notes Between His Majesty's Government in the Union of South Africa and the Portuguese Government Respecting the Boundary Between the Mandated Territory of South-West Africa and Angola. See Ashton et al., 2005.

Treaty. 1931(b). Territory Boundary: Caprivi Zipfel; Northern Rhodesia and Portugal. See Ashton et al., 2005.

Treaty. 1933. Exchange of Notes Between the Government of the Union of South Africa and the Government of Northern Rhodesia Regarding the Boundary Between the Caprivi Zipfel and Northern Rhodesia and the Grant of Privileges to Natives on Northern Rhodesia on Islands Belonging to the Caprivi Zipfel. See Ashton et al., 2005.

Treaty. 1943. Exchange of Notes Between the Governments of the Union of South Africa and Portugal Respecting the Boundary Between the Mandated Territory of South-West Africa and Angola. See Ashton et al., 2005.

Treaty. 1964. Agreement Between the Government of the Republic of South Africa and the Government of Portugal in regard to Rivers of Mutual Interest and the Cunene River Scheme. Signed in Lisbon on 13 October 1964, in Republic of South Africa Treaty Series. Pretoria: Government Printer.

Treaty. 1969(a). Agreement Between the Republic of South Africa and the Government of Portugal in Regard to the First Phase Development of the Water Resources of the Kunene [sic] River Basin. South Africa Treaty Series, Vol. 1; 1969.

Treaty. 1969(b). Agreement Between the Governments of the Republic of South Africa and Portugal Relative to the Cabora Bassa Project. South Africa Treaty Series, Vol. 7; 1969.

Treaty. 1971. Agreement Between the Government of the Republic of South Africa and the Government of the Republic of Portugal in Regard to Rivers of Mutual Interest, 1964 Massingir Dam. South Africa Treaty Series, No. 5; 1971.

Treaty. 1980. Agreement in Respect of a Servitude to be Granted by Swaziland to South Africa for the Inundation of 3 800 Acres (1 540 hectares) in Swaziland by the Pongolapoort Dam and the Instruments of Ratification Thereto.

Treaty. 1983(a). Agreement Between the Government of the Republic of South Africa the Government of the Kingdom of Swaziland and the Government of the People's Republic of Mozambique Relative to the Establishment of a Tripartite Permanent Technical Committee. Signed at Pretoria on 17 February 1983, in South Africa Treaty Series, No. 12; 1986.

Treaty. 1983(b). Agreement Between the Government of the Republic of South Africa and the Government of the Kingdom of Swaziland with Regard to Financial and Technical Assistance for the Construction of a Railway Link in the Kingdom of Swaziland.

Treaty. 1983(c). Agreement Between the Government of the Republic of South Africa and the Government of the Kingdom of Swaziland with Regard to Financial and Technical Assistance for the Construction of a Railway Link in the Kingdom of Swaziland.

Treaty. 1984(a). Agreement of non-Aggression and Good Neighbourliness Between the Government of the Republic of South Africa and the Government of the People's Republic of Mozambique. South Africa Treaty Series, Vol. 14; 1986.

Treaty. 1984(b). Agreement Between the Governments of the Republic of South Africa, the People's Republic of Mozambique and the Republic of Portugal Relating to the Cahora Bassa Project. South Africa Treaty Series, Vol. 15; 1986.

Treaty. 1986(a). Treaty on the Lesotho Highlands Water Project between the Government of the Republic of South Africa and the Government of the Kingdom of Lesotho. 85 pp.

Treaty. 1986(b). Exchange of Notes Regarding the Privileges and Immunities Accorded to the Members of the Joint Permanent Technical Commission.

Treaty. 1986(c). Agreement Between the Government of the Republic of Botswana, the Government of the People's Republic of Mozambique, the Government of the Republic of South Africa and the Government of the Republic of Zimbabwe Relative to the Establishment of the Limpopo Basin Permanent Technical Committee. Signatory Document, signed by Representatives of Four Governments. Harare, 5 June 1986. 3 pp.

Treaty. 1987. Samewerkingsooreenkoms Tussen die Regering van die Republiek van Suid-Afrika en die Oorgangsregering van Nasionale Eenheid van Suid-Wes Afrika / Namibië Betreffende die Beheer, Ontwikkeling en Benutting van die Water van die Oranjerivier (Cooperation Agreement Between the Government of the Republic of South Africa and the Transitional Government of National Unity of South-West Africa / Namibia Regarding the Control, Development and Utilization of the water from the Orange River). Signatory Document, signed by Representatives of Two Governments. Mbabane, 13 November 1987. 5 pp.

Treaty. 1988. Agreement Relating to the Supply of Water from the Molatedi Dam in the Marico River Between the Department of Water Affairs of the Republic of Bophuthatswana and the Water Utilities Corporation in the Republic of Botswana and the Department of Water Affairs of the Republic of South Africa. See Ashton et al., 2005.

Treaty. 1992(a). Treaty on the Development and Utilization of the Water Resources of the Komati River Basin between the Government of the Republic of South Africa and the Government of the Kingdom of Swaziland. Signatory Document, signed by Representatives of Two Governments. Mbabane, 13 March 1992. 51 pp.

Treaty. 1992(b). Treaty on the Establishment and Functioning of the Joint Water Commission between the Government of the Kingdom of Swaziland and the Government of the Republic of South Africa. Signatory Document, signed by Representatives of Two Governments. Mbabane, 13 March 1992. 17 pp.

Treaty. 1992(c). Agreement Between the Government of the Republic of South Africa and the Government of the Republic of Namibia on the Establishment of a Permanent Water Commission. 10 pp.

Treaty. 1992(d). Agreement on the Viooolsdrift and Noordoewer Joint Irrigation Scheme Between the Government of the Republic of South Africa and the Government of the Republic of Namibia. 33 pp.

Treaty. 1992(e). Declaration Treaty and Protocol of Southern African Development Community. Signatory Document, signed by Representatives of Ten Governments. Windhoek, 17 August 1992. 29 pp.

Treaty. 1995. Protocol on Shared Watercourse Systems in the Southern African Development Community (SADC) Region. Signatory Document signed by Representatives of the Ten Member States in Johannesburg on 28 August 1995.

Treaty. 2000. Agreement Between the Governments of the Republic of Botswana, the Kingdom of Lesotho, the Republic of Namibia, and the Republic of South Africa on the Establishment of the Orange-Senqu River Commission. 13 pp.

Treaty. 2002. Tripartite Interim Agreement Between the Republic of Mozambique and the Republic of South Africa and the Kingdom of Swaziland for Cooperation on the Protection and Sustainable Utilization of the Water Resources of the Incomati and Maputo Watercourses. Signatory Document, signed by Representatives of the Three Governments. Johannesburg, 29 August 2002. 48 pp.

Treaty. 2003. Agreement Between the Republic of Botswana, the Republic of Mozambique, the Republic of South Africa and the Republic of Botswana on the Establishment of the Limpopo Watercourse Commission. Signatory Document signed by Representatives of the Four Governments, Maputo, 27 November 2003. 14 pp.

Trolldalen, J.M. 1992. International River Systems, in International Environmental Conflict Resolution: The Role of the United Nations. Washington, DC: World Foundation for

Environment and Development. *Reprinted in* Wolf, A. (Ed.) 2002. Conflict Prevention and Resolution in Water Systems. Cheltenham: Edward Elgar. (Pages 114-147).

Turner, J.W. 1998. Continent Ablaze: The Insurgency Wars in Africa 1960 to the Present. Johannesburg: Jonathan Ball Publishers.

Turton, A.R. 2003. Environmental Security: A Southern African Perspective on Transboundary Water Resource Management. *In:* Environmental Change and Security Project Report. The Woodrow Wilson Centre. Issue 9 (Summer 2003). Washington, DC: Woodrow Wilson International Center for Scholars. (Page 75-87).

Turton, A.R. 2005(a). A Critical Assessment of the River Basins at Risk in the Southern African Hydropolitical Complex. Paper presented at the Workshop on the Management of International Rivers and Lakes, hosted by the Third World Centre for Water Management and the Helsinki University of Technology. 17-19 August 2005. Helsinki, Finland. Forthcoming chapter in a book as yet untitled.

Turton, A.R. 2005(b). Hydro Hegemony in the Context of the Orange River Basin. Paper presented at the Workshop on Hydro Hegemony hosted by Kings College and the School of Oriental and African Studies (SOAS), 20-21 May 2005, London. Forthcoming chapter in a book as yet untitled.

Turton, A.R. 2006. Shaking Hands with Billy: *The Private Memoirs of Anthony Richard Turton.* Limited edition pubication for collectors of Africana. Krugersdorp: AR Turton Publisher.

Turton. A.R. and Meissner, R. 2002. The Hydro-Social Contract and its Manifestation in Society: A South African Case Study. *In:* Hydropolitics in the Developing World: A Southern African Perspective. A.R. Turton and R. Henwood (Eds.) 2002. Pretoria: African Water Issues Research Unit (AWIRU). (Page 37-60).

Turton, A.R., Meissner, R., Mampane, P.M. and Seremo, O. 2004. A Hydropolitical History of South Africa's International River Basins. Report to the Water Research Commission. Pretoria: Water Research Commission.

Turton, A.R., Schultz, C., Buckle, H.; Kgomongoe, M., Malongani, T. & Drackner, M. 2006. Gold, Scorched Earth and Water: The Hydropolitics of Johannesburg. In *Water Resources Development,* vol. 22, no. 2; 313-335.

Van Rensburg, T. 1980. Camp Diary of Henrietta E.C. Armstrong: Experiences of a Boer Nurse in the Irene Concentration Camp 6 April–11 October 1901. Pretoria: Human Sciences Research Council (HSRC).

Vas, A.C. and Pereira, A.L. 1998. The Incomati and Limpopo International River Basins: A View from Downstream. *In:* The Management of Shared River Basins. Experiences from SADC and EU. H.G. Savenije, and P. van der Zaag, (Eds.) Page 112-124. The Hague: Ministry of Foreign Affairs.

Wolf, A.T., Yoffe, S.B. and Giordano, M. 2003. International Waters: Identifying Basins at Risk. Water Policy, 5(1), 29-60.

7

Competition for Limited Water Resources in Botswana

Umoh T. Umoh[1], Santosh Kumar[2], Piet K. Kenabatho[3] and Imoh J. Ekpoh[4]

[1]University of Botswana, Dept. of Environmental Science,
Private Bag UB00704, Gaborone, Botswana
E-mail: umoht@mopipi.ub.bw
[2]University of Melbourne, Department of Mathematics and Statistics,
Parkville, Victoria 3010, Australia.
E-mail: skumar@ms.unimelb.edu.au
[3]University of Botswana, Department of Environmental Science,
Private Bag UB00704, Gaborone, Botswana.
[4]University of Calabar, Department of Geography, Calabar, Nigeria

INTRODUCTION

Botswana, with an area of approximately 582,000 square kilometers is a land-locked country and lies between latitudes 17°30′S and 29°S, and between longitudes 20°E and 29°E. The country is bordered to the north by Zambia, to the north-west by Namibia, to the north-east by Zimbabwe and to the east and south by South Africa. The major factor that controls the climate of the country is the fact that Botswana lies on the equator-ward side of Southern Hemisphere's Subtropical High Pressure Belt. The high pressure belt blocks the traveling mid latitude depressions and anticyclones, so that their influence is not exercized directly over Botswana. As a result Botswana's climate is basically arid to semi-arid and hot (Bhalotra, 1984).The amount of rainfall amount is low and highly variable over space and from one year to another.

The overall surface water available to Botswana consists of two components, one composed of runoff originating from rainfall within Botswana. The other component is composed of international river systems, which have the whole or major proportion of their catchments outside the country. The major international river systems are Okavango, Kwando/ Linyanti/Chobe, Limpopo and Molopo/Nosob, but contribute less runoff to the entire drainage system of Botswana. All rivers originating within Botswana are ephemeral and these are concentrated in the eastern part of the country. The total amount of surface water runoff in Botswana is estimated to be in order of 1.2 mm/a. This translates to a total impoundment of 340 million cubic meters (MCM) of surface water storage spread between the northeastern and south eastern side of the country. However these resources are rarely replenished to their full capacity owing to insufficient rainfall, escalating demand as well as evaporation losses. For this reason the Government of Botswana initiated a water transfer scheme which connects one of the dams in northern Botswana to those in the south, particularly Gaborone dam (141 MCM) which is located in the capital city of Gaborone. The volume of groundwater in storage across the country is estimated to be in order of 100000 MCM with recharge generally low and averaging 2.7 mm/a (Botswana National Atlas, 2001). Taking the surface runoff and comparing it with other countries, it does become apparently clear that Botswana is one of the most water-scarce countries in the world.

The responsibility for water in Botswana is shared between a number of government ministries and departments, parastatals and other bodies, national and local. However, the overall responsibility for policy, planning and development of Botswana's meagre water resources falls within the portfolio responsibilities of the Ministry of Mineral, Energy and Water Resources (MMEWA). Water is a public good in Botswana and the main body of laws directly concerned with water comprises four statutes, namely: The Water Act of 1968, The Borehole Act of 1956, The Water Works Act of 1962 and Water Utilities Corporation Act of 1970. Some of these statutes have remained unchanged since coming into force while some have been amended to cover new and changing circumstances. However, the current review of the National Water Master Plan is expected to recommend institutional as well as legal reforms in order to improve water resources management in Botswana.

Water is an extremely scarce resource in Botswana. This poses one of the major development constraints in many parts of the country particularly in the southwest and western parts of the country. Conflicts over current and future allocations and competition amongst users of water resources exist in Botswana. These conflicts and competition typically involve water demand among settlement, agriculture, mining, energy and wildlife. The settlement component comprises domestic, commercial, industrial and institutional while agriculture is split into livestock, irrigation and forestry. The total demand also includes distribution and treatment losses. The present total demand is estimated at about 240 million cubic meters per annum. This huge demand outweighs the presently available water resources.

Historically, increased demand for water has been met by developing additional supplies to avoid conflicts amongst various water-demanding sectors. This chapter examines the major water-consuming sectors in Botswana. It outlines integrated water management strategies as a way of overcoming the current conflicts among water-demanding sectors and thus meeting Botswana's present and future water needs.

RELIEF AND DRAINAGE

The geology of Botswana is made up of old igneous and metamorphic rocks known as the Basement complex rocks. Younger sedimentary rocks covering about 70% of the country overlie these rocks. There are surface outcrops of Basement complex rocks in the eastern parts of the country and in the Ghanzi Ridge. Elsewhere in Botswana the Basement complex rocks is covered mainly by Kalahari sands and in small areas in the east by Karoo rocks. Salt pans occur in parts of the sedimentary rock areas of Botswana of which Makgadikgadi pans covering an area of more than 7000 square kilometres are the largest. Botswana is located in Kalahari Basin, which is ringed by areas of high-land including the Drakensberg, the Namibian Highlands and the Bie Plateau in Angola. The topography of Botswana consists mainly of a gently undulating plateau with a mean elevation of about 1000 meters above sea level. The elevations vary between 1491 meters at Otse Hill, the highest point in Botswana and less than 800 meters in the Limpopo Basin in the extreme east.

Botswana can be divided into six drainage basins or catchments. These are the Molopo/Nossop, Limpopo, Makgadikgadi, Kwando/Linyanti/ Chobe and the Okavango basins. Their respective location and areal extent are listed on Table 1.

Table 1. Major Drainage Basins in Botswana

Drainage Basin	Region	Area (km²)
Limpopo	East	80000
Makgadikgadi	North	30000
Okavango	North-west	97000
Kwando/Linyanti/Chobe North	North	26000
Molopo/Nossop	South	71000
Uncoordinated	Central	259000

Source: Botswana National Atlas, 2001.

Most of the rivers and streams in Botswana are ephemeral because of the semi-arid climate. The few perennial rivers such as the Okavango, the Zambezi-Chobe and the Limpopo have their sources outside the country. The Okavango River and its delta in the north-western part of the country constitute a major drainage system in Botswana accounting for about 95% of total surface water in Botswana. However, this resource remains unused for portable water supply owing to its pristine nature as it is protected under the RAMSA sites, and it is shared resources between three countries –Angola, Namibia and Botswana–under the Okavango Commission known as OKACOM. The Makgadikgadi pans, an inland drainage has some rivers such as Boteti, Nata and Mosetse flowing into them during the wet season (Figure 1). Most of these valleys are usually dry except during the rains.

Fig. 1. Botswana drainage

CLIMATE

The climate of Botswana is continental semi-arid. Due to the country's location in the sub-tropical high pressure belt of southern hemisphere in the interior of Southern Africa far away from oceanic influences, rainfall is low and the seasonal and diurnal ranges of temperature are high. Rainfall is highly variable in terms of both space and time. Annual rainfall ranges between a maximum of 650 mm in Kasane (in the extreme northeast), and a minimum of less than 250 mm in the south west (Bhalotra, 1987) (Figure 2). The rains fall mostly in the summer months between November and March. The extreme northeastern parts receive the highest amount of rainfall because when the I.T.C.Z. retreats northward at the end of the rainy season, it causes rain to fall again. Very little, if any, rain falls in the winter period from May to September. Drought is common since the

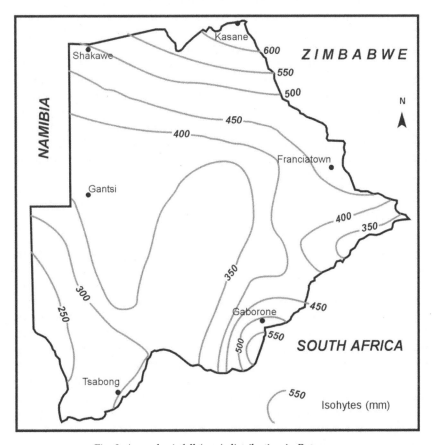

Fig. 2. Annual rainfall (mm) distribution in Botswana

country is located at the edge of the sub-tropical high pressure belt. Relative humidity (RH) varies according to the rainfall pattern, and therefore ranges between 90% over the extreme northeastern parts during the rainy season, and below 10% over the dry southwestern parts of the country during the dry winter season.

Temperatures are generally high because of the country's geographical position. The highest mean monthly maximum temperatures range between 32°C and 35°C in the northern half of the country, during the months October and January. On the other hand, the lowest mean monthly minimum temperatures vary between 2°C and 7°C in July, with the lowest temperatures recorded in the southwest. Yearly potential evaporation averages reach a maximum of 3.1 to 3.3 m in the dry south western parts, where there is generally a lot of sunshine . Maximum evaporation occurs in October and November, prior to the onset of rains (Pike, 1971). Annual potential evaporation reach a minimum in July, with values not exceeding 2.5 m, over the extreme northeastern parts of the country. This is because annual average sunshine is minimal over these areas due to clouds and rainfall which occur more than in any part of the country.

Drought is a recurrent phenomena in Botswana and it is evidently the most important climatic issue. Susceptibility to drought is high given the poor, erratic and unreliable nature of its rainfall and generally poor quality of its soil. The degree of susceptibility vary from a low level on the islands in the Okavango Delta or the low-lying lands along the Limpopo River to a very high level on the areas of deep Kalahari sand in the south-west of the country where rainfall is only about 250 millimeters per annum and coefficient of variability is 80% (Cooke, 1979).

WATER RESOURCES AND CLIMATIC LIMITATIONS

Water is a scarce resource in Botswana. Due to low and variable rainfall coupled with high evaporation rates, the country has limited water resources. The few perennial rivers that flow into the country have external sources, while the rest of the streams that cross the country are ephemeral. Temporary surface water are held in natural pans for a short period of time following the rains. Due to limited surface water resources, a large part of the country depends on groundwater sources for its supply. Surface water sources account for 35% of the total water supply in the country (Arntzen et al., 1999). Although surface water resources account for only a third of the total supply, they provide 90% of the water used in urban areas.

In contrast the majority of rural villages obtain their water supplies from groundwater sources. This implies that two thirds of the country's water supply comes from groundwater sources whose recharge rate is uncertain. The depths of boreholes range from 30 meters to 600 meters and the quality also varies from east to west where saline water is found. Water reuse is another potential source which could be developed particularly as demands increase. The estimated return flows from urban areas are 10.5 MCM/year (1990) growing to 64.4 Mm^3/year in 2020 (Khupe, 1994).

The problem of limited water resources in Botswana is compounded by drought occurrences. There is no doubt that Botswana has been experiencing increasing incidence of drought spells since the beginning of the 20th century. Rainfall regimes of the country are generally in the form of alternating wet and dry epochs. Years of increased rainfall were 1922, 1941, 1959, 1972, 1988, 1995 and 2000. Periods of marked decrease were observed for 1911–1914; 1932–1934; 1945–1946; 1951–1953; 1965–1969; 1982–1987, 1992; 2001–2002 and 2004. Long-term fluctuations in climate have occurred in the past in Botswana. A great lake once existed in the basin now occupied by Makgadikgadi pans in northern Botswana, fed by a number of large rivers from the south and the west whose valleys are now totally dry.

Drought is a periodic reduction in moisture availability below average conditions. Or simply defined as the non-availability of adequate amount of water for man, animals and plants. Drought results in the depletion or exhaustion of soil and shallow groundwater and administers shocks to the ecological system. Botswana experiences considerable distress during drought occurrence. Disasters caused by drought are also strongly affected by such diverse factors as poor agricultural practice, increase in population density and the country's inability to provide alternative supplies of food, water and employment.

Major drought in Botswana in the 20th century occurred in 1913–15, 1940–41, 1948–49, 1967–69, 1972–73, 1983–85, 1987 and 1992–94 (Cooke, 1979; Tyson, 1987; Umoh, 2003). The causes of these climatic phenomena and the impact of such events on hydrology and water resources as well as socio-economic activities on water related sectors have been documented in the literature (e.g. Tyson, 1987; Darkoh, 1998; Vogel, 1998, 2000).

Drought in the early 1980's has been identified as one of the most severe on record (Dent et al., 1987). The annual total of rainfall falls short of the long-term mean by as much as 61% during the 1983/84 drought. A look at the mean total annual rainfall and mean annual temperature figures for different years in different parts of the country, the data set show a trend towards decreasing amounts of rainfall at all the stations. Yet the annual mean temperature has shown no decline. Indeed, the mean annual temperature values tend to be highest in years of lowest rainfall. Table 2 shows the chronological order of the drought disaster between 1965 and 2000 in Botswana. In all drought situations, humans and animals are affected because of widespread water shortages and crop failure.

Table 2. Drought disaster in Botswana (1965–2000)

Year	Drought disaster	Population Affected
1965	Drought	60,000
1968	Drought	60,000
1969	Drought	87,600
1982	Drought	409,770
1983	Drought	409,770
1984	Drought	1,037,300
1985	Drought and locust infestation	880,000
1986	Drought	648,000
1987	Drought	671,000
1992	Drought	100,000

Source: The OFDA (Office of U.S. Foreign Disaster Assistance) International Disaster Database

Drought exerts remarkable effects on river flows, groundwater recharge and other biophysical components of the water resource base, and demands for that resource. Changes in water resources and demand will impact on water supply, flood risk, power generation, navigation, pollution control, recreation, habitats and ecosystems services. Surface water resources in the Botswana consist of rivers, most of which are ephemeral and wetlands, notably the Okavango Delta and the various salt pans. Most of the rivers have low flows that are highly variable and erratic. High rates of evaporation accounts for significant reduction in sustainable yields from reservoirs.

It is observed that there was high frequency of droughts in the 1980's. Annual drought frequency probability for three stations in Botswana has been calculated by Pike (1971) as follows: Kasane, 1 year in 20; Gaborone, 1 year in 15 and Ghanzi, 1 year in 7. Drought index analysis (Umoh, 2003) shows persistence of negative indices for virtually all stations from 1965 to 2000. The annual mean temperature shows no decline. The product moment coefficient of correlation (r) calculated for the data set is -0.91 with student t test of 8.3816 that is significant at the 0.001 probability level given a degree of freedom of 11. This implies that potential evapotranspiration would be increasing.

The drought of 2004/05 has impacted seriously on dams within the country resulting in decrease in water storage in the dams. Table 3 indicates the status of water supplies and recommended management strategies by Water Utilities Corporation (WUC) in major dams of Botswana. In all the dams, there is a great reduction in water levels and as a result, limits their optimal performance. Water levels at the Gaborone dam, as an example dropped steadily to the lowest recorded level of 22% (see Table 3). Gaborone

Table 3: Status of water supplies in dams

Dam	Capacity MCM	% Level	Av % drop per month	Months of supply without inflow	Drought status	Recommendations
Gaborone	141.4	22.0	2.6	4	Severe	Water Conservation Campaigns, Water Restriction, Water Rationing
Bokaa	18.5	34.0	4.3	5	Severe	Water Conservation Campaigns, Water Restriction
Nnywane	2.2	64.3	4.3	10	Severe	Water Conservation Campaigns, Water Restriction
Letsibogo	100.0	52.7	3.0	15	Moderate	Water Conservation Campaigns
Shashe	85.3	80.2	3.3	18	Moderate	Water Conservation Campaigns

Source: Water Utilities Corporation, 2005

dam, at this water level has reached the critical drought phase and is estimated to sustain water supply for about four months assuming there is no further inflow. No significant rainfall was expected till the next rainy season, which is October if the onset of rains is early. The estimated average monthly drop in water level was 2.6%. Considering that evaporation rate increases during the hot months (September to February) the rate of drop in water level will increase. Other issues to consider when the dam reaches such low levels are the raw water pump suction pressure, the true extent of siltation and how it is positioned relative to draw off structures. As a result of the drought situation as indicated in Table 3 above, the WUC implemented some of the recommendations including water restrictions and water conservation campaigns targeting a reduction in demand by 25% in Gaborone and surrounding villages. Recent reports by the same department indicate that there was a decrease in demand by 35%, which is higher than the set target. This alone implies that with more conservation measures in place, the demand in major towns of the country can be controlled so that water use is rationally distributed across other sectors of the economy.

COMPETITION FOR WATER BY VARIOUS SECTORS

The Botswana National Water Master Plan (BNWMP) categorizes five main sectors responsible for water consumption. These include human settlements, mining and energy, livestock, irrigation and forestry, and wildlife (SMEC et al., 1991). Under human settlement water is used for domestic, commercial and industrial, and institutional purposes. Agricultural uses are taken into account by sectors of livestock, irrigation and forestry. Botswana water supply consists of two segments, a formal water supply sector that provides for urban areas, large villages, and small villages, and the privately controlled water sector that supplies water to livestock areas, land areas, and other forms of settlement outside villages. There is competition for the limited available water resources of the country by various water-consuming sectors as follows:

Household Sector

The 2001 census estimated the population of Botswana to be 1,680,863. The total number of households and the average household size were estimated to be 282,318 and 5.1 respectively. The portion of the population which lived in urban centres and urban villages was estimated to be 46%, whereas the rest of the population (54%) lived in rural villages and smaller settlements. The household sector is a major consumer of water in Botswana. The Botswana National Water Master Plan revealed that the sector consumed 17.3 million m^3 of water in 2000, accounting for 14% of total water consumption. The consumption of water in this sector is projected to reach 67.6 million m^3 in the year 2020, which will be 26% of the total water consumption.

It is estimated that 83% of the households have access to piped water in Botswana. A growth in household incomes resulted in an increase in private water connections. The Income and Expenditure Survey revealed that the proportion of households with private water connections increased from 52% in 1993/94 to 71% in 2000/01 in urban areas. The proportion of those in urban villages and rural areas increased from 45% and 9% in 1994/95 to 58% and 16% in 2000/01 (Central Statistics Office, 2002). The rapid increase in the demand for water is attributed to the rapid growth in population, higher incomes and greater access to private water connections.

Industrial and commercial Sectors

Industrial and commercial sectors which are large consumers of water in Botswana are manufacturing; water and electricity; construction; trade, hotels and restaurants; transport and communications; banks, insurance and business services. A few establishments such as Botswana Meat Commission, breweries, hotels and restaurants, and the Botswana Power Corporation are responsible for the bulk of water consumption. Table 4 shows statistics of water consumption in Lobatse for the period 1996-1998. It shows that consumption by the BMC accounts for one third of the total water consumption in the town. The Power station in Selebi-Phikwe consumed 595,000 m³ of water in 1994/95. The Kgalagadi Breweries in Gaborone was estimated to consume 2,750 m³ of water per day in 1990, which was 46% of the total water consumption in the industrial and commercial sector in this urban centre (Arntzen et al., 1999).

Table 4. Water consumption in Lobatse, 1996-1998.

Year	Sector	Water consumption (000 m³)	% of total
1996/97	Domestic	367	18.5
	Government	467	23.5
	Commercial and industrial	282	14.2
	Botswana Meat Commission	671	33.8
	Town Council	196	9.9
	Total	**1983**	**100.0**
1997/98	Domestic	483	18.1
	Government	603	22.6
	Commercial and industrial	571	21.4
	Botswana Meat Commission	792	29.7
	Town Council	222	8.3
	Total	**2671**	**100.0**

Source: Water Utilities Corporation, 1998.

Government Sector

Statistics on water consumption by government sector is outlined in Table 5. The demand is highest in the capital city of Gaborone, which has the largest government sector, and smallest in Jwaneng, which has the smallest government sector.

Table 5. Water consumption by the government sector in urban areas (000 m³; 1994-1997)

Town/city	1994/95	1995/96	1996/97
Gaborone	5926	5287	4541
Lobatse	442	439	463
Francistown	1362	1374	1387
Silebi-Phikwe	544	561	519
Jwaneng	44	43	45

Source: Water Utilities Corporation, 1997

In terms of the contribution of the government sector to the total water demand in the urban areas, the capital city of Gaborone has the highest. In 1995/96, the contribution of the government sector to the total water demand in Gaborone was 35%, whereas the contributions of this sector to the total water demand in Lobatse, Jwaneng and Selebi-Phikwe were as low as 23%, 3.4%, ans 26%, respectively (Arntzen et al., 1999).

Agriculture Sector

In Botswana, the agricultural sector is an important source of informal sector employment where more than 80% of the population is involved. It plays a very important role in the economy of the country. The National Water Master Plan estimated the total water consumption for agricultural sector in 1990 to be 54.2 million m^3, which was 46% of the total water demand in the country. The main water-consuming sub-sectors in the agricultural sector are livestock, irrigation, and forestry.

The number of cattle was estimated to be 2.7 million in 1990. The total water consumption in this sub-sector was estimated to be 34.7 million m^3, accounting for 41% of total water consumption in the country. Water consumption in this sector is projected to reach 44.1 million m^3 in the year 2020 (SMEC et al., 1991).

The estimated irrigation potential of Botswana is 20,000 ha but the actual total area irrigated is 1,380 ha. The irrigation sub-sector consumed 18.8 million m^3, which accounted for 16% of the total water consumption in the country (Arntzen et al., 1999). The eighth National Development Plan (1997/98-2002/2003) states that the Government intends to construct a number of dams in the country, which will be used for irrigation. It also states that intensive and efficient irrigation system, such as drip irrigation will be promoted.

The forestry sector uses water for watering trees in woodlots and plantations. The consumption of water in this sub-sector was estimated to be as low as 0.1 million m^3, or 0.08% of the national water consumption. Projection of the Botswana Water Master Plan of water consumption in the sub-sectors of irrigation and forestry is estimated to reach 28.9 million m^3 in 2000, and 46.9 million m^3 by the year 2020.

Mining Sector

The main activities in the mining sector include diamond, copper-nickel, coal, soda ash, salt, and gold. The mining sector has grown very rapidly since the 1970s, and its contribution to GDP increased from 17% in 1975/76 to 34% in 1993/94. The mining sector is a major consumer of water resources in Botswana. The Botswana National Water Master Plan estimated the consumption of water in this sector to be 20.7 million m^3 in 1990, which was 17.7% of total water consumption.

FUTURE WATER DEMAND IN BOTSWANA

The demand for water is increasing rapidly in Botswana. This is due to the following factors:

i)   high rate of increase in population,
ii)   very rapid growth of Gaborone and other major urban centres
iii)   high rate of internal migration from rural areas to the urban areas and major villages where per capita consumption is appreciably high,
iv)   The development of industry and new mines.

These trends are expected to continue. The water forecasts for different sectors for the period 1990–2020 are shown in Table 6. The forecast shows that the total water consumption will rise from 116.9 million m^3 in the year 1990 to reach 323.5 million m^3 by the year 2020, an increase of 177% (Arntzen et al., 1999). The largest increases are expected to occur in the settlement and mining sectors. The demand for water in the settlements will increase very rapidly as a result of the escalated growth in population and incomes, reaching 38%, 46% and 52% of the total water demand in 2000, 2010 and 2020, respectively.

Table 6. Trends in water demand in Botswana

Demand category	Estimated Demand (10^6m^3/a)			
	1990	2000	2010	2020
Settlements	33.8	68.8	109.9	167.8
Mining and Energy	22.9	33.6	52.2	58.7
Livestock	35.3	44.8	48.3	44.1
Irrigation and Forestry	18.9	28.9	38.5	46.9
Wildlife	6.0	6.0	6.0	6.0
Total	**116.9**	**182.1**	**240.9**	**323.5**

The diamond industry consumes more than 60% of the total water requirements of the mining sector and it is expected to maintain this position for a long time to come.

FUTURE WATER SUPPLY

Various alternative sources of water supply aimed at meeting the demand up to 2020 in Botswana were considered by BNWMP. The Lower Shashe Dam expected to be completed in 2005, with the catchment of 7,810 square kilometres has an estimated mean annual runoff of about 120 million cubic meters. The estimated capacity and annual yield stand at 400 and 73 million cubic meters respectively and when completed, this dam will be the largest in the country.

Beyond 2020, development of resources along the international river system will become the only viable option. In eastern Botswana possible dam sites include Riversdale and Seleka along the Limpopo River. In the north, the Chobe and Zambezi systems are possibilities. Botswana currently imports 7.2 million cubic meters of water from the Molatedi dam in South Africa. Thus an increase in the amount of water imported from the neighbouring countries is also a possibility for future supply. In other areas of the country groundwater will continue to be the only viable source of future supply, thus necessitating more groundwater exploration work in these areas and the need to enhance natural groundwater recharge process through artificial recharge.

MEETING BOTSWANA'S FUTURE LONG-TERM WATER REQUIREMENTS

Botswana's water scarcity problems and population pressure on water is one of the severest in the world. Both the surface water and groundwater resources are meagre and therefore, the question in the longer term, is not whether the country can construct more dams and develop new well fields, but on using the limited resources Botswana has on a sustainable basis.

Demand management offers a cost-effective alternative to new resource development by slowing the rate of demand growth. In an effort to encourage people to use water sparingly and efficiently, the underlisted water-demand management and conservation strategies are employed:

i)   a pricing policy in which, after meeting the minimum basic human requirements, the more water you use the more you have to pay for every cubic meter,

ii)   supply restrictions particularly in times of drought such as in the mid-1980's and in times of low dam water level as in 2004/2005,

iii)   encouragement of the use of water saving devices and technologies, particularly in public institutions,

iv)   public education and mobilization in water conservation.

CONCLUSION AND RECOMMENDATIONS

Water is a finite and very scarce resource in Botswana, due to the semi-arid climatic condition of the country, characterized by recurrence of droughts and very high evaporation rates in surface water resources. As a result, water-demand management and conservation measures are introduced in order to reduce pressure for the development of new water resources. Competition for limited or diminishing water resources by different users in the country is increasing. The major water consuming sectors are settlement, mining and energy, livestock, irrigation and forestry, and wildlife. This is further exacerbated by the erratic rains and unpredictable climatic conditions. The forecast has shown that water consumption will rise from 116.9 million m³ in 1990 to reach 323.5 million m³ by the year 2020, an increase of 177% whereas water development and supplies have limitations.

In Botswana, it is evident that water demand by far outweighs available water supply, thus a more realistic water- demand management is essential. Tate (1999) in his definition of water demand management emphasized

i)   the need for demand management to be socially beneficial
ii) the need to reduce average or peak water demands
iii) the need to integrate both ground and surface water
iv) the importance of water quality considerations.

From our discussions, the major objective of water- demand management in Botswana is to reduce the need for continuing expansion of conventional water supply systems such as boreholes and dams. The aim is to meet water demand on a sustainable basis. This strategy includes three main types:

- Reduction of water consumption
- Increasing water supply from non-conventional sources
- Reducing losses of conventional sources.

Demand management measures recommended, therefore, should include:

i)   Water restriction and policing
ii) Water rationing
iii) Loss reduction programmes and
iv) Public education and awareness campaigns.

It is also recommended that the government should design instruments which will increase the attraction of water-demand management measure through a combination of:

- Legislative instruments (eg. water use restriction, building and product specifications) which prescribe the desired behaviour of water users.
- Economic instrument (water charges, product charge, WDM subsidies) which remove the difference between private and social net benefits.
- Consultative instrument (e.g. negotiated agreements with major water

consuming sectors) which are aimed at a compromise solution close to the optimal situation for the society.

The preferred suggestions and recommendations, if effectively implemented by various stakeholders will manage to curb water demand crises in Botswana.

REFERENCES

Arntzen, J., Kgathi, D.L., Segosebe, E. and Chigodora, K. 1999. 'Water Demand Management – Botswana Country Study. Paper prepared for the World Conservation Union IUCN Regional programme for Southern Africa. Gaborone, June 1999. [On-line: http://www.iucn.org/places/rosa/wdm/countries/botswana.html.

Bhalotra, Y. P. R. 1984. Climate of Botswana Part I: Climatic Controls. Botswana Meteorological Services, Gaborone, Botswana.

Bhalotra, Y.P.R. 1987. Climate of Botswana Part II: Elements of Climate. Botswana Meteorological Services, Gaborone, Botswana.

Central Statistics Office. 1995. Household Income and Expenditure Survey; 1993/94. Government Printer, Gaborone, Botswana.

Central Statistics Office. 2002. 2001 Population and Housing Census: Population of towns, villages and associated localities in AUGUST, 2001. Government Printer, Gaborone, Botswana.

Cooke, H. J. 1979. The problems of drought in Botswana. *In:* Proceedings of Symposium on Drought in Botswana. M.T. Hinchey (Ed.). The Botswana Society; University Press of New England, pp 7-20.

Darkoh, M.B.K. 1998. The nature, causes and consequences of desertification in the drylands of Africa Land Degradation and Development, 9, 1–20.

Dent, M.C., Schulze, R.E., Wills, H.N.M. and Lynch, S.D. 1987. Spatial and temporal analysis of drought in the summer rainfall region of southern Africa. Water South Africa, 13 (37), 42.

Government of Botswana. 2001. Botswana National Atlas. Department of Survey and Mapping, Gaborone, Botswana.

Government of Botswana. 2003. National Development Plan 9, 2003/04-2008/09, MFDP, Gaborone, Botswana.

Khupe, B.B. 1994. Integrated water resource management in Botswana. *In:* Proceedings of Integrated Water Resource Management Workshop. Kanye, A. Gieske and J. Gould (Eds). Botswana, pp. 1-10.

Pike, J.G. 1971. Rainfall and evaporation in Botswana. Tech. Doc. No. 1, FAO/UNDP/SF Project 359, Rome.

SMEC, Snowy Mountains Engineering Corporation Ltd., WLPU Consultants and Swedish Geological AB. 1990. Botswana National Water Master Plan, 5, Hydrogeology. Ministry of Mineral Resources and Water Affairs, Gaborone, Botswana.

SMEC, Snowy Mountains Engineering Corporation Ltd., WLPU Consultants and Swedish Geological AB. 1991. Botswana National Water Master Plan Study; Final Report: 3: Economics, Demography and Water Demands. Ministry of Mineral Resources and Water Affairs, Gaborone, Botswana.

SMEC, Snowy Mountains Engineering Corporation Ltd., WLPU Consultants and Swedish Geological AB. 1991. Botswana National Water Master Plan Study; 6: Ministry of Mineral Resources and Water Affairs, Gaborone, Botswana.

Tate, D.M. 1999. Comments on Botswana draft paper on water Demand Management. Kanata, Ontario; GeoEconomics Associates.

Tyson, P.D. 1987. Climate Change and Variability in Southern Africa, Oxford University Press, Cape Town, South Africa.

Umoh, U.T. 2003. Effects of drought spells on water resources development in Botswana. Paper presented at 4th WATERNET/WARFSA Annual Symposium, 15–17 October, 2003, Gaborone, Botswana.

Umoh U.T., S. Kumar, and A.A. Oladimeji. 2004a. Climate change and water resources management in semi-arid Southern Africa. *In:* Climate Change: Five Years after Kyoto, Velma I. Grover (Ed.). Science Publishers, Inc, Plymouth, UK, 10, 211:229.

Umoh U.T., A. A. Oladimeji and S. Kumar. 2004b. Adaptation to effects of climate change in Southern Africa. *In:* Climate Change: Five Years after Kyoto. Velma I. Grover (Ed), Science Publishers, Inc, Plymouth, UK, 11: 231-249.

Vogel, C.H. 1998. Disaster management in South Africa, South African Journal of Science, 94 (3): 98-100.

Vogel, C.H. 2000. Climate and Climate Change: Causes and Consequences. *In:* The Geography of South Africa in a Changing World, R. Fox and K. Rowntree (Eds). Oxford University Press, Oxford, pp. 284–303.

Water Utilities Corporation, 1997. Annual Report, Gaborone, Botswana.

Water Utilities Corporation, 2004 and 2005 Reports, Gaborone, Botswana.

8

Water Conflict in West Africa: The Niger River Basin Experience

Umoh T. Umoh[a], Imoh J. Ekpoh[b] and Santosh Kumar[c]

[a]University of Botswana, Department of Environmental Science,
Private Bag UB00704, Gaborone, Botswana.
E-mail: umoht@mopipi.ub.bw
[b]University of Calabar, Department of Geography, Calabar, Nigeria.
E-mail: imohjekpoh@yahoo.com
[c]University of Melbourne, Department of Mathematics and Statistics,
Parkville, Victoria 3010, Australia.
E-mail: skumar@ms.unimelb.edu.au

INTRODUCTION

Water is one of the most important elements in nature and in human society. It is everywhere around us in the atmosphere, in the lithosphere, in the plants, in animals and in ourselves. About 70 percent of the human body, by weight, is made of water, upon which the functioning of the body depends. In our daily lives we require water to drink, prepare our food, bath ourselves, wash our clothes, irrigate farms, manufacture goods, generate electricity, enjoy water sports and recreation and above all, obtain fishes and other aquatic organisms. Not only is water essential in all life on the planet, its presence in the atmosphere protects the earth from excessive cooling or heating, therefore ensuring a more habitable planet for mankind.

Despite its importance, it is a well known fact that water is not always available in the right quantity and quality in all places and at all times. Roughly, 97.5 percent of the earth's water resources are in the oceans as saline water. Only 2.5 percent of the world's water is fresh, and 77.5 percent of this water is 'locked up' in ice caps and glaciers, leaving a tiny fraction for human use. The problem as is often the case with natural resources is that water is not evenly distributed around the globe. Many societies spend their time fighting floods while others suffer from desiccating droughts. Perhaps more than we realize, our lifestyles depend on the availability of freshwater. Think of a situation where our taps should go dry, our household routines would collapse, our health would be at risk, industries would grind to a halt, and agriculture would be in dire straits. Indeed, the entire fabric of our societies would begin to crumble.

The problem is not simply one of unequal spatial distribution alone, rather there is the added dimension of trans-boundary water resources. Increasingly, the demand for water and the services it can render is moving rapidly in many societies and there is a paradox in many instances, in that, it is often the regions which need the water most are the ones which lack it most. The River Niger system presents a classic example of this paradox. The river traverses some four major West African countries, namely, Guinea, Mali, Niger and Nigeria. However, of the four Niger River countries, it is Nigeria with a population of over 150 million that would naturally require more water, but unfortunately, she is at the tail end of the river system, often receiving 'left over' from countries located in the upper and middle courses of River Niger.

Conflicts over water allocation in the Niger River Basin of West Africa have arisen between agricultural, urban and environmental uses. The frequent droughts of the Sahelian region of the sub-continent exacerbate conflict between competing agricultural and urban demands. Lakes created by dam construction displace people living upstream while those living downstream of the dam, find the change in river regime disadvantageous. Flood no longer reach lands where crops depended on moisture left in the soil when water subsided. Dams interfere with fish migration. Kainji lake for example displaced 50,000 people who were resettled in specially built villages. Many of these people found the location of their new settlements and houses unsatisfactory. These aggravate conflicts.

This chapter discusses the enduring conflict in the management of waters in the River Niger, a river which traverses four independent West African countries, often with conflicting interest on developmental needs. It identifies three potential elements of conflicts in the River Basin namely: geography, environmental change and resource dependence.

THE PHYSICAL CHARACTERISTICS OF RIVER NIGER

The River Niger, with a length of 4,200 km, is the third longest river in Africa, after the Nile and Congo (Iloeje, 2003). With this impressive length, it is by far the most important river in the region. The river originates from the Fouta Djallon mountains in Guinea which is the high water mark of West Africa. The Tembi river considered to be the main source of the Niger. Inside Guinea, the River Niger is fed by the Mafou, Niandan, Milo and the

Tinkissi tributaries. From Guinea the river runs through the interior plateau in a north easterly direction towards the inland delta in Mali. As it crosses the Guinea-Mali border, the Niger is joined by the Fie River and thereafter by the Sangarani River near Kangare. From there, the river flows in a north-easterly direction towards the inland delta in Mali where it is joined by one of its very important tributaries, the Bani River, which itself has a length of 1,120 km. The inland delta at Mopti in Mali is quite large, extending some 390 km from Djena to Timbuktu, with Lake Debo at its centre (Figure 1). The delta area is swampy and the soil sandy, which probably accounts for the loss of nearly two-thirds of the waters of the Niger between Segou and Timbuktu due largely to seepage and also to intense evaporation especially along the fringes of the Sahara Desert.

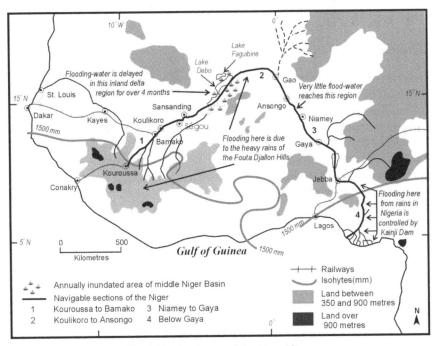

Fig. 1. The landscape of the River Niger

From the inland delta in Mali, the river proceeds in a north-easterly direction before turning to the south-east to form the famous Niger Elbow. It then flows through the arid areas in the surroundings of the ancient city of Gao before entering the Republic of Niger. Inside the Republic of Niger, the river is joined by three major tributaries namely: the Faroul River from Upper Volta; the Dargol River and the Sirba River. Further south of Niamey, River Niger is joined by the Rivers Garoubi and Tapoa. The Niger then flows for a while along the boundary between the Republics of Niger and Benin, thus becoming a continuous river. This section receives an important tributary from Benin Republic, the Mekrou River. Thereafter, the river enters the Federal Republic of Nigeria, where it is joined by numerous tributaries, the most important of which are the Sokoto, Malendo, Kaduna, Oyo, Oro, Kampi, Obi and Moshi Rivers. The most important tributary of the Niger, however, is the Benue which merges with the river at Lokoja in Nigeria. The Benue itself, which rises in Cameroon, is joined in that country by such tributaries as the Faro, Mayo-Kebbi and Lissaka Rivers, and inside Nigeria by the Gongola, Dongo, Ankwa, Taraba, Okwa and others. From the confluence with the Benue, the Niger heads southwards and empties into the Atlantic Ocean in the Gulf of Guinea through a network of outlets that constitute its maritime delta, the Niger Delta (Fig. 1).

Landlords of the River Niger

From the description of the flow from source to destination (mouth), it is clear that many countries have a stake in the waters of the Niger (Figure 2). Guinea has possession of the upper course of the river, Niger and Mali Republics control the middle course, with Mali controlling a length of 1,160 km while Niger controls 500 km of the river length. The Federal Republic of Nigeria controls the lower course of the river. These four countries are the major stake holders in the River Niger Project, although Benin Republic is often claimed to be a part of the River Niger Commission since a section of the main stream forms its boundary with the Republic of Niger.

However, the bottom line is that Benin Republic, Cote d' Voire, Upper Volta and Cameroon are Basin States of the River Niger by virtue of the numerous tributaries and sub-tributaries originating in those countries or even passing through their territories (Godana, 1985). So, in all, about eight countries have a stake in the River Niger and can conveniently be said to constitute the Basin States of the River Niger.

Rainfall Climatology of the River Niger Basin

The River Niger, together with all its tributaries, is a river unparalleled in the West African sub-region. Indeed, it traverses the sub-region from one end to the other. Such a large basin area, which spans all the major vegetation and climatic zones of West Africa, is bound to experience widely varied hydrology. West African rainfall is largely accounted for by the anti-cyclone monsoon winds, which blows in a southeasterly direction, which change direction to southwesterly on crossing the Equator. These are moisture-laden winds because they form over the Atlantic Ocean and

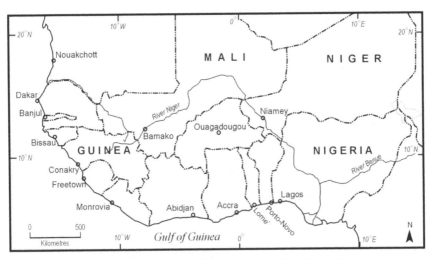

Fig. 2. West Africa showing the basin states of the River Niger.

constitute the main source of rain over the River Niger Basin (Hayward and Oguntoyinbo, 1987).

In West Africa, rainfall decreases from the coast hinterland. The southern coastal areas receive more rain because of its proximity to the source region of the southwest, tropical maritime winds. Rainfall decreases hinterland in a north, northeastward fashion. The wettest parts, with over 2000 mm of rainfall are Guinea, South-West Sierra Leone, Liberia and South-East Nigeria, especially the Oban Hill region where rainfall of over 4000 mm have been recorded. Individual falls of rain are very heavy, and may last several days. Rainfall at the beginning and end of the rainy season is usually accompanied by thunderstorms, some of which may be very violent, causing damage to buildings, trees, electric poles and washing away soil and plants. In the cities, flash flooding and gullying are some of the environmental hazards often caused by tropical thunderstorms. In Freetown and Conakry, which is the source region of the River Niger, annual rainfall averages about 3500 mm, and comes mostly in 5 to 7 months (Hayward and Oguntoyinbo, 1987). This rain is like that on the south-west coast of India and Mynamar and is appropriately called the 'monsoon'. In the rest of West Africa the isohyets appear almost as parallel straight lines running from west to east, with the 250 mm isohyets recorded in Timbuktu (Figures 3 and 4).

The Niger delta, where the River Niger enters the Atlantic Ocean enjoys rainfall well in excess of 3000 mm, and there is rain in every month of the year. Indeed the entire southern coast of West Africa from Cote d' Voire to Cameroon receives 9-12 months of rainfall. Stations hinterland receives less rainfall. For instance, Dakar, Bamako, Ouagadougou and Kano are

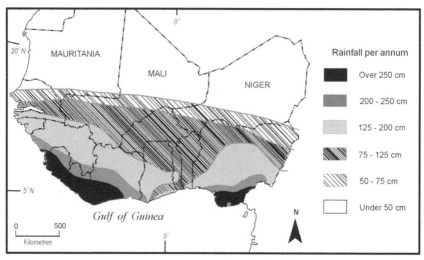

Fig. 3. Annual distribution of rainfall in West Africa.

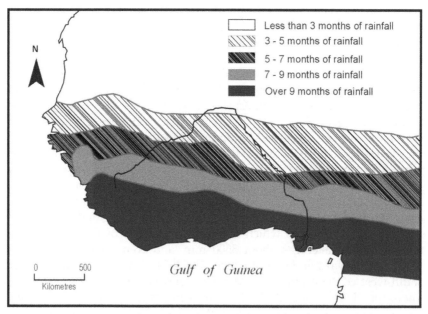

Fig. 4. The length of the wet season in West Africa.

found within the very much drier southern Savanna region running east to west. In the coastal stations generally, rainfall follows a pattern which shows that the heavy rains usually begin about April and rises to a maximum in June, before decreasing slightly in August (the little dry season), and then ascending again to another peak in September, and finally diminishing with the advance of the dry season in October or November. This is why many coastal rainfall isobars display double maxima. Other stations hinterlands have only one peak rainfall, usually about August.

Descriptive statistics for long-term annual rainfall series in the Niger River in Nigeria are shown in Table 1. Generally, the temporal pattern has fluctuated with a tendency to decline. During the 80-year study period, Izom registered the highest mean annual rainfall of 1513.1 mm while Lokoja had the least being 11221 mm. The standard deviation ranged between 179.9 for Konagora and 323.3 for Kachia (Umoh, 2004). It can also be seen from Table 1 that annual variability in rainfall over Niger River Basin differ from station to station. The coefficient of variation lies between 12 percent at Izom and 23.5 percent at Kaduna. Even though the annual variability for all the stations are relatively low, the year to year variability is lowest in Izom and highest in Kaduna. Other stations fall in between these extreme values. The distribution for the stations is significantly and positively skewed. This means that the bulk of the values are less than the mean. Kaduna has highest skewness of 1.90 while Kachia has the least, 0.35.

Table 1. Statistical properties of annual rainfall at six stations in Niger River Basin

Station	Mean	Standard deviation	Coefficient of variation (%)	Skewness
Izom	1,513.1	181.8	12.0	0.45
Kachia	1,438.2	323.3	22.5	0.35
Kaduna	1356.3	319.0	23.5	1.90
Kontagora	1,126.2	176.9	15.7	0.58
Lokoja	1,122.1	225.4	20.1	0.42
Minna	1,230.1	203	16.5	0.70

Results of analysis of basic statistical parameters of Kaduna, Minna and Lokoja reveals that the mean annual rainfall of Kaduna over 80-years period is 1356.3 mm with standard deviation (SD) of 319 mm. Of this total 1254.1 mm or 92.5 percent (with a SD of 244.5 mm) fell in the months of May to October. Mean annual rainfall and rainy season coefficient of variation (CV) are 23.5 percent and 19.5 percent respectively. For Minna and Lokoja, the annual rainfalls are 1230.1 mm and 1122.1 mm respectively; with annual coefficient of variation of 16.5 percent for Minna and 20.1 percent for Lokoja. The percentages of mean annual rainfall and rainy season coefficient of variations range between 16.5 percent - 23.5 percent (Umoh, 2004). The low values of annual coefficient of variation for the stations suggest that rainfall is less variable from year to year and that its variability is greater in the individual months.

The Hydrology of the Niger River Basin

In its upper reaches, the River Niger is a torrential stream fed by abundant rainfall in the Guinea Highlands of the Fouta Djallon where the mean annual rainfall is over 2000 mm. The rainfall decreases sharply as the river flows through increasing dry countryside towards the inland delta. At the inland delta area, the mean annual precipitation records mere 254 mm.

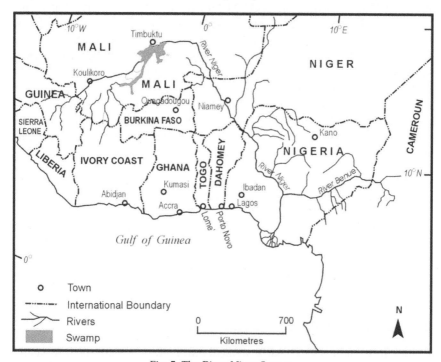

Fig. 5. The River Niger System

However, as the river leaves the inland lake region and turns south, rainfall increase again until it records over 3000 mm in the Niger Delta Region (Figure 5).

The most important and prominent tributary of the River Niger, is River Benue. The upper Niger rises in June and decreases in December. The middle Niger, on the other hand, reaches its maximum near Timbuktu only in January. April to July is the low water season. Below the River Benue confluence, The Niger is at its lowest in April and May. In June, the river is subject to great fluctuations and may overflow its banks. It begins to rise in level about the middle of August, to attain its maximum in September. Between the high and the low water marks, the difference is as much as 35 ft. At the mouth, the River Niger carries an estimated annual volume of 175,000.00 cu.m (Myeni, 1980). The whole river from source to destination has a variable flow, attaining anything from 5,000 cu.m/sec at peak flow to 50 cu.m/sec at lowest flow (Yakemtchouk, 1971).

Rainfall–runoff relations over three drainage basins of the Niger River in Nigeria indicates that areal rainfall is lowest in Kaduna basin and highest in Gurara basin with the mean annual rainfall ranging from 1220 mm in Kaduna (Wuya) to 1474 mm in Gurara (Jere). Surface runoff in the drainage basin varies between 127 mm in Kaduna and 228 mm in Gurara. Runoff as percentage of rainfall is between 10.2 percent and 10.9 percent for Kaduna basin; 12.1 percent for Gbako basin and 12.4 percent-15.4 percent for Gurara basin. The 30-year average of areal rainfall over the Niger basin is 1326.4 mm; while surface runoff is 163.6 mm constituting about 12.2 percent of the rainfall. Rainfall has dominant effect on peak flow regime variations (Umoh, 2004)

The result of annual areal volume of some water balance parameters for the three drainage basins are presented on Table 2. In the drainage basin there is a wide difference between Potential Evapotranspiration (PET) and Actual Evapotranspiration (AET) which indicates a distinct climatic variation in the area. Soil moisture deficit (SMD) starts in December and ends in April. Of the three basins, soil moisture deficit is highest in the Kaduna basin and lowest in Gurara.

Table 2. Annual areal volume of some water balance parameters in three sub-catchment basins of The Niger River ($\times 10^6 \text{m}^3/\text{year}$)

	Kaduna (65,150 km²)	Gbako (7,540 km²)	Gurara (23,730 km²)
Rainfall	81,632.9	9,688.9	33,863.6
Surface Runoff	8,404.3	1,146.1	4,840.9
Infiltration	42,086.9	4,878.3	17,607.6
AET	35,246.1	4,991.4	16,705.9
SMD	30,750.8	2,915.6	8,281.7

Table 3 summarizes the computation of annual averages of water balance components of parts of the Niger River Basin. The mean annual areal rainfall of the catchment basin is 1321.7 mm. Potential evapotranspiration (PET) is 1072.6 mm while actual evapotranspiration (AET) is 635.6 mm. The soil moisture capacity (RAS) is 170 mm on the basis of 1.0 mm effective root zone, while infiltration or recharge to groundwater is 678.3 mm. Soil moisture deficit (SMD) and surface runoff are 404.3 mm and 161.6 mm respectively (Umoh, 2004).

Table 3. Annual averages of water balance of the Niger River Basin (1970-2000)

	Kaduna	Gbako	Gurara	Mean
Rainfall (mm)	1253	1285	1427	1321.7
P.E.T (mm)	1112	1053	1053	1072.6
R.A.S (mm)	150	180	180	170.0
Infiltration(mm)	646	647	742	678.3
AET (mm)	541	662	704	635.6
SMD (mm)	472	392	349	404.3
Surface runoff (mm)	129	152	204	161.3

The Niger River Basin experiences pronounced climatic variations with their accompanying climatic events and hydrometeorologic consequences of these events. Rainfall (amount, intensity, rainy season length) and stream-flow (flood and low flows) have fluctuated over the years with a declining trend. The decline in rainfall is reflected by overall decline in annual peak flood and minimum water level. Spatial and temporal climatic variability and alterations in the hydrological parameters have serious implications on agriculture, water needs/supply and environmental hazards. These are also the potential causes of water conflict within the drainage basin.

The Distribution of the Niger River Basin among Basin States

As with most resources, the Niger River Basin is unevenly distributed among the countries that have a stake in it. Consequently, the level of dependence on the basin's water resources varies from one basin to the other, partly as a factor of the ratio between the quality of its available water and the total water resources in the country, and partly as a factor of the proportion of the river system within each country.

The headwaters of the River Niger are located in Guinea where some 95,000 km³ of that country's total surface of 245,857 km³ fall within the river drainage basin (Myeni, 1980). In this region, the high mountain of Fouta Djallon generates relief rainfall which endows the country with plenty of water resources. After Guinea, the next Basin State is Mali. This country has the largest share of the total drainage basin area of River Niger. Roughly 620,000 km³ of the country's total surface area (12,400,000 km³) fall within the Niger basin, accounting for about 50% of the land area.

Next to Mali is the Republic of Niger with a total basin area of 490,000 km², amounting to about 23.3 percent of the total basin area of the river and about 39 percent of Mali's total surface area, Incidentally, most parts of the Republic of Niger is desert and very dry so that the Niger River serves as the main source of reliable water supply for a greater part of the population, a factor which explains the concentration of human habitation along the river valley. The proper harnessing of the river water resources seems to be the only long-term plan for the controlled development of the country's largely nomadic population.

The Federal Republic of Nigeria is the last basin state of the River Niger. The Niger Basin occupies some 580,000 km^2 or about 63 percent of the total land area of Nigeria's 923,768 km^2. Nigeria, being the lowest downstream beneficiary of the River Niger and as such, is usually affected by the on-channel uses of the river by upstream basin states.

Cote d'Voire is considered a Niger River Basin State courtesy of its being the source of two sub-tributaries which empty into the Bani River, a tributary of the Niger which joins the latter in Mali. Of all the basin states, Cote d'Voire has the smallest share of the drainage basin area, with some 25,000 km^2 of basin area. The Republic of Benin is a contiguous basin area to that part of River Niger mainstream which forms part of its boundary with the Republic of Niger. It also contributes more than three tributaries to the main stream of the river, and 44 percent of the country's total surface area of 112,622 km^2 lies within the drainage basin of the River Niger system.

Cameroon is also considered a basin State because it has the headwaters of the River Benue, which is, of course, the most important tributary of the River Niger. About 90,000 km^2 or 19 percent of the country surface area of 475,442 km^2 is covered by the Niger River basin. Since the Benue is a major tributary of the River Niger, on-channel and off-channel uses of River Benue by Cameroonian authorities have substantially affected the flow of the river in Nigeria. Upper Volta is another basin state of the River Niger by virtue of its being the source of the tributaries that join the River Niger in Mali and in Republic of Niger. The area covered by the Niger River Basin in Upper Volta is roughly 81,700 km^2. Again, on-channel and off-channel uses of the tributaries can affect the volume of the water in the River Niger, especially in Mali and Niger Republic.

Finally, it must be pointed out that the Niger River is concern not only to basin states, but to other African States, which, though not geographically located within the basin, may be regarded as complementary zones. These are States within the zone of influence of the hydrological basin of the Niger River, such as Liberia, Sierra Leone, Ghana, Togo and Chad. For instance, crops from irrigation schemes in the Republic of Niger are exported to other West African States. Similarly, most of the fish caught in the Niger River is marketed in countries such as Ghana and Liberia.

Conflicts in using the Water Resources of the River Niger

Increasingly, water resources development involves the use of river systems or groundwater shared by more than one user, region, state or country

(Barrow, 1987; Ekpoh, 2004). In the case of the River Niger, each of the basin States has its own independent schemes and projects for harnessing the water resources within its national boundary. This has often brought conflicts between basin states, especially between those upstream and downstream. The principal objective of most basin States is the production of electricity through hydro-power generation. Nigeria as an example, constructed three hydro-electric power stations in the 1960s, during the 1962–1968 development plan. The projects involved damming River Niger at Jebba and Kainji, and its tributary, the River Kaduna at Shiroro. The Kainji Dam on River Niger was commissioned in 1968 with the capacity to produce 760 Mega Watts of electricity. The Jebba Dam on the River Niger produces 540 Mega Watts of electricity, while the Shiroro Dam on River Kaduna, a tributary of the River Niger produces 600 Mega Watts of electricity.

Since the 1980's and after the droughts of 1972/1973 and 1983/1984 in the Sahel of Africa, the Nigerian authorities have been worried by the decreased inflow of water from outside Nigeria into the River Niger. The Black Flood from the upper Niger and the White Flood from the Sokoto-Rima Basin within Nigeria have changed their pattern of inflows thereby affecting drastically, the power generating capacity of the Dams on the Niger River. The causes of the decreased inflows have been attributed to drought in the Sahel zone of Mali and Niger Republics through which the River Niger flows before entering Nigeria. As an illustration, the total yield of the upper and middle Niger into the Kainji reservoir in Nigeria steadily decreased from $51 \times 10^9\,\mathrm{m}^3$ in 1970 to an all time low of 24.3×10^9 m^3 at the height of drought in 1973 (Oyebande, 1975). Periodic blockage of the River Niger during low flows at Niamey has also been cited as another factor. The construction of dams as well as development of large irrigation schemes along the upstream course and tributaries of the River Niger have also contributed to the reduced inflow of water into the Kainji Lake. The Jebba and the Shiroro reservoirs also suffer from reduced inflow of water, just like the Kainji Lake.

One of the most important environmental changes brought about by dam construction within the Niger River Basin which usually results in conflict is large scale flooding of quality agricultural lands and valley floor land. In many parts of the river basin reservoirs have wiped out the much needed vegetation such as forests, woodlands and grazing lands. In Kano State (Nigeria) some 3,000 km² of forest reserve and farmland now lies under artificial lakes. Dams are used to control flooding downstream but this has turned out to be a catastrophe as far as agriculture is concerned. In Nigeria, for example, the flood plains downstream of the dams had traditionally been put to flood-farming for production of paddy rice. It is estimated that with the construction of the Kainji Dam, about 77% of the floodplain between Baro and Kacha became unsuitable for rice cultivation. Beyond Gapan, the loss was as high as 94% (NEST, 1991).

Dams and irrigation projects displace people creating resettlement problems. The Kainji Lake displaced 50,000 peasant farmers, the Tiga Lake, 12,000; Bakolori Lake, 14,000 and Goronyo Lake, 20,000 (NEST, 1991). The dam construction within the river basin has led to changes in the grazing and migratory patterns of cattle and humans thereby creating greater demand on any of the available grazing lands in the lake region. The various restrictions of the Fulani's annual migration in Northern Nigeria have resulted in a series of clashes between farmers and the cattle rearers.

Following the incessant blockage and overexploitation of the River Niger upstream in Niger and elsewhere, the Nigerian Government went into international cooperation with its neighbours in the water resource sector. To avoid confrontation with the Niger authorities over incessant blockage of the flow of River Niger, the Nigerian Government went into agreement with their Niger counterparts to supply her electricity generated at Kainji Dam, rather than Niger developing its own hydro-electric power station. The cooperation has worked well and has substantially averted serious conflict between the Nigerian and Niger Government over water resources of the River Niger.

CONCLUSION

The Niger River is an international river and does not observe political boundaries as it traverses many independent states. Therefore, water resources development of such region will have to be jointly managed to avoid conflict, and minimize harm to both the water resources and the environment. As River Niger runs through several countries in West Africa, each country wants to maximize its share of benefits. The Niger River Basin is well endowed with water resources, both surface and underground. The present water conflicts arise from two major factors. One is the poor distribution of water in time and space in relation to the needs. The other is lack of adequate planning and management of water resources to ensure a temporal and spatial convergence between water demands and supplies. There is need therefore to control the hydrological cycle to adapt itself to space, time, quantity and quality requirements.

It is suggested that, for the River Niger, the different basin states should not pursue diverse projects, rather, the ultimate goal should be that of harnessing the water resources of the River Niger for the socio-economic growth of the peoples and nations of the region through the increasing use of the river basin management approach. This will involve the setting up of a suitable management team of experts and administrators (the administrators representing the Basin States) to pursue a comprehensive or integrated planning and management for attainment of broader regional prosperity. If this is done, the future of the Niger River Basin promises great improvement in power generation, domestic and industrial water supplies, irrigation, for livestock, for waste disposal, for fishing, for recreation and for navigation. Above all, water conflicts will be minimized in the region.

REFERENCES

Barrow, C. 1987. Water Resources and Agricultural Development in the Tropics. Longman Scientific and Technical, Essex, England.

Ekpoh, I.J. 2004. Environmental Hazard: Events and Valuation Techniques. St. Paul's Publishing and Printing Co. Calabar, Nigeria.

Godana, B.A. 1985. Africa's Shared Water Resources. Frances Printer (Pub) Ltd. London, UK.

Hayward, D. and Oguntoyinbo, J. 1987. Climatology of West Africa. Hutchinson, London, UK.

Iloeje, N.P. 2003. A New Geography of West Africa. Longman, Essex, England.

Myeni, M.B. 1980. Intra-African Attempts at Cooperation in Common Goods Case Study: The River Niger Commission. Geneva, Switzerland.

Nigeria Environmental Study/Action Team (NEST) 1991. Nigeria Threatened Environment: A National Profile. NEST Publication, Lagos, Nigeria.

Oyebande, L. 1975. Water Resource Problems in Africa. *In:* Problems and Perspectives, African Environment Special Report 1 by International African Institute, London, UK. pp. 119-127.

Umoh U.T. 2004. Hydroclimatological approach to sustainable water resources management in semi-arid regions of Africa. *In:* Water Resources in Arid Areas. D. Stephenson, E.M. Shemang, and T.R. Chaoka (Eds), A.A. Balkema Publishers, London, UK. pp. 389 – 395.

Yakemtchouk, R. 1971. L'Afrque en droit international, Paris, France.

9

River Development and Bilateral Cooperation: Lesotho Highlands Water Project Case Study

Naho Mirumachi

University of Tokyo, Tokyo, Japan, King's College London,
University of London, London, U.K.

INTRODUCTION

Developing international rivers would not be complex if there were no upstream influences on downstream water and if rivers flowed accordingly to our man-made boundaries. However, it is these two distinct characteristics of international rivers that have made water a volatile subject for basin states. Water scarcity has often been considered to cause instability and even conflict in the form of war between nations.[1] Starr (1991) raised the issue of 'water wars' in the Middle East. Gleick (1993) claimed that securing water supply can become a prioritized political or military objective and how water can be the potential instrument for war. International river development for water allocation among basin states can easily become a contentious issue.

Even though currently there is a growing consensus that water scarcity is not the major and sole factor that prompts war, river development can still cause negative sentiment and rivalry. Many basin states are in neither open conflict situations nor cooperative situations regarding river

---

This is an earlier version of a paper submitted to Water International (in review).

[1] Concerning the various discussions of the link between environment and security, in general refer to the following: Brock, L. (1997). The environment and security: Conceptual and theoretical issues. In N. P. Gleditsch (Ed.), *Conflict and the Environment* (pp.17-34). Dordrecht: Kluwer Academic Publishers. Deudney, D. (1991). Environment and security: Muddled thinking. Bulletin of Atomic Scientists, 47, 23-28. Gleick, P.H. (1991). Environment and security: The clear connections. Bulletin of Atomic Scientists, 47, pp.17-21. Mathews, J.T. (1989). Redefining security. Foreign Affairs, 68 (2), 162-177.

development. The overarching question of this chapter stems from this: When and how can basin states move from non-cooperation to cooperation for river development? Instead of perceiving international river development as a 'problem' that requires solutions for conflict resolution or prevention, this chapter emphasizes how it can be an opportunity for states. In other words, if states can perceive international river development as a way of enhancing their benefits, cooperation would be a win-win situation for all the basin states.

The chapter first gives a brief review of the arguments why water scarcity or the existence of shared rivers does not necessarily lead to conflict. It then explores the case study of the Lesotho Highlands Water Project (LHWP) from the perspective of cooperative development. There have been attempts to transfer water in the Zambezi River, Okavango River and Mekong River. However, none of them has achieved the same level of success as this Southern African case.[2] The case study highlights how states can negotiate the costs and benefits of cooperation so that the project can provide non-water sector benefits. The existence of such benefits is considered the key in shifting from non-cooperation to cooperation.

UNDERSTANDING CONFLICT AND NON-COOPERATION

It seems highly unlikely for states to go into war over resource scarcity in general. Homer-Dixon (1999) states that environmental scarcity itself does not cause violence, for it is the combination of other factors like political, economic or social conditions that are influential. According to him, scarcity is "mainly an *indirect* [emphasis in original] cause of violence, and this violence is mainly *internal* [emphasis in original] to countries" (Homer-Dixon, 1999). In regards to international rivers, acute conflict between the upstream and downstream states is only in limited situations:

> [T]he downstream country must be highly dependent on the water for its national well-being; the upstream country must be threatening to restrict substantially the river's flow; there must be a history of antagonism between the two countries; and, most importantly, the downstream country must believe it is militarily stronger than the upstream country (Homer-Dixon, 1999).

---

[2] For detailed accounts of the cases, refer to the following respectively: Scudder, T. (1993). Pipe dreams - Can the Zambezi River supply the region's water needs? Cultural Survival Quarterly, Summer 1993, 45-47. The IUCN Review of the Southern Okavango Integrated Water Development. Available at http://www.iucn.org/themes/wetlands/okavango.htm. Ojendal, J. (1995). Mainland Southeast Asia: Co-operation or Conflict over Water? In Hydropolitics: Conflicts over Water as a Development Constraint, (pp.149-177). London: Zed Books.

Wolf's (1998) Transboundary Freshwater Dispute Database analyzed 412 international crises during 1918 to 1994. His findings show that *"there has never been a single war fought over water* (emphasis in original)" (Wolf, 1998). He states that "the actual history of armed water conflict is somewhat less dramatic than the water wars literature would lead one to believe: a total of seven incidents, in three of which no shots were fired" (Wolf, 1998). His argument does not contradict Homer-Dixon's findings of water conflict rarity at the international level.

Toset et al. (2000) conducted a statistical study concerning the relationship between resource competition over water and armed conflict. By using a dataset of 1274 dyads with shared rivers during 1816-1992, they sought the probability of militarized dispute. Their findings show that though water scarcity in both the states of the dyad does contribute to dispute, it cannot be singled out as the reason of conflict.

From six case studies, Elhance (1999) makes "contingent generalizations" that states often choose against water wars but there exists the lack of cooperation, which can increase the potential of interstate conflict from water scarcity. He uses the term 'lack of cooperation' loosely and synonymous to 'non-cooperation'. In this chapter, non-cooperation includes two situations: (1) no explicit institutions or agreements for bilateral/multilateral cooperation (2) defective institutions or agreements. To give an example, between India and Nepal, though there are several treaties concerning the Ganges River, the riparian relationship cannot be considered as cooperative. The Mahakali Treaty signed in 1996 was initially hailed as a landmark in Indo-Nepal integrated water development on the Mahakali tributary of the Ganges River. However, this benefit-sharing treaty has not been implemented. There are differences in the interpretation of the treaty between the two states and a major component of the treaty, Pancheswar Multipurpose Project, has been delayed. As Gyawali and Dixit(2000) note, the 'Mahakali impasse' has brought the two basin states into a non-cooperative situation.

Elhance (1999) listed several impediments of cooperation, which keep the basin in non-cooperation: erosion of state sovereignty through water sharing, conflicting domestic political interests for interstate cooperation, difficulty in determining the economic value of water and lack of data and technical expertise. The following case study will show especially how bilateral negotiations can overcome basin states' apprehension towards sovereignty erosion.

LESOTHO HIGHLANDS WATER PROJECT: BILATERAL AGREEMENT OF WATER TRANSFER

The Lesotho Highlands Water Project (LHWP) is an inter-basin water transfer scheme on the Senqu/Orange River. The Senqu/Orange River is shared by four states (Figure 1). Originating in the highlands of Lesotho, the Senqu River changes its name to the Orange upon entering South African territory. Though there are some differences between sources, Turton (2003), for example, measures that the river basin spans over a total of 964,000 km² in which Lesotho accounts for 4% of the basin area, but contributes to 41% of the mean annual runoff. Lesotho and South Africa signed the Treaty on the Lesotho Highlands Water Project between the Government of the Republic of South Africa and the Government of the Kingdom of Lesotho in 1986. According to Article 4(1) of the Treaty, it aims to "enhance the use of water of the Senqu/Orange River by storing, regulating, diverting and controlling the flow of the Senqu/Orange River

Fig. 1 Map of the Orange River basin. (*Source:* South Africa Department of Water Affairs and Forestry, Orange River Project http://www.dwaf.gov.za/orange/default.htm)

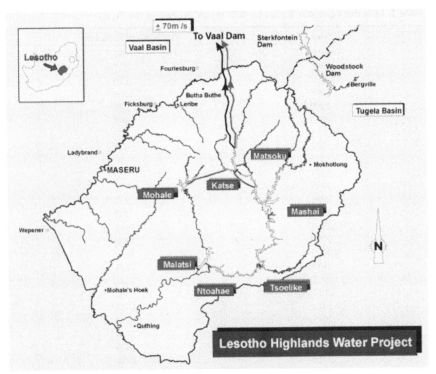

Fig. 2 Schematic map of water transfer in the Lesotho Highlands Water Project.
(*Source:* South Africa Department of Water Affairs and Forestry, Orange River Project
http://www.dwaf.gov.za/orange/default.htm)

and its affluents" to deliver water and generate hydroelectricity. The flow of the Senqu River is diverted northwards into the South African Vaal River (Figure 2). Using the altitude of the highlands as an advantage, water stored in dams is transferred by gravity. Thus pumping is not required in this scheme.

The project includes the construction of five large dams, underground transfer tunnels through the Maluti Mountains, a hydropower station and other supplementary infrastructure. Almost 95% of this construction takes place in Lesotho territory (Wallis, 1992). The development is divided into four phases. Currently, Phase 1 has been completed. The first half of this phase, 1A, consisted of the 185 m high concrete Katse Dam, 72 MW Muela Hydropower Station and a network of transfer and delivery tunnels. Phase 1B developed the Mohale Dam, Matsoku Weir and additional transfer tunnels.

On completion of all phases, 70 m^3/sec of water will be transferred (LHDA, n.d. b). Until then, a minimum quantity for water delivery is determined by the two countries in the Treaty (Table 1). South Africa

Table 1. Minimum quantity for water delivery as defined in the Treaty

Calendar year	Million cubic meters
1995	57
1996	123
1997	190
1998	258
1999	327
2000	398
2001	470
2002	543
2003	618
2004	695
2005	772
2006	852
2007	932
2008	1014
2009	1098
2010	1183
2011	1271
2012	1361
2013	1452
2014	1545
2015	1640
2016	1736
2017	1835
2018	1934
2019	2036
2020	2139
After 2020	2208

Source: Treaty (1986, Annexure II)

currently receives 30.2 m³/sec from Phase 1 (Wallis, 2000). The transferred water is primarily used for industrial activity and potable water supply in the Gauteng Province, which encompasses large cities such as Johannesburg and Pretoria.

There are three institutions—one bilateral and two implementing agencies—for the governance structure of the LHWP. The Lesotho Highlands Development Authority (LHDA) oversees 'project implementation, operation and maintenance' in Lesotho (Treaty, 1986, Article 7(1)). Specifically, it is responsible for the delivery of water, implementation of future phases and development of domestic environmental and socio-economic policies concerning the local communities affected by the project (LHDA, n.d. b). South Africa's Trans-Caledon Tunnel Authority (TCTA) undertakes the implementation of the project in its territory and financing for the water transfer on behalf of the Department of Water Affairs and Forestry. Since infrastructure for the delivery of water is now complete in South African territory, it oversees "operations and maintenance of the Delivery Tunnel North, Ash River Outfall and the appurtenances" (TCTA-LHDA, 2003). Governing the two national organizations is the Lesotho Highlands Water Commission (LHWC). This body is an upgraded version of the Joint Permanent Technical Commission (JPTC), which had been set up from the initial stages of the project. LHWC is formed by three representatives from both countries so that all agreements are made bilaterally. Its new role emphasizes not only total responsibility for the implementation but also as "the single conduit for communication with the Project authorities" of the two countries (Wallis, 2000).

COSTS AND BENEFITS

The Treaty of the Lesotho Highlands Water Project determines that South Africa be responsible for all costs of the water transfer (Treaty, Article 10(1)). Phase 1A cost close to US$ 1800 millon (Tromp, 2006). For Phase 1B, costs are estimated to be around US$ 846 million (Tromp, 2006).

In terms of the royalties, South Africa pays fixed and variable royalties to Lesotho. The fixed royalties are calculated from the amount South Africa would save compared to the next cheapest water scheme created unilaterally outside of Lesotho territory. This Orange-Vaal Transfer Scheme (OVTS) was planned to pump water out of the Orange River into the Vaal River through a network of canals and pipelines outside the Lesotho border. Water would be delivered from a distance of 500 km (Wallis, 1992). The OVTS costs amounted more than that of the LHWP for two main reasons: expensive coal-fueled pumping and higher capital cost for maintenance and operation. In addition, the water quality would not be as good due to the long transfer distance (Mohamed, 2003). Specifically, it was estimated that compared to the OVTS, the LHWP would be 56% of the cost, allowing South Africa to save 44%. Hence, it was decided that 56% of the difference in cost would be fixed royalties. The fixed annuity has been made since 1995 and is expected to continue "until all project costs have been redeemed and Lesotho's share of the net benefit has been paid for in full" (Wallis, 1995). Variable royalties are calculated for the actual amount of transferred water. Since the delivery of water from the Katse dam in 1998, it is reported that a total of US$ 230.95 royalty payment has been made between 1996 to 2004 (Tromp, 2006).

For South Africa, the largest material benefit is the procurement of cheap water supply. The LHWP supplies water to major cities that are the heart of economic activity of the country. Around 70% of the workforce and around 40% of the South African GDP is generated in the Gauteng region (TCTA-LHDA, 2003). In addition, the region has experienced a long history of drought. This has lead to the creation of numerous dams on the surrounding Vaal, Ash, and Wilge Rivers. Therefore, it is easy to surmise the importance of securing stable water supply at the minimal cost.

As for Lesotho's burden of the project, the Treaty places the responsibility of the implementation, operation and maintenance of the Muela Hydropower Plant (Treaty, Article 10(2)). This is equal to 5% of the LHWP construction cost (http://www.tcta.co.za/). Any other development envisaged is also the responsibility of Lesotho. This includes costs of resettlement and compensation for the locals affected with by the infrastructure construction. Environmental conservation is undertaken by LHDA. With the push from both international and domestic NGOs, several programs covering health and sanitation, environment and heritage, skill development, community participation and education have been conducted.

The most notable benefit for Lesotho is the revenue or royalties that the project brings into the country. It was agreed that Lesotho would receive roughly US$55 million in the form of royalties every year during Phase 1, which accounts for "25% of Lesotho's total annual export revenues and 14% of the Government's public revenues over a 50-year period" (LHDA, n.d. b). It is interesting to note that the LHDA sources (LHDA 2003, 2004) and LHWC Source (Tromp, 2006) present different figures for water deliveries and royalty revenue. The following table shows both figures from the two organization for reference (Table 2).

Table 2. Water deliveries and royalty revenue

Year	Planned deliveries (million m³)		Actual deliveries (million m³)		Royalty payments (million Maluti, LHDA) (million US$ LHWC)	
	(LHDA)	(LHWC)	(LHDA)	(LHWC)		
1999/2000	538	327	540	539.6	146.93	22.18
2000/2001	573	398	574	570.3	158.05	24.31
2001/2002	591	470	584	436.8	182.95	28.15
2002/2003	615	543	585	554.2	205.91	32.38
2003/2004	695	618	687	692.6	207.85	31.49

Source: Adapted from LHDA (2003), LHDA (2004) and Tromp (2006)

Another large benefit is the creation of the 72 megawatt Muela Hydropower Station. The significance of this plant is that it allows the country to be self-sufficient in electricity generation. Previously, Lesotho had been importing electricity from South Africa's state-owned power supply company, Electricity Supply Commission (ESKOM). The new plant "conceived primarily as a strategic investment" will reduce dependency on South Africa (LHDA, n.d. a)

In addition, Lesotho has been selling a small percentage of the electricity through Lesotho Electric Corporation (LEC). The percentage of exports is still low due to the high competition in regional electricity supply with ESKOM dominating the market. Furthermore, because the infrastructure is large, it is projected that cost recovery will take several years. However, once it has been recovered, long-term benefits can be foreseen. The Muela Hydropower Station *only* became plausible with the creation of the Muela Dam. This represents Lesotho maximizing benefits from the bilateral project through strategic investment.

CHARACTERISTICS OF BILATERAL PROJECT NEGOTIATIONS

The event that directly influenced the signing of the treaty between the two countries in 1986 is the regime change in Lesotho. Contrary to the previous ruling party, the new military government by Lesotho Paramilitary Force leader, Major-General Justin Metsing Lekhanya took an eager position for project implementation. Domestic support for Lekhanya's attitude towards economic development through utilizing its water was strong. In addition, South Africa found the government "more compliable" to conduct negotiations compared to the previous regime (Meissner and Turton, 2003). The reasons for such a change of attitude reflects how highly political issues influenced the process of bilateral negotiations of the project. Specifically, contention over South Africa's apartheid policy was the biggest issue. It took three decades of negotiation before the signing of the treaty. In the following section, three keywords that highlight the characteristics of negotiations are discussed in detail: resource need, mistrust and security.

Resource Need

Initially, the concept of the LHWP was conceived as the Oxbow Scheme in 1956 (Meissner and Turton, 2003). A South African engineer, Ninham Shand, proposed a water transfer scheme using the Elands River. This scheme would tunnel through the Maluti Mountains, where it would be possible to develop a hydropower station as well. Basutoland (as Lesotho was called before its independence) would supply water discharged from the power station to the Orange Free State Goldfields at an 'attractive' price (Ninham Shand Consulting Engineers, n.d.). In addition, Basutoland would sell all generated electricity to South Africa (Meissner and Turton, 2003). However, this plan was not considered by the South African Department of Water Affairs, despite there being high demand for water in the Free State Goldfields. Instead, water supply from the Vaal River was considered as a potential (Ninham Shand Consulting Engineers, n.d.).

According to Meissner and Turton (2003), this original scheme was not realized largely due to South Africa's reluctance towards buying water and electricity from Lesotho. Another factor was that, as Lesotho did not have the economic capacity to build the infrastructure by itself it would make South Africa's participation indispensable for project implementation (Meissuer and Turton 2003). It can be said that South Africa perceived the commitment to a capital-consuming bilateral development as a risk.

However, increased water demand prompted South Africa to reconsider the Oxbow Scheme in the mid 1960s. Severe drought strained domestic water supplies, elevating Lesotho's water source to a potential option for long-term, though still supplementary, water source. Meissner and Turton (2003) note that there were several political issues limiting the scheme to be only a secondary consideration. These included the issue of potential political confrontation over apartheid through the project and South Africa's 'unwillingness' to be reliant on Lesotho (Meissner and Turton 2003). Even though in the face of domestic water strain, the capacity to control the water was an important aspect for South Africa in the negotiations. Once again, dependence on foreign resources for its economic heartland was considered to weaken the sovereignty of the state.

Eventually in the 1970s, British and South African consultants and the World Bank created preliminary feasibility reports of the LHWP (Ninham Shand Consulting Engineers, n.d.). Furthermore, in 1978, the first meeting of the Joint Technical Committee (JTC) was held between the two countries to conduct a joint preliminary feasibility investigation (Turton et al. 2004). These events mark how resource need was constantly high throughout the three decades despite the sovereignty dilemma of South Africa. Upon entering the 1980s, the Department of Water Affairs started imposing higher water consumption rates on residents (*Rand Daily Mail,* January 28, 1984:3). Tighter controls on water use and a completely revised water tariff system were debated as well (*Rand Daily Mail,* March 17, 1984:2, March 21, 1981:3, March 31, 1984:1). This caused a rise in interest for water transfer at the local level. There were voices to continue with the LHWP plans for it could generate "greater cooperation and development on the sub-continent" (*Rand Daily Mail,* October 20, 1984:3).

Mistrust and Apartheid

From its independence from British colonial annexation in 1966, Lesotho prioritized seeking new means to secure revenue. The LHWP held much potential for this objective, especially since Lesotho offers few other profitable natural resources to be 'exploited' large scale except water. Nevertheless, there was some concern that South Africa will impose apartheid through economic influences if it committed to the scheme (Meissner and Turton, 2003). The immediate period after independence raised "the question... [of] what kind of position a politically independent

Lesotho was going to occupy in relation to its powerful neighbor" (Scott, 1985), for the Basotho people have had a "history of struggle to preserve their national identity against great odds" (Weisfelder, 1979). For Lesotho, developing the economy was an urgent agenda but retaining its integrity was equally important.

The basic stance of the ruling Basotho National Party (BNP) was to maintain a peaceful relationship with its big neighbor *and* to pose as a model for other African states opposing apartheid. The leader of state, Chief Leabua Jonathan believed that his "realistic' BNP policies of sustaining and augmenting 'bread and butter' relationships with South Africa" was its political stronghold (Weisfelder, 1979). However, in reality, this would eventually cause domestic opposition and South African disapproval. Already from the initial relations with South Africa, Lesotho had to struggle with how dependent it can be on a politically contradicting neighbor. Baffoe expressed this as the double-edged situation where Lesotho can "use the devil [South Africa, for economic gains] but have to let it penetrate [Lesotho] a little" (personal communication, March 8, 2004).

Lesotho drastically changed its attitude towards apartheid in the 1970s, causing political relations to sour rapidly. Realizing that domestic opposition towards his cooperation with South Africa could result in political failure, Jonathan converted to openly criticizing the apartheid regime. Tension mounted as South Africa accused its neighbor for harboring the then banned African National Congress (ANC) members. According to Barber and Baratt (1990) (as cited in Meissner and Turton, 2003), Lesotho was seen as "an extremist state". Hostility and mistrust between the two countries prevented any negotiation from 1976 to 1978 (Turton et al., 2004).

Water and Security

The few years preceding the signing of the Treaty experienced peaked political tensions from the apartheid controversy. There was acute conflict; in December 1982, South Africa launched an attack against ANC members in Lesotho that involved civilian casualties (Hayashi, 1999). Jonathan accused South Africa of backing the anti-governmental Lesotho National Liberation Army (LNLA) and killing Basotho politicians. South Africa strengthened its hard-line policies in retaliation. It proposed to create a peace pact to ensure security in the area. When Lesotho did not comply, South African technicians withdrew from the LHWP negotiations. This was considered the "most serious threat so far to the success of the two countries" (*Rand Daily Mail*, August 25, 1984:2). There had been no technical obstacles in realizing the project. However, Minister of Foreign Affairs Pik Botha displayed discontent with Lesotho's aggressive attitude in propagating negative sentiment and accusations. He threatened that South Africa would discard the project plans unless security situations were improved by Lesotho (*Rand Daily Mail*, August 27, 1984:2).

The project was politicized to force Lesotho into signing such a pact. Lesotho, insisted on separating the water scheme from the security pact issue (Turton et al. 2004). A security pact would be unnecessary as Lesotho was not a threat to the region. Lesotho held prospects of the LHWP as a chance to "heighten economic co-operation" (*Rand Daily Mail*, August 31, 1984:3). Though delegates of both countries confirmed to resume talks in the end (*Rand Daily Mail*, September 1, 1984:3), South Africa did not abandon its claim that participation was "dependent on satisfactory arrangements to secure the project from sabotage" (*Rand Daily Mail*, October 9, 1984: 4).

The accusation of Lesotho harboring ANC members and counter accusations of South Africa destabilizing Lesotho by use of LNLA forces climaxed into South Africa blocking the border with Lesotho (*The Star*, January 4, 1986:2). South Africa closed the border to protect its people by controlling border traffic against ANC 'terrorists' operating from Lesotho. Furthermore, South Africa was "increasingly edgy" about growing communist influences with the large presence of Soviet diplomats and the opening of the Cuban embassy in Lesotho's capital (*The Star*, January 15, 1986: 4). Turton et al. (2004) judge that communist threats were the "last straw" for South Africa during this security-conscious period.

Confrontation on apartheid escalated into security issues, with the Cold War setting adding to the complexity of neighbor relations between South Africa and Lesotho. This resulted in domestic unrest in Lesotho for quick resolve in border closure. The new government was expected to act swiftly to terminate the blockade.

Non-water Sector Benefits

For South Africa, securing ways to obtain water for its economic heartland was a persisting issue during the thirty years. At the same time, increased opposition to apartheid and heightened national security concerns due to the Cold War also prompted South Africa to conceive the LWHP as a way to prevent, or at least reduce, diplomatic stress between Lesotho. By strategically halting and reopening negotiations over the LHWP, it sought to induce Lesotho into taking a less aggressive attitude over apartheid. At the time of signing the project, South Africa benefited from having not only additional water for urban needs in Ganteng but also additional reassurance that the upstream state would not jeopardize the scheme. Hence the sovereignty of the state would not be weakened by engaging in a 'risky' project concerning vital resources.

For Lesotho, the role of the LHWP as an economic driver was a major motive for engaging in talks with South Africa. However, domestic politics influenced how the Basotho government should position the project in regards to state integrity. The Jonathan regime was ambivalent about its dependence on South Africa that it eventually caused political unrest. The new regime of 1986 was more pragmatic about the LHWP; the project would offer tangible benefits for the state and heighten economic development under the new government. It can be said that the new government sought symbolic significance in the LHWP. It was a project that would bring in visible material benefits as a "crucial step towards 'economic self-reliance'" (Ciccolo, 1992). This led to stronger domestic support for the new government.

SIGNIFICANCE OF NON-WATER SECTOR BENEFITS

The new Lesotho regime of 1986 was initially popular domestically and in South Africa. There was a certain degree of opportunism for economic development within the new government (Baumbach, I., personal communication, March 3, 2004). This domestic approval of the project signals of Lesotho's national prestige to take on a project of such magnitude and reap benefits from it. The LWHP provided Lesotho with new ways to generate its economy in the long term with its *own* resources while maintaining economic ties with South Africa.

The benefit of having a politically amiable neighbor for South Africa has great importance in the context of regional development. Especially after the establishment of the Southern African Development Community (SADC) in 1993, reginoal integration has become prominent. Socio-economic development is a main issue. Translating the LHWP into this context of SADC's mission, there are high hopes of uplifting the region through the construction and operation of the project. South African President, Thabo Mbeki stated the following in a speech at the inauguration ceremony of the Mohale Dam (http://www.lhwp.org.ls/news/apr04/speechbypresident.htm):

> This project sparkles like a jewel in the crown of the Southern African Development Community (SADC) and the African Union, proving that we can, as Africans, accomplish sustainable development, to the mutual benefit of neighboring countries and as an example of projects that are needed all over our continent to achieve our renaissance.

This reflects how South Africa views the project as an impetus for regional growth. Hence, in order to fulfill the 'moral obligation' to the development of the whole Southern African region (van Niekerk, P., personal communication, March 15, 2004) and achieve "regional and social benefits outside South Africa" (Maartens, W., personal communication, March 10, 2004), cordial relations with neighboring countries is an important step for long-term vision, and not just for overcoming differences in political ideology of apartheid and Cold War issues.

CONCLUSIONS

The Lesotho Highlands Water Project illustrates how basin states can make the transition from non-cooperation to cooperation in international river development. The key lies in the non-water sector benefits. From the outset of the negotiations of the projects, the benefits directly occurring in the water sector were clear. For Lesotho, it would be able to generate hydropower with its own water resources in its own territory to be self-sufficient. For South Africa, there would be stable supply of water for its economic heartland. However, 'high politics' can influence the climate of negotiations, making such material benefits insufficient for bilateral cooperation. Lesotho negotiated with South Africa in a way that would separate project agreement with issues of contentious apartheid. It meant that Lesotho would be able to posit the project as a sign of economic development through its water resources while maintaining political integrity as a state opposing apartheid. In this case study, having this benefit unrelated to the water sector weighed heavily in tipping the bilateral situation to cooperation. Considering the disparate differences of power with South Africa, the influence of the project on Lesotho's national prestige cannot be undermined. South Africa also perceived the project as a way to emphasize its role in regional development. Hence, ameliorated relation with its neighbor was the first step from moving away from non-cooperation to cooperation.

Since it is difficult to quantitatively share water in an ever-changing river environment, benefits outside the water sector benefits function as a way to enforce cooperation. Inevitably, this results in preventing cooperation from eroding sovereignty. It emphasizes less on sharing the actual water resource or the diminishing of the pie. International river development can be opportunities for win-win situations if states can determine how it can benefit the not only the water sector but also in other spheres such as the political and economic.

Acknowledgements

The research carried out for this chapter was partly funded by the New Research Initiatives in Humanities and Social Sciences of the Japan Society for the Promotion of Science (JSPS) and by the Core Research for Evolutional Science and Technology (CREST) of the Japan Science and Technology Agency (JST)

REFERENCES

Ciccolo, A. 1992. Environmental responsibility under severe economic constraints: Re-examining the Lesotho Highlands Water Project. The Georgetown International Environmental Law Review, 4, 447-467.

Elhance, A.P. 1999. Hydropolitics in the 3rd world: Conflict and cooperation in international river basins. United States Institute of Peace Press. Washington D.C., USA.

Gleick, P.H. 1993. Water and conflict: Fresh water resources and international security. International Security, 18(1), 79-112.

Government of the Republic of South Africa and Government of the Kingdom of Lesotho. 1986. Treaty on the Lesotho Highlands Water Project between the Government of the Republic of South Africa and the Government of the Kingdom of Lesotho.

Gyawali, D. and Dixit, A. 2000. Mahakali impasse: A futile paradigm's bequested travails. *In:* Domestic Conflict and Crisis of Governability in Nepal, Kumar (Ed). pp. 236-301. Centre for Nepal and Asian Studies, Tribhuvan University, Kathmandu.

Hayashi, K. 1999. *Nanbu afurica seiji keizai ron.* Institute of Developing Economies, Japan External Trade Organization, Tokyo, Japan.

Homer-Dixon, T.F. 1999. Environment, Scarcity, and Violence. Princeton University Press, Princeton, USA.

Lesotho Highlands Development Authority (LHDA) 2003. Annual report 2002/2003 [Brochure]. Maseru: Author.

Lesotho Highlands Development Authority (LHDA) (2004). Annual report 2003/2004 [Brochure]. Maseru: Author.

Lesotho Highlands Development Authority LHDA (n.d. a). The Lesotho Highlands Development Authority: Muela Hydropower Plant Project [Brochure]. Maseru: Author.

Lesotho Highlands Development Authority LHDA (n.d. b). The Lesotho Highlands Water Project commemorative journal 1986-1996 [Brochure]. Maseru: Author.

Meissner, R. and Turton, A.R. 2003. The hydrosocial contract theory and the Lesotho Highlands Water Project. Water Policy, 5 (2), 115-126.

Mohamed, A.E. 2003. Joint development and cooperation in international water resources. *In:* International Waters in Southern Africa. M. Nakayama (Ed.). United Nations University Press, Tokyo, Japan, pp.209-248.

Ninham Shand Consulting Engineers (n.d.) The Lesotho Highlands Water Project, Inauguration of Phase 1A: The role of Ninham Shand Consulting Engineers [Brochure]. Cape Town: Author.

Scott, T.S. 1985. Lesotho: The politics of dependence. Kurimoto Gakuen Souritsu Gojyushunen Kinen Nagoya Shouka Daigaku Ronshu 30: 739-756.

Starr, J.R. 1991. Water wars. Foreign Policy, 82: 17-36.

Toset, H.P.W., Gleditsch, N.P. and Hegre, H. 2000. Shared rivers and interstate conflict. Political Geography, 19: 971-996.

Trans-Caledon Tunnel Authority(TCTA) – Lesotho Highlands Development Authority (LHDA) (2003). Sustainable development: Lesotho Highlands Water Project [Brochure]. Pretoria: Author.

Tromp, L. 2006. Lesotho Highlands: The socio-economics of exporting water. Proceedings of the Institution of Civil Engineers - Civil Engineering, 159, 44-49.

Turton, A.R. 2003. An overview of the hydropolitical dynamics pf the Orange River basin. *In:* International Waters in Southern Africa. M. Nakayama (Ed.) pp. 136-163. United Nations University Press, Tokyo, Japan.

Turton, A.R., Meissner, R., Mampane, P.M. and Seremo, O. 2004. A hydropolitical history of South Africa's international river basins. Gezina, South Africa: Water Research Commission.

Wallis, S. 1992. Lesotho Highlands Water Project, Volume 1. Laserline, Surrey, UK.

Wallis, S. 1995. Lesotho Highlands Water Project, Volume 3. Laserline, Surrey, UK.

Wallis, S. 2000. Lesotho Highlands Water Project, Volume 5. Laserline, Surrey, UK.

Weisfelder, R. 1979. Lesotho: Changing patterns of dependence. *In:* Southern Africa: The Continuing Crisis, G.M. Carter and P. O'Meara (Eds.). Indiana University Press, Bloomington, USA. pp. 249-268.

Wolf, A.T. 1998. Conflict and cooperation along international waterways. Water Policy, 1, 251–265.

Australia

10

Water: An Essential Commodity and Yet a Potential Source of Conflict with Special Reference to Australia

Syed U. Hussainy[1] and Santosh Kumar[2]*

[1]Faculty of Engineering, Science and Technology, Victoria University
PO Box 14428, Melbourne City, MC8001, Australia
[2]Department of Mathematics and Statistics, University of Melbourne
Parkville, Vic. 3108, Australia
*School of Computer Science and Mathematics, Victoria University,
P.O. Box 14428, Melbourn City, MC8001, Australia

INTRODUCTION

Water, a public good and an essential commodity to sustain life is not only essential for biota but its unparallel properties directly influence the Earth's environment and its interaction with various forms. Life on Earth would not have been possible without its unique properties. The Earth is the only planet, that we know of, where water exists in liquid form in such quantities. For a conflict/controversy to arise, some form of disagreement must arise between two rival parties. In the case of water, there are many stakeholders often with conflicting interests. The Oxford Dictionary describes conflict as a clash, a disagreement, a quarrel, or a controversy. Water as a public good may create any one or more of the above between the stakeholders, be they neighbouring farmers, neighbouring states within a country or neighbouring countries. There are many international incidences of conflict related to use of water, damming of the waterways for generating hydroelectric power and/or agriculture and the discharge of pollutants in the river downstream before it enters the neighbouring country. The conflicts may also arise due to vested interest of the parties involved; these may include the politicians, farmers, industries, ecologists and bureaucrats and possibly the general public. Conflicts can be powerful, they are not just negative: they can bring about positive transformations in the society.

Sections 2 and 3 discuss the availability, utilization and supply of water. Section 4 presents briefly water-related conflicts arising between some nations and Section 5 focuses on the conflicts on the most populated part of the world, Asia. Section 6 presents flows in the river management system as a source of conflict. The Australian scenario has been presented in detail in Section 7. Finally, Section 8 focuses on a Hollywood film producer who selected water as a scarce commodity as theme for movie production. These discussions are indicative of potential conflicts that can arise due to water.

WATER AVAILABILITY AND ITS UTILIZATION

Although water is in 'abundance' in streams, lakes and rivers as freshwater and in oceans as saline water, it has been a source of conflict over the centuries, right through the biblical times (Hatami and Glerck, 1994). In some parts of the world, water is present in abundant quantities and in other parts water is a scarce commodity. The total amount of water is more than 1404 million km³ and covers 70% of the Earth surface (Cunningham, 1999). In recent years it has been shown that the ice comets that enter the Earth atmosphere contribute about 40 tons of water in the atmosphere. The cosmic rain over the past billion years may have been a major source of water on Earth (Cunningham, 1990). A gentle misty rainfall can refresh plants and animals and even converts dead brown soil into green meadows, whereas a violent thunderstorm and floods may result in washing away stream banks creating potential problems downstream.

Rain falls unevenly on the planet. In some places it rains more or less constantly while in some other areas there is no precipitation of any kind. For instance, at Iquique in the Chilean desert, no rain has fallen in recorded history, whereas Cherrapunji, in India, records 22 metres of rain in a year. Very heavy rainfall is typical of tropical areas especially where monsoon winds carry moisture-laden sea air on shore. Each year approximately 496,000 km³ of water evaporates from the Earth's surface. Ninety per cent of the evaporated water falls back on to oceans as rainwater. The remaining 10% is carried by winds over the continent, where it combines with the water evaporated from inland resources to provide a total continental precipitation of about 111,000 km³. A large proportion of the precipitation seeps into the soil and is stored as groundwater. The 40,000 km³ flows back to the ocean each year as surface runoff or underground flows represent the renewable supply available to sustain freshwater-dependant ecosystems including human needs. The global water budget is thus balanced by a circulation system in land, in the atmosphere, and in the oceans that move water from areas of excess to deficit.

Since the advent of agriculture, over 33% of the world's forests have been cleared and converted for other uses. The United States of America and Africa have lost about half of their forests while Brazil, the Philippines and Europe have lost 40%, 50% and 70% respectively. This increases the demand for good quality water for various anthropogenic purposes including agriculture.

At any single moment 97% of the Earth's water is in the oceans, the remaining 3% is freshwater (non-saline), of this 97% is locked up in polar ice, glaciers and in deeper inaccessible aquifers leaving only 0.003% available for the bioactivities including anthropogenic requirements. This small quantity of water available as a scarce resource has a potential to create conflict among its users especially in the industrialized world. The population of humans and animals including the domestic animals have expanded over the years destroying forests and stripping away protective vegetation from what was once a fertile land to exposing the base soil for erosion. For all life activities, clean freshwater is a necessity and through historic times, the availability of water determines where and when the biota inhabited the Earth.

WATER A SCARCE COMMODITY

South America and Asia receive about a quarter of the total global runoff and have about 12% of the total land area of the world. Most of the rainfall and runoff in South America occurs in the rainforests of the American basin where there is not much human habitation. Only 27% of the runoff occurs in areas inhabited by about 6% of world's population. In Asia, however, much of the runoff does occur in areas suitable for agriculture and caters for nearly 60% of the world's population.

In terms of per capita water supply, Iceland has the highest renewable water supply 670,000 m^3 per-person-per-year. Whereas countries like Kuwait and Bahrain have no renewable water supply, they rely entirely only on desalinated seawater. Egypt, in spite of the River Nile as a major resource, has only 30 m^3 of renewable water per-capita-per-year. Rainfall is never uniform in geographical distribution or yearly amounts. Water shortages have their most severe effect in semi-arid zones where moisture availability determines the distribution of plants and animals. The cycle of drought and wet years is cyclical and has the potential to affect the country in economic and social terms. During drought years, wasteful farming practices combined with the dry weather create 'dust bowls'. In February 1983 wind stripped topsoil from million hectares of land from Western Victoria, Australia and the billowing dust clouds turned day into night. The topsoil was carried as far as New Zealand. For the past six years

Victoria, Australia has been experiencing below average rainfall and many towns and cities in Victoria are experiencing water restrictions. However, in early February 2005 Melbourne got 175 mm of rain in a 36-hour period causing major floods in the western creeks and waterways. There is also a great worry that the climate change brought about by the Green House effect will make drought and floods more severe and more frequent than in the past.

WAR-MONGERING NATIONS AND WATER CONFLICTS

Water has been the basis of hostilities between neighbouring countries since biblical times in the most arid part of the world – the Middle East. In 1200 BC, when Moses and Israelites found themselves trapped between the Pharoah's army and the Red Sea, Moses miraculously parted the water of the Red Sea, allowing his followers to escape. In 695 BC in quelling the rebellious Assyrians, Sennacherib razed Babylon and diverted the principal canal so that its water was retained over the ruins (Hatami and Gleick, 1994). In 1503 Leonardo and Machiavelli planned to divert the Arno River away from Pisa during a conflict between Pisa and Florence (Honan, 1996). Water resources and water storage dams have always been a military target during the war periods. In recent times, the USA bombed dikes in the Red River delta, rivers and canals during a massacre bombing campaign in North Vietnam.

In 1975 South African troops moved into Angola to occupy and defend the Ruacana hydropower complex, including the Gove Dam on the Kunene River in order to defend the water resources of south-western Africa and Namibia (Meissener, 2000). During 1967, the Syrian attempt to divert water from the Jordan River ignited the six-day war between the Arab countries and their neighbours. As a consequence Israel was able to control the two essential water resources: the Golden Heights (the water shed for the Jordan River), and the mountain aquifer under the occupied West Bank Anwar Sadat (1979) declared the vital importance of water in his statement, "The only matter that could take Egypt to war again is water" (Gleick, 1991, 1994).

Similar potential conflicts have occurred between Egypt and its neighbors over the only water source, the Nile. During 1975 amassed its forces at the border when Syria reduced the flow in the Euphrates. During the 1991 Gulf War, Turkey threatened to completely cut off water flowing to Iraq and Iraq in turn threatened to blow up the Ataturk Dam if any moves were made to reduce its water supply.

Similar statements have been echoed by Boutrous–Ghali in 1988 when he stated, " The next war in our region will be over the waters of the Nile, not politics (Gleick, 1994). Gleick (2003) provides a chronological conspectus of conflicts wherein either water has been the source of conflict or water was used as the target during the conflict between different factions.

WATER AS A SOURCE OF CONFLICT IN
DEVELOPING ASIAN COUNTRIES

In recent times China has been going through a rapid economic development to provide the basic resources like water and electricity for its economic development. China has been constructing dams on major rivers such as Yangtze, Mekong and Yellow River. The origin of these major rivers is close to each other on the Tibetan plateau. The rivers, due to overexploitation, are in poor state and are known to cause floods and kill hundreds of people each year. The Yellow River, which has been historically relied upon in northern China often dries up in some parts altogether. The Mekong River raises complex issues for China and its South-East Asian neighbours such as Cambodia and Vietnam. The many dams, constructed to provide water and hydro-electricity, are responsible for the falling water levels when these rivers cross the international borders. These are the very same countries to whom China is currently trying to prove that its new status as a regional power does not present a threat. In recent months, the Chinese State Environment Protection Agency (SEPA) ordered 30 major projects to be stopped because they had failed to carry out environmental impact assessment. Two of these were the power stations for the giant three Gorges Dam and the third one was Xiluodu Dam in the mountains of Yunnan province. Even the state news agency Xinhua admitted that this dam would flood some of the world's most spectacular canyons and gorges (Spencer, 2005). The construction of these major dams and their potential to cause floods or falling of water levels in the neighbouring countries raises complex issues even for China and even these countries and can be the potential source of conflicts between them. One of the concerns cited by Spencer (2005) is that there are people in the government who support building dams in upper stretches of the river and some who oppose it. A few who doubt those party leaders are generally concerned about the environment, but for many other analysts the economic growth for the province over powers the losses. There is a whole new debate now as to how to judge officials. One is even talking about moving away measuring GDP growth to include things like the environment. Mr. Li Peng, the former Prime Minister of China played a major role in pushing the three Georges Dams through in the face of huge opposition. The people around Li Peng are still there and still powerful, and these people might not want to take on his son. Two leading power companies are controlled by children of Mr. Li Peng. However, China in the height of huge opposition is changing its policies related to power generation (Spencer, 2005).

Australia is already a crucial energy supplier for the booming Chinese economy and in 2004 signed a US $25 billion deal for liquefied natural gas from Australia. China has nine operating nuclear power plants, two in the pipelines and 10 in the planning stage (McPhedran, 2005).

FLAWS IN RIVER MANAGEMENT AS A SOURCE OF CONFLICT

Major rivers in Continental Europe, Asia, America and Africa face problems due to conflict of interests. Water management is a bigger issue than is generally thought of and one that requires a complete change in attitude and infrastructure. The major rivers like the Rhine, Mississippi, Nile and the major gorges cross many international borders as they meander through between their origins to their ultimate destination. As the water flows through the international borders, each country draws water to meet its requirements and often discharges the wastes both industrial and agricultural, sometimes inadequately treated into the main river or its tributary downstream safely ignoring that the same river water could be drawn for people's daily needs in the downstream countries. It is said that, "molecules of water in a tap in Rotterdam must have already been drunk by five different people as the Rhine River passes through cities and towns on its way from the Alps to the North Sea."

On the Indian subcontinent civil unrest took place in Karachi over severe water shortages, riots resulting in death and injuries as some of the ethnic groups accuse the government of favouring the populus Punjab (Pakistan) over Sindh Province in water distribution (Nadeem, 2001). In Gujarat (India) riots were reported during 2000 in the Jamnagar district during a protest against the authorities failure to provide adequate supply of water from Kankavati Dam. Inter-state disputes between the two South Indian states of Karnataka and Tamil Nadu have been taking place over many decades on sharing of Kaveri River water and the control of water resources. A tribunal was established in 1990 in order to settle the disputes, yet the conflict continues to take lives in both the states. The tribunal has not been able to settle the conflict (Butts, 1997). Similar disputes are not uncommon in the Middle East between Israel and its Arab neighbours since the 1967 war. This mainly involved the Jordan River, Golan Heights and the newly built East Ghor canal (Gleick, 2003).

WATER AS A SOURCE OF CONFLICT IN AUSTRALIA

Geographic Profile of Australia

Australia is the world's largest island as well as the smallest continent. It stretches the latitude of 10:41 S to 43:39 S. The distance between the northern and southern extremities is 3680 km, covering an area of 768, 2300 sq. km. Given its size it is not surprising that Australia is a land of great climate and topographic diversity. It ranges from dense tropical rain forests and mangrove swamps of North Queensland through the broad grassland of Eastern Australia, the alpine zone of snowy South Wales of northern Victoria to the vast desert of Central Australia, which extends westward to the central coast line of Western Australia.

Climate

The Australian continent is characterized by great climate variations. Its size and longitudinal range produce climates, which range from steamy tropics to cold, dry continentality. The tropic stretches across the north of the continent, resulting in subtropical and temperate conditions along the Eastern Coast. The snowy mountains in Southern New South Wales and the highlands of Tasmania experience alpine conditions while the vast central area is arid desert. Australia is a very dry continent. It is second only to Antarctica where the annual precipitation is less than 50 millimeters. More than half of the continent receives less than 300 millimeters of rainfall in a year. The extreme dryness is exacerbated by the domestic effect of continentally.

In spite of the regular rains in the north, most of Australia is vulnerable to extended drought conditions. The problem is that moist air masses rarely reach the continent centre: consequently there is dryness in the continent. The dryness in the continent is also extended by El-Nino. It produces an extended period of dryness over the northern part of the continent. This has a flow-on effect because of the rain, which would normally flow into the vast Murray-Darling River System do not arrive and consequently large areas of Western New South Wales and Queensland are adversely affected. Most of the densely populated Southern Australia experiences a modified Mediterranean climate, which is produced by movement of anticyclones or high-pressure systems crossing the southern part of the continent. While the southern east line experiences wet winters and dry summers further north of the coast is under the influence of tropical cyclones.

Tasmania on the other hand is most unique, with one of the most extraordinary landscapes in the world – rugged, ancient and diverse with temperate to alpine forests. Its geographical structure is completely different to the mainland (Millwood, 2003). During 1950 to 1980, the Hydroelectric Commission of Tasmania was damming rivers and creating massive hydroelectric schemes in the name of progress, with a vision of Tasmania becoming an industrial centre through cheaper electricity. During 1960s an ancient lake, Lake Pedar in the southwest highland mountains was drowned to create Gordon Dam to produce hydroelectricity in spite of a severe conflict with the naturalists, academics and others who dislike the destruction of endemic fauna and flora that inhibits the lake environment. Similar attempts were made to drown Franklin River for a similar hydroelectric project by the technocrats. The opposition to the project and the public pressure was so vehement that conflict between the government and the public throughout Australia reached to such levels that the government of Tasmania decided to withdraw the project. The Franklin and its surrounding environs have been described by the naturalists as the 'Mecca' of Australian landscape (Millwood, 2003).

Australian Water Resources and its Demand

Water supply to the major Australian cities to meet the daily requirements is predominantly via impounding or storage resources. Nearly 60% of Australians (Total population 20.1 million) live in New South Wales and Victoria, an area covering 12% of the continent. To put these figures in perspective that there are 2.3 Australians per square kilometer and that only a few countries have a lower population densities than this. It is also worth noting that there are fewer people in the whole of Australia than the in the greater metropolitan area of some of the major cities in the world e.g. Tokyo and Seoul. During 2000/2001 a total population of 13.3 million Australians were supplied with potable water and 12.3 million by Waste Water Services (WSAA Facts, 2001). Some 90% of the water and sewer connections supply residential households. Non-households include industrial, commercial, municipal consumers. In rural areas and remote residences the main source of water supply is from the rainwater collected into a tank from the roof of houses.

The major cities of Sydney and Melbourne rely on surface water impounded in reservoirs with adequate treatment as required to meet the NHMRC standards. Although the Brisbane city council pumps water directly from the Brisbane River, the river flow is secured by the reservoir managed by South East Queensland Water Corporation, the water wholesalers for the Brisbane region. Groundwater is a significant ongoing source for Perth (WA), New Castle (NWS) and Geelong (Victoria) (WSAA Facts, 2001).

In Perth groundwater usage continued at just over 50% for 2000/01. Hunter Valley (NSW) continued to supply 25% to 30% of its potable water needs from groundwater. It has recently commissioned an upgrade of groundwater storage system to allow increased capacity to source and substitute water quality problems which occur in 'surface shortages' (WSAA, 2001). Barwon water (Victoria), on the other hand, due to increase in rainfall in Geelong area has reversed its usage of groundwater with the usage falling to 40% compared to 52% in 1999/2000.

Although Australia is poorly endowed with surface water resources, it is fortunate in having approximately 60% of its area underlain by groundwater basins. Most of these are artesian, and the longest, the Great Artesian Basin covers approximately 20% of the area of the continent. One major problem/conflict with exploitation of the Australia artesian groundwater is in fact that it is salty and generally contains between 800 and 5000 mg of total dissolved solids per litre. Despite these restrictions and the fact that many bores in the Great Artesian Basin have to be sunk to depths in excess of 600 metres, the pastoral industry in the arid zone owes its existence to the presence of these groundwater resources.

Urbanization also introduces a conflict between the hydrological cycle, surface runoff deterioration of water quality in waterways and receiving waters. This will also introduce a conflict between the availability of water, its quality and quantity to meet all the requirements for potable, industrial and agricultural.

The unpredictability and unreliability of precipitation have an important bearing on the characteristics of Australian surface runoff across the continent. These include:
1. The average annual flow of Australian rivers is small by comparison with equivalent river systems around the world.
2. A considerable variation in stream flow throughout the continent.
3. A marked seasonal variation in stream flow.
4. A marked variation in stream flows from year to year.
5. A different seasonal variation in stream flow exhibited in various parts of the country.

The high annual and seasonal variability and the large flood flows lead to problem of storing and developing water resources, which are particular to Australia.

Water Resources Administration

One of the most important aspects of water allocation and management in Australia is the legislative system adopted by the Australian states in the late 19th century. The Irrigation Act of 1886 stated that all streams are public property. It abolished all riparian rights and nationalized all surface water in Victoria. Other states have since followed the lead given by Victoria from much of the legal arguments over water rights.

Water resources development and management are therefore the responsibility of the individual states. In most states this responsibility is shared between two governing bodies. One agency has the prime responsibility for urban supply to the capital city and its environs, while the other deals with assessment and development of all water resources. In addition to the states activities, the Commonwealth Government has played a leading role in establishing co-operative organizations to manage river systems and development of projects, which are of importance to more than one state and may pose potential conflicts among these states. For example, as a result concern that irrigation development along the River Murray was in serious conflict with the navigation on the river, the Murray River Water Agreement was signed in 1915 by the Commonwealth, New South Wales, Victoria and South Australian governments. The agreement established the Murray Rivers Commission to regulate the flow in the river to meet the requirements of each state. The Commission acts as a regulatory agency to ensure an equitable distribution in the river system and storages. Another example of interstate and commonwealth co-operation in the Snowy Mountain Authority created in 1949 to supervise the construction of the Snowy Mountain hydroelectricity to New South Wales and Victoria, the scheme diverts water for use in irrigation projects further west.

Australian Water Resources and their Development

Much of Australia's history and the pattern of European settlement has been centred on the search and demanded for reliable water supply. Some of the names of the rivers illustrate the importance of the river as a sustainable water resource for their settlement. One such river in Victoria is Werribee, meaning, "back bone" in their native language. Most of the Australian population is concentrated along the north-east, north-west, south-east and south-west coastal regions as these are the areas relatively endowed with surface water resource. Water is mainly used for three major purposes in this country.

1.   The first use is to meet urban water demand of the major metropolitan areas.

2.   The second major use is for large-scale irrigation and stock supply schemes, which have been developed primarily in the Murray-Darling Basin.

3.   The third usage is the generation of hydroelectric power in the Snowy Mountains and Tasmania.

Hydroelectric power development is largely confined to the Snowy Mountains and Tasmania, although a number of major hydroelectric schemes are to be found incorporated with other major reservoirs or as a major system located along the eastern side of the Great Dividing Ranges. Tasmania has led Australia in developing its water resources for hydroelectric power generation. Although this resource has formed the basis for much of the industrial development in Tasmania, in recent times it has been the basis for conflict between the community and the civic authorities. This has been discussed in one of the earlier sections above. It clearly demonstrates the public concern for the development of water resources and its impact on the environmental and social wellbeing.

The Snowy Mountain Schemes is a multipurpose scheme designed to provide water conservation as well as hydroelectric power generation and to divert the headwaters of Snowy Rivers into the upper reaches of Murrumbidgce River and Murray River System. In recent years due to a prolonged drought and low runoff from the catchment into the stream and its storages conflict arose as to how much water could be used to generate hydroelectric power and how much water be released to sustain the ecology of the river system. There have been several debates between the various states governments, Federal Government and the Murray-Darling Basin Authority.

It can be seen, therefore for Australian rivers, there is a conflict between the degree of regulation and the variability of stream flow and that, because of the evaporative losses, larger storages does not necessarily lead to higher yield or better efficiencies on economic of scale.

One such river, wherein many governments have an interest, is the Murray-Darling River Complex. The catchment for the streams together forms the largest river system covering four states (Queensland, New South Wales, Victoria and South Australia) and one territory (The ACT). A total of 33 different government departments in the various states have responsibility for it and about 256 local governments have a stake in it. The region of the Murray-Darling catchment covers only about 14% of Australia's irrigated land and grazes, about 50% of sheep and a 25% the cattle. The region also produces about 40% of rice and 80% of canned fruit. Many small towns and farmers draw their water from and dispose of their waste into the river of the system. Adelaide, although outside the catchment, relies on the Murray River for about 50% of its water requirements. About three quarters of all the water used in the country comes from the rivers in the region (Annon 1994).

The conflict is that users upstream can affect the quality and quantity of the water available for those further downstream. The main contaminant of the river water is salt. The river carries about 2.5 tonnes of salt into the South Australia every minute. Most of this comes from inflows of saline groundwater. These inflows are caused by rising water tables, which in turn are brought about by irrigation and clearance of native vegetation and its replacement with shallow rooted crops. For sustainable development ecology should be regarded as a part of economics. However, the myopic view of technocrats, see the two as being antagonistic rather than synergistic in achieving the desired sustainable end. With this viewpoint recharge rates have increased by up to 20 millimeters a year as against 0.1 millimetre in the case of native vegetation. The groundwater is therefore rising, bringing dissolved salt to the surface. Irrigated areas around the Murray River, especially in Victoria are now badly affected by irrigation salinity. As a consequence 50% of the river salt loads comes from anthropogenic activities in the catchment.

Due to overutilization of water resources in the basin, the river during the summer months has been subjected to low flows and even been infested with blooms of cyanobacteria in the early 1990's. Quite often, a sand bar blocks the mouth of the river and the river discharge to the ocean is often reduced to a trickle. Due to increase in salinity, many of the river red gum trees in the area are dying from increase in salinity or lack of quality of water. The survey commissioned by the Murray-Darling Basin Commission showed that in just 18 months the number of river red gums and black box trees counted as stressed, dying or dead had increased from 51.4% to 75.4%. The decline in the tree population was universal across hundred of kilometers of flooded plains in Victoria and South Australia. As a result of dams, irrigation dams and other uses, the Murray's flood plains are no longer frequently flooded, a decent flood to occur once every three years. They now occur once in ten years and are unable to flush away the accumulated salts. The stress due to infrequent flooding has been compounded by the extensive irrigation and associated elevated saline groundwater leading to a permanent loss of trees in parts of the flood plains and no capacity for recovery. Hence there is a conflict between the river flows to maintain the ecology of the river and the flood plains and the harvesting of the water to meet the increase in demand for industry and other needs.

Short-term Management Steps to Alleviate the Problem

As a temporary measure to alleviate the problem the Governments involved agreed as a first step to release 500 billion litres to Murray, but the process has been delayed by bureaucratic formalities (Fyle, 2004).

A worst case scenario would suggest that the remaining trees, not currently displaying signs of stress may show such signs within the next 15 to 18 months without a significant change in environmental condition such as increase in rainfall or extensive flooding. The loss of trees would have a compounding effect on the fauna that depends on these trees for food and habitat. The decline in the quality of the ecosystem would have an impact on the aquatic biota including the native fish. The conflict of interest in the overutilization of water resources and the river ecology eventually leads to a decline in environmental quality reduction in the economic output, export potential of goods and services and eventually between the various states that depend on the river system to grab a lion's share to sustain their economy and quality of life in that realizing that there is no conflict between economy and ecology support economy and a better quality of life. Water management is a serious issue that requires a complete change in attitude and infrastructure.

ROLE OF WATER IN THE CONFLICT BETWEEN THE RICH AND THE POOR

It is a known fact that some observations can lead to major events in one's life. For example, a tale from the Indian epics describes an observation and its consequence well. It is on the life of Prince Siddhartha, who experienced a major change in his own life by observing a hungry person, a sick person and finally the body of a dead person. He went in isolation, and later on became the world famous 'Lord Buddha', whose legacy has stood the test of time.

Similarly, a water-related observation was made by an Indian born Hollywood filmmaker, Shekhar Kapur while travelling from Mumbai Airport to his hotel in Colaba, an upper class suburb with modern hotels and facilities comparable to the West. He observed that on the surface roads were wide and smooth but under the overpasses, India's most populous slums existed in Mumbai. It is estimated that the population of 600,000 is confined under these overpasses on an estimated land of 175 hectares, which is half the size of New York City's Central Park. Kapur observed that the only freshwater supply they had is the water that seeps out of cracks in the pipes that carry water to steel-and-glass blocks in the downtown Mumbai. He observed that these people living in these slums, known as *'Dharavi'* can adapt to small space, live shoulder to shoulder, run miniature factories, and keep even animals in the same space they have. He concluded, "Water will soon be the World's most valuable commodity, and places like *'Dharavi'* will have none. He decided to undertake this subject as the theme for his movie project, called 'Paani', which means water in Hindi. This film is his obsession; so much so that he left some unfinished projects to undertake this one on 'Water', and has moved to India leaving behind the world of Hollywood. The plot for this production as given by Perry (2005) is:

"There is an upper city like Vegas and lower city like the colony of *Dharavis*. All the water has been sucked by the upper city and skinny girls of *Dharavi* are going to be 'water rats' and the heroes in the film. These heroes live in dry pipes and crawl to the upper city and steal water. Corrugated iron shacks are going to be the lower city, where they make everything for the upper city. No laws, no health and safety no restriction on population, a total free enterprise. In the vision of Kapur, these two cities meet only once when upper city ventures in the lower city's Paradise Club, where any thing goes. In the movie the divide between the rich and the poor leads to a revolution, mainly due to water. Kapur has made his point well in the films he has directed before and in this one his target is the importance of water. He believes that money can dazzle the rich and blind them to know the plight of the poor. Water being an essential commodity is used as a major part of the play.

CONCLUDING REMARKS

Thus, the water conflicts, are not only drawing the attention of various governments and planners as discussed in this chapter but has now come to the attention of the entertainment industry. Some producers have had remarkable success in teaching the general community those social issues, which were impossible tasks for many governments. Our belief is that based on his past records, Kapur is likely to demonstrate water conflicts and water scarcity well in the form of a simple story. It is time when users of water do understand how precious is this commodity, which they have taken for granted for so long.

REFERENCES

Annon, 1994. Environmental Science, Australian Academy of Sciences.

Butts, K. (ed.), 1997. Environmental change and regional security, US Army War College.

Chivas, D.D. 1994. Environmental Science Action for Sustainable Future, The Benjamin/ Cummings Publishing Co. Inc.

Cunningham, W.P. and Saigo, B.W. 1999. Environmental Science: A Global Concern, 5th Ed. The McGraw-Hill Company Inc. NY, USA.

Fyle, M. 2004. Flood Murray gums dying of thirst, The Age, Melbourne, November 14.

Gleick, P.H. 1991. Environment and Society: The clear connection, Bulletin of the Atomic Scientists.

Gleick, P.H. 1994. Water, war and peace in the Middle East, Environment, 36, 6-9.

Gelick, P.H. 2003. Water conflict chronology, www.worldwater.org/conflict.htm

Hatami, H. and Gleick, P.H. 1994. Chronology of conflicts over water in legends, Myths and History of the Ancient Middle East, Environment.

Hindu, The, 2002. Ryots on the rampage in Madya, The Hindu (Indian National Newpaper), October 31.

Honan, W.H. 1996. Scholar sees Leonardo's influence on Machiavelli, The New York Times, December 8, 18 pp.

Hussainy, S.U. and Kumar, S. 2006. Fresh Water A Scarce Resource: Its Global Utilization and Sustainable Work Practices in Melbourne, Australia, Water: Global Common Global Problems, V. Grover (ed.), Science Publishers, USA.

Hussainy, S.U. and Kumar, S. 2006. Liquid waste management and the utilization of the treated effluent with reference to Melbourne, Australia, Water: Global Common Global Problems. V. Grover (ed.), Science Publishers, USA.

McPhedran, I. 2005. China to buy uranium, Herald Sun, Melbourne, February 8.

Nadeem, A. 2001. Bombs in Karachi killed one, Associated Press, June 13.

Millwood, S. 2003. Wilderness: The vision, Wildlife Australia, 40(3), 24-27.

Perry, A. 2005. The Numbers Man, Time – An Australian weekly, April 4, pp. 62-63.

Spencer, R. 2005. China moves to Protect great rivers, The Age, Melbourne, February 4th.

WSAA Facts. 2001. The Australian Water Industry Water Services Association of Australia.

Comparisons between Continents

11

Enhancing Sustainability in River Basin Management through Conflict Resolution: Comparative Analysis from the U.S. and South Korea[*]

Young-Doo Wang,[1] William James Smith, Jr.,[2] John Byrne,[1] Michael Scozzafava[3] and Joon-Hee Lee[4]

[1] Center for Energy and Environmental Policy, University of Delaware
[2] Department of Environmental Studies, University of Nevada, Las Vegas
[3] U.S. Environmental Protection Agency
[4] Ministry of Environment, South Korea

INTRODUCTION

Humans often alter river basins to such an extent that they are arguably as much cultural, as physical, constructs. Cross-basin transfers of water, engineering of stream channels, and effluents of massive quantities of point and non-point pollution from pipe, earth and sky merely manifest a few of the more obvious examples of how the modern basin is being impacted by human activities. The scale of such endeavors across time and space is truly astounding. From the Three Gorges Dam in China, to the often small and forgotten relics from times gone by that are broken dams blocking fish passages in the U.S. – humans alter landscapes and hydrology at multiple scales to place it in the 'service of man' (McPhee, 1989).

These complex activities can take place in river basins that form a setting of intense competition – competition for water for economic development, for maintaining ecology that provides for livelihoods and food, for recreation, and for distribution amongst people of varying political and economic statures for direct consumption, and more. Maintenance of any balance struck between such diverse stakeholders and competing

[*] Wang and Smith Jr. are co-lead authors (E-mail/Wang: youngdoo@udel.edu; Smith Jr.: Bill.Smith@ccmail.nevada.edu)

needs must consider both short- and long-term views of conflict resolution for multi-generational equity and sustainability to be achieved.

Conflict is inherent in river basin management,[1] wherein diverse 'stakes' are held, multi-purpose resources are shared, political leaders answer to many forces both within and beyond the basin, and heterogeneous visions of the past, present and future often collide. Thus, conflict mitigation is a core element of a sustainable basin management. Issues causing conflicts typically include, but are not limited to, diminished access to resources caused by impairment, denial of use for economic purposes, property rights, and concerns regarding environmental health.

The purpose of this chapter is to offer insight regarding how to mitigate conflict caused by multi-stakeholder competition for water in river basin settings. A lot of interesting work on water conflict has been in international in scope (Yoffe et al., 2004; Espey and Towfique, 2004; Just and Netanyahu, 2004; Mostert, 2003; Beach et al., 2000; and Murphy and Sabadell 1986). This chapter provides a comparative analysis of river basins that are situated in the U.S. and South Korea. The core argument of this chapter is that sustainable water quality and quantity (W) in a given river basin can be enhanced by conflict resolution through balanced consideration of socio-political equity (E), ecological viability (E) and economic development (E), or **WE³**.

LITERATURE ON WATER CONFLICT RESOLUTION

Science clearly takes a 'back seat' in conflict mitigation analysis. In fact, although science provides important background data and information for making decisions, such as water quality measurements, conflict and its mitigation in the search for 'equity' are both very much cultural constructs.[2] Therefore, unlike some aspects of water management, it is not possible to ascribe very specific universal techniques to address conflict. Rather, only broad principles can be offered with any certainty and universality. The importance of 'place' results in incubating diverse conflicts, and necessitates the creation of equally diverse ways to address them.

It may be no longer necessary to plead a case for the worth of holistic basin management in academic literature. However, if the general public cannot make the connections associated with holistic basin management, improvements in governance including conflict mitigation are unlikely. This is one of the reasons that cultivating the notion of shared space is so

---

[1] Flack and Summers (1971) point out as early as the 1970s that computer-aided conflict resolution can also help facilitate making problems clearer in complex multi-user scenarios.

[2] As McGinnis et al. (1999) note, "...like science itself, watershed planning is carried out in a context that is conditioned by culture and society."

important when trying to help community-based organizations and the public adopts a basin-based identity, in addition to their many political ones. This shared identity provides a rationale for cooperation, rather than conflict (Smith Jr., 2003). Fostering a basin-based identity can help make clear to stakeholders just how vital cooperation is for positive change.

McGinnis et al. (1999) take a bioregional perspective regarding choosing the basin as a unit of analysis. Interestingly, they combine activism and ecology, two features that reveal themselves as key elements in the case studies that follow. They state that:

> Generally, bio-regionalists are involved in a process of cultural change at two levels – as a conservation and restoration strategy, and as a political movement that calls for devolution of power to ecologically and culturally defined regions and watersheds (1999).

The basin and bioregion is selected as a unit for conflict resolution because it lends itself to 'integrated' and 'comprehensive' approaches. But the authors also note that basin-based approaches nurture community-based organizations that can foster conflict resolution, support ecology, and maintain local control over resources (Thomas, 1997). To this end, they note that even the U.S. federal government (especially the Environmental Protection Agency) has funded many such groups.[3]

There are many studies that have been suggested for understanding basin conflict. This is likely a result of what Dinar (2004) notes as a trending upwards of interest in conflict and cooperation in the water resources arena in recent years. According to Dinar, one of two possible explanations is viable. First, water is a source of increased conflict and cooperation. Or, communication about conflict and cooperation has been significantly enhanced. Core issues, such as economic 'development' upstream resulting in water quality impairment and endangering public health downstream, are not exclusively national or international in nature. Nevertheless, much interesting work has been international in scope.

Yoffe et al. (2004) explore a large Trans-boundary Freshwater Dispute Database that provides a framework for quantitative explorations of the relationships between freshwater resources and international cooperation and conflict.[4] Espey and Towfique (2004) use a logistic model to determine factors that have influenced the formation of bilateral international water treaties over the last 60 years. They report that the larger a water basin in

---

[3] As urbanization expands across the globe, the traditional connection with the natural world and the understanding of river's essential place in the sustainability of life diminishes without purposeful intervention. For example, it is difficult to see a basin between skyscrapers and through pavement covering once viable streams – so basin-based arguments find little public support without extra efforts to intervene by government, NGOs, or both.

[4] Mostert (2003) also provides an interesting analysis of conflict and cooperation at a global scale.

terms of percent of a country's size, the more likely the country is to form a management treaty. Notably, the more control over the watershed a given country has, the less likely it will sign a treaty.

Just and Netanyahu (2004) use the case of the Palestinian-Israeli shared aquifer to explore relationships between interconnected game theory and conflict mitigation:

> … application of interconnected game theory to modeling of bilateral agreements for sharing common pool resources under conditions of unequal access. Linking negotiations to issues with reciprocal benefits through interconnected game theory has been proposed in other settings to achieve international cooperation because it can avoid outcomes that are politically unacceptable due to the "victim pays" principle (2004).

Murphy and Sabadell (1986) suggest that decision-makers (in various countries) have at their disposal legal bases for agreements buoyed by mainly hydrologic and economic models that are expected to produce 'equitable water allocations' that may mitigate conflict. The authors state that the under-appreciated gap that needs to be filled is a tool for testing the political aspects of proposed solutions to conflicts. Murphy and Sabadell propose a theoretical model to address this issue by objectively measuring the impact of an individual country's political decisions upon the negotiations between them.

Beach et al. (2000), in their discussion of 'general theory conflict resolution,' point to authors such as Kaufman and Bingham who have investigated the potential for consensus building in its various related forms, such as 'joint problem solving.' The authors, in addressing international river basin negotiations, note that being locked into extreme positions in negotiations leaves little space for bargaining. This does not leave room for the necessary acknowledgment of shared sovereignty necessary to compromise.

McGinnis et al. (1999) propose that "long-term watershed planning requires rebuilding a community-based infrastructure that can support important social and bioregional networks and partnerships" (1999). These authors concede the importance of a formalized process of collaborative decision-making and conflict resolution. They state that the collaborative decision-making model focuses on three issues: 1) Reliance upon scientific information; 2) Neutral facilitation and mediation (including issue audits); and 3) Public participation.[5]

---

[5] In addition, seven general characteristics of "less formal" collaborative decision-making include: representation for weaker parties; equal access to scientific information; participation; accountability and legitimacy; commitment to the process through implementation (resources may be offered); sustaining cultural values; and creating *adaptive* decision-making.

Sneddon (2002) points out that some models based on co-management aspire to constant cooperation and 'win-win'-like scenarios. The author examines this theory in practice in a 'medium-sized, altered' river basin in NE Thailand, and explores obstacles and co-management techniques for mitigating social conflicts occurring in the study area. He points out in the course of this analysis the mismatch in scale of governance between resource management agencies with natural unit boundaries and governmental ones with boundaries dictated by politics. Sneddon notes how this mismatch in scale can impact decision-makers and make them feel insecure regarding the questioning of their logic and authority. Thus, it is the institution itself that may begin to wage conflict on those who wish to co-manage.[6]

Steinberg and Clark (1999) point out that many basin conflicts have less to do with abhorrence towards others, and more to do with 'positive attempts' to hold on to or reclaim a place. At times, what the conflicts are about on the surface is merely a representation of the tug-of-war regarding values occurring under the surface. Recognizing this as being the case is worthwhile as the shareholders position to begin negotiations. In addition, this exposes the fact that conflict analysis that merely paints a picture of upstream-downstream clashes over resources may lack proper context.

Smutko et al. (2002), based on original data from stakeholder assessments conducted in North Carolina, indicate how individual interests or preferences affect stakeholder participation. A factor such as level of uncertainty can increase the need for collaboration, but can, at the same time, decrease the willingness to engage. Conversely, the need for collaboration can decrease as clarity increases, while willingness to engage may increase as the problem becomes clearer. Other factors like risk and urgency of decision show a positive relationship between need for collaboration and willingness to engage. The authors conclude that citizens are more likely to get involved when they perceive an issue to poise some type of threat to their welfare.[7]

Leach and Pelkey (2001) explore how to make watershed partnerships successful and mitigate conflict by reviewing the empirical literature on

---

[6] The author goes on to point out the similarity between co-management regimes and conceptual models for "common-pool resources" managed by diverse groups. The idea is that collective actions for "sustainable watershed management" must ensure that all stakeholders in a basin are included in the building of management schemes and third parties develop cooperative avenues.

[7] The authors used an issue attribute approach to evaluate the need for collaboration and stakeholder willingness to engage for specific issues based on a number of fundamental explanatory characteristics. This short list of seven attributes is based in part on other studies (Yoder, 1999), and includes: 1) Level of uncertainty; 2) Balance of information; 3) Risk; 4) Time-horizon effects; 5) Urgency of the decision; 6) Distribution effects; and 7) Clarity of the problem.

the subject in order to generalize what has been learned and suggest future research paths. After reviewing a compilation of 37 studies on the issue, they were able to form 210 conclusions that were further simplified into 28 themes. Through a factor analysis utilizing Sorensen's similarity index,[8] they were able to explain 95% of variance in themes based on four broad factors: 1) Resources and scope; 2) Flexibility and informality; 3) Alternative dispute resolution; and 4) Institutional analysis and development.[9]

Demand-side management can also play a role in avoiding conflict over scarce resources. If water pricing is done properly, sustainable water use can be fostered as significant conservation is achieved without overburdening low-income customers. This occurs while reducing water intakes from streams so as to mitigate any stresses in stream ecosystems and increasing water volume for competing downstream users (Wang et al., 2005).

BUILDING ON THE LITERATURE

In summary, water conflict literature reviewed in this chapter highlights the following:

- Reliance on scientific information and hydrologic and economic models (Smutko et al., 2002; McGinnis et al., 1999; Yoder, 1999; Murphy and Sabadell, 1986);
- Notion of shared space of bio-region and co-management (Smith Jr. 2003; Sneddon, 2002; Leach and Pelkey, 2001; Beach et al., 2000; McGinnis et al., 1999);
- Politically acceptable and socially equitable resolution (Just and Netanyahu, 2004; Murphy and Sabadell, 1986);
- Local control and public participation (McGinnis et al., 1999; Thomas, 1997);
- Uncertainty, risk, and values under the surface (Smutko et al., 2002; Steinberg and Clark, 1999);

---

[8] A formula is used in biodiversity studies, it calculates similarity based on the number of themes two factors have in common.

[9] Factor 1 explains 24% of variance and suggests odds of a successful outcome are enhanced when a partnership has adequate time, abundant resources, and a limited scope. Factor 2 explains 21% of variance and stresses the fact that a partnership's strength is its ability to provide a flexible, informal alternative to traditional forms of resource management. Factor 3 explains 33% of variance (including funding) and describes methods that partnerships can employ when participating in negotiation or resolution processes. Factor 4 explains 17% of variance and uses a "rational actor" model of collective action to explain why some communities overcome the desire to "free ride" and effectively manage common resources.

- Conservation and restoration, economic development and environmental health (Wang et al., 2005; Dinar, 2004; McGinnis et al., 1999; Thomas et al., 1997);
- Property rights and percent of control over the watershed (Espey and Towfigue, 2004; Just and Netanyahu, 2004);
- Neutral third party facilitation and medication (Sneddon, 2002; Beach et al., 2000; McGinnis et al., 1999);
- Mismatch in scale of governance between resource management agencies (Sneddon, 2002);
- Joint problem solving and common ground (Beach et al. 2000);

The literature points to major values of finding politically, socially, environmentally, and economically justifiable common grounds in resolving water conflicts. In fact, it is pointed out that 'Negotiated Rulemaking' was established as an official tool in a U.S. Congressional Act in 1990. Again, the dimensions of water conflict resolution such as third parties to facilitate,[10] public participation,[11] and reliable information,[12] and shared space and values[13] are underscored.

On the basis of these findings in the above literature, we investigate how conflict resolution can enhance opportunities to achieve sustainability in river basin management. Conflict resolution, for the purpose of this chapter, is characterized as a method of mitigating conflicts through transparent, democratic and participatory 'social transactions' or 'trade-offs.' Sustainability is defined here as a measure of the potential to enhance current and future water quality and quantity (W) in a given river basin through balanced consideration of social-political equity (E), ecological

---

[10] "Efficiency" for some may result from limiting the 'interference' of third parties or NGOs. However, not carving out a role for civil society in the process will likely reduce democratic practice, spur public resistance, rather than engage the public in pushing forward progressive ideas, and likely damage opportunities for sustainable resolutions.

[11] It can be argued that a better role for government to play in regard to civil society and resource management is to empower it. Thereby allowing the public to become a positive force in pushing positive action in a basin and encouraging institutional support (Smith Jr., 2003). This stands in contrast to assuming a position wherein "giving away" power makes a government feel weakened or threatened. This is, at times, an alternative to bringing issues to court. Also, there is a difference in approach in that empowering is often *proactive* in nature, and this has the potential to mitigate conflict rather than waiting for periods of "crisis."

[12] Information *transaction* occurs between all parties, regardless of whether there is agreement regarding that information. This facilitates negotiation in the public sphere and governmental scales regarding what options are equitable and environmentally sound enough in the long and short-term to be considered sustainable (Fraser, 1993).

[13] The scope of analysis may have to expand beyond the basin scale to reveal forces underlying the stresses that are manifest in basin conflicts. Commodification of water resources can come packaged with a utilitarian logic that might not make recognizing other values (i.e. cultural) possible, and this can have harmful impacts, reducing opportunities for conflict mitigation. This is one reason that conflict management and basin management in general must be multi-scale, socially and ecologically adaptive for long-term success.

viability (E) and economic development (E), or **WE³**. Conflict is considered to be resolved when **WE³** is balanced.

Conflicts in regional river basin management can arise among coalitions representing varied economic, environmental and socio-political interests as shown in Figure 1. Coalitions can be formed in private and/or public entities when their members share interests. Conflict resolution is a process of addressing differences among stakeholders aims (**WE³**) so that the outcome is sustainable water. 'Sustainable water' is defined here as socio-politically equitable, economically efficient, and ecologically viable water management outcomes.

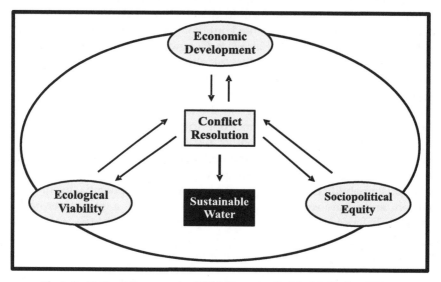

Fig. 1. Institutional Framework of WE³: Toward a Sustainable Regional River Basin Management

In order to explore the viability of the **WE³** approach, conflict resolution cases are examined using two U.S. and two South Korean river basins. Before introducing the Delaware (established in 1961) and Susquehanna River Basin groups (established in 1973), a brief review of compacts, a prime tool for conflict management in the U.S., is undertaken. This is followed by the South Korean cases of Nakdong and Han River Basin being evaluated. Comparative evaluations of both U.S. and South Korean cases are then considered in relation to the **WE³** framework.

U.S. Water Compacts

Sherk (2004) has compiled a bibliography of interstate water compacts. A number of important interstate water compacts were established in the

U.S. during the early 1950's. Literature during this period recognized the phenomenon, noting the significance of such agreements for water allocation and natural resource conservation (Zimmerman and Wendell, 1952; Lepawsky, 1950). While the interstate compact is actually older than the Constitution, use of such agreements in natural resource issues had been limited before this time. Interstate compacts were most often used to settle disputes directly related to boundary areas. However, with the 1950's came the emergence of more interstate compacts than in any time of U.S. history (Zimmerman and Wendell, 1952).

In the United States, the present status of interstate water compacts indicates a significant resource management divide between the eastern and western states. While some western compacts address issues other than water allocation, in most cases they do not (McCormick, 1994). This is in significant contrast to the eastern states, where these agreements delegate significant power to multi-state regulatory commissions, including the power to allocate water and approve specific water resource projects in their jurisdiction. This difference may be, in part, a result of the profound contrast between Eastern (Riparian) and Western (Prior Appropriation) water law.[14]

Federal-interstate compacts are widely recognized as more effective than interstate compacts when it comes to managing interstate water resources and resolving multi-stakeholder conflicts (Sherk, 1994; McCormick, 1994). Federal-interstate compacts use federal-interstate basin commissions as a means to provide for joint exercise of sovereign powers over water resources in river basins. As the preferred institutional arrangement for water resource planning and management in multi-state watersheds, the National Water Commission first recommended in 1973 that federal-interstate compacts be utilized to resolve interstate water conflicts.[15]

Often, water allocation and diversion disputes between states are addressed using different methods like litigation, legislation, and market mechanisms. While these methods have their strengths, most agree the establishment of a federal-interstate compact is the most effective strategy

---

[14] Riparian law ensures no user impedes another adjacent user's water rights, while Prior Appropriation guarantees access for only those with primary right to the resource ("first in time"). Eastern water law allows upstream residents to use the resource as needed, so long as their consumption doesn't hinder the rights of downstream users. Since western water law developed according to a first-in-right basis, water is like a property right. Any management scheme that takes away that property would be met with resistance (McCormick 1994). Thus, while law clearly designates user rights in the west, allocation in the east needs to be regulated.

[15] However, it seems this recommendation has not been followed, as the number of compacts approved since that time has decreased when compared to those approved before 1973 (Sherk, 1994). (Of course, this may partially be due to the fact that the most urgent scenarios may have been addressed earlier.)

for resolving interstate water disputes in the U.S. This is especially true when the compact leads to the establishment of a regional management entity (Sherk, 1994).[16] Four federal-interstate compacts have been enacted in the United States: the Delaware River Basin Compact (1961); the Susquehanna River Basin Compact (1970); the Alabama-Coosa-Tallapoosa River Basins Compact (1997); and the Apalachicola-Chattahoochee-Flint River Basins Compact (1997).[17]

THE CASES: TOWARD SUSTAINABILITY

The Delaware River Basin

The Delaware River, 330 miles (531 km) in length from Hancock, New York to the Delaware Bay, is the longest un-dammed river east of the Mississippi River. The Delaware River directly drains an area of over 13,539 square miles (35,066 sq. km) that includes the states of Delaware, New Jersey, New York and Pennsylvania along the east coast of the United States (Figure 2). Due to the massive region that is supplied water by the river, approximately 15 million people rely upon the Delaware River Basin to provide water for drinking and industrial use (DRBC website 2005). Consequently, the regional influence of the Delaware River across political boundaries requires management so as to integrate all concerned federal, state, and local organizations and governments.

After several decades of litigation between Delaware, New Jersey, New York and Pennsylvania over the Delaware River, the Delaware River Basin Compact was envisioned (Featherstone, 2001). The Compact, established on October 27, 1961 recognized for the first time since the nation's birth an equal partnership of the federal government and a group of states for river basin planning. This resulted in the development of the regulatory agency named the Delaware River Basin Commission (DRBC). Since 1961, the DRBC, represented by the governors of the four participatory states and a federal designee, has been charged with regulatory, management, planning and coordination powers.

Within the selected powers are the duties of water quality protection, water supply allocation, regulatory review (permitting), water conservation initiatives, watershed planning, drought management, flood control, and

---

[16] Such an organization can be effective in implementing policies and addressing complex intergovernmental problems spanning state boundaries and agency functions. This effectiveness, however, is dependent on the political and financial support of all parties involved with the organization (Featherstone, 2001).

[17] A fifth compact, the Interstate Compact on the Potomac River Basin (1940), has federal membership, but the federal government is not a signatory party to the Compact (Featherstone, 1999).

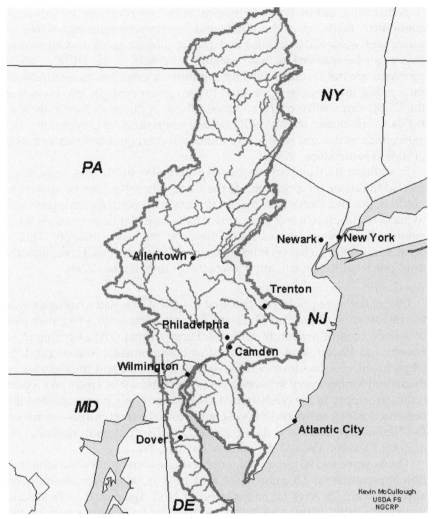

Fig. 2. Major tributaries of the Delaware River Basin
Source: Delaware River Basin Collaborative Environmental Monitoring and Research
Initiative, 2003.

recreation. These duties are accomplished through business meetings and
hearings on policy matters and water resource projects under regulatory
review, as well as meetings of advisory committees, all of which are open
to the public (DRBC website 2005). Given the scale of the area and wide
variety of rural, urban and mixed stakeholder interests that exist, it is not
uncommon for the DRBC to have to cope with conflicting citizens and
advocacy groups.

Addressing issues typically begins with consideration by advisory committees made up of the public and government officials, as well as water and wastewater utilities and public interest groups. Often most issues can be resolved without making proposals to the DRBC. When problems are not readily solved by committees, conferences, or seminars, then public meetings are sponsored to gain greater insight into ways that the DRBC can resolve conflict. Among those opinions that are critical to the decisions chosen by the DRBC include environmental groups, industry, agriculture, water and wastewater utilities, recreation, academia and civic groups (Featherstone, 1999).

In addition, the Compact requires all proposed policies, rules, regulations and additions or modifications of the Comprehensive Plan be subject to public notice and formal hearing. Finally, when the commission is prepared to make its decision, it still is driven to make sure that there is a consensus among all members, regardless of the need for just a majority. This is because the DRBC relies on funding from each signatory and consequently does not want to alienate any viewpoints (Featherstone, 1999).

The Conflict

One of the most public conflicts that the DRBC has had to mitigate was the construction of the Point Pleasant Pumping Station that began as part of a flood control project. In 1966, the Pennsylvania (PA) Department of Forests and Waters (now Department of Environmental Resources), U.S. Department of Agriculture-Soil Conservation Service and the counties of Bucks and Montgomery in Pennsylvania prepared a joint study and report on water supply in the Neshaminy Creek Basin. The report evaluated the construction of a series of 10 flood-control and/or multipurpose dams on the Neshaminy Creek and its tributaries, and two pumping stations, one at Point Pleasant, PA and the other at Yardley, PA.

The water would be pumped through a single underground transmission line approximately 2.5 miles (4.02 km) to a storage reservoir built on approximately 28 acres (11.33 hectares) of land designated as Bradshaw Reservoir. Also, water was to be pumped into the headwaters of the Neshaminy Creek where some of the withdrawal would be used for public water supply at the North Branch Water Treatment Plant in Chalfont, PA. The project was preliminarily approved by the DRBC and added to their Comprehensive Plan on October 26, 1966 (Delaware Water Emergency Group et al. 1981).

Controversy was to surround the construction. In 1974, the Nuclear Regulatory Commission granted a construction permit to the Philadelphia Electric Company (PECO) for the Limerick nuclear generation plant. The permit contained a provision for withdrawal of water from the Delaware River at Point Pleasant, and transportation of water through the use of transmission lines and the waterways of the Perkiomen Creek Watershed to the plant as additional cooling water. As a matter of protocol, environmental impact assessments (EIA) were prepared (as had been completed for the prior projects), and the withdrawal for the Limerick plant was approved. The validity of the construction permit was challenged through court proceedings where specific objections were raised as to the adequacy of the past EIAs in relation to the Limerick project. The Court of Appeals for the Third Circuit affirmed the issuance of the construction license in Environmental Coalition of Nuclear Power et al. v. Nuclear Regulatory Commission and Philadelphia Electric Co. (Delaware Water Emergency Group et al. 1981).

Work on the Chalfont water treatment plant began in the mid-1970s, but was then suspended when studies projected a smaller demand for public water in Bucks County due to lower projections of the future population. Subsequently, the plant was down-scaled from 80 million gallons per day (mgd) to 40 mgd, reducing the approximately 150 mgd maximum withdrawal to a 95 mgd withdrawal. In 1980 the DRBC made a decision on the basis of past environmental assessments by DRBC and the U.S. Department of Agriculture-Soil Conservation Service. This statement had concluded that the projects would be 'beneficial' to the communities of both creeks, and not detrimental to the Delaware River (provided various express conditions as to the control and use of water were enforced), and decided to recommend a 'negative declaration.'

A 'negative declaration' would do away with the need to prepare another EIA for the projects due to the reduction in maximum withdrawal. Public notice of intent to issue a negative declaration and of the preparation of the environmental assessment was given and public hearings were held. Consequently, the DRBC published a "Final Environmental Assessment For the Neshaminy Water Supply System" project sponsored by the Neshaminy Water Resources Authority (NWRA) and PECO, without an additional assessment of the decreased withdrawal. In 1981 the DRBC approved the Neshaminy Water Resources Authority's and PECO's application for the projects, with final approval for the construction of the pumping station, the conduits and the water-treatment facilities, as originally contemplated in 1966 (Delaware Water Emergency Group et al. 1981).

These issues in the 1970s were motivation for the formation of an environmental coalition (including many conservation and ecology groups) to rally against the proposed projects. The proposed discharge of river water into the relatively clean headwaters of two Bucks County streams caused outrage, as did the use of precious water resources for cooling dangerous and 'unnecessary' nuclear facilities. One of the most contested aspects of the projects was the use of the 'negative declaration' to accept the project after the overall withdrawal of water was decreased from 150 mgd to 95 mgd. This was underscored in 1981 through the case of Delaware Water Emergency Group, et al. v. Gerald Hansler (head of DRBC), Neshaminy Water Resources Authority and Philadelphia Electric Company.

Litigation was used to challenge the validity of approvals granted by the DRBC to PECO and NWRA to construct facilities for the withdrawal, diversion and use of water from the Delaware River by means of the pumping station at Point Pleasant, PA. The concerned citizens that became the plaintiffs in the case wanted to challenge the DRBC's approvals because they believed that the 'negative declaration' was a failure of DRBC to have a new, updated environmental impact statement (EIS) prepared. They also believed that there was a failure to adequately consider various potential environmental effects of the projects.

From the point of view of the defendants, all environmental issues were fully considered in prior environmental impact statements, and present facilities authorized by DRBC approvals merely down-scaled the size of previously approved projects. As the smaller size would have less adverse impacts on the environment, the defendants asserted that the 'negative declaration' was fully justified. The defendants asserted that every environmental impact had been fully studied and was carefully considered by the DRBC in the environmental assessment prepared for the present applications and in the prior analysis. And, that there was more than adequate public notice and participation, all appropriate governmental agencies had been notified, and the responses of the agencies were carefully considered prior to the approvals.

The judge overseeing the case concluded the proceedings by stating that, "Under the circumstances of this case, the decision of DRBC approving the applications of NWRA and PECO by way of a 'negative declaration' and without preparing another EIS was a reasonable determination based upon the facts presented to it" (Delaware Water Emergency Group, et al. 1981). Thus, in 1983, construction of the Point Pleasant pump began after two decades.

Following the beginning of construction, the opposition only grew more intense. In 1983, 'Del-Aware Unlimited' emerged as the non-governmental organization (NGO) leader of opposition to the project. They were able to obtain a non-binding referendum question on the ballot in Bucks County (whose taxpayers would provide the money to build the system) as to whether the constituents supported the construction. A 56% to 44% vote politically killed the Point Pleasant Pumping Station. Also through the election, the Bucks County electorate voted out the county commissioners who supported the project and replaced them with a majority of officials who had spent years fighting to stop the pump through the courts (Carluccio, 1987).

Consequently, the new commissioners appointed anti-pump members to the NWRA of Bucks County, which had originally contracted to build the pump, and the executive director of Del-Aware Unlimited, Tracy Carluccio, became executive secretary of the NWRA. In 1984, the NWRA shut down the project, but proponents for the project filed suit to have the construction contract enforced and the project reopened. The suit was settled in 1987 when a judge ruled the contracts valid and ordered the construction to resume (Stevens, 1987).

The Outcomes

Backed by litigation efforts, the Point Pleasant Pumping Station project was legally determined to be environmentally safe, and thus, the decision of the DRBC was upheld to allow the construction of the project. In 1987 construction of the Point Pleasant project resumed after three years, along with the arrest of more than 200 demonstrators violating court injunctions prohibiting interference at the construction sites. Ten objectors, ranging in ages from 20-67, were jailed for contempt, two of them fasting and only drinking water. 'Dump the Pump' became the slogan of anti-pump demonstrators who felt that forestland would be destroyed, and were against the inter-basin transfer of water over land (Carluccio, 1987). As the project was completed in 1994, the Bucks County Commissioners sold the finished Point Pleasant Pumping Station to North Penn and North Wales Water Authorities in Montgomery County for $55.2 million (Lazar and King, 1994).

Sustainability Implications

While some aspects of the **WE³** model have been resolved in the Point Pleasant pumping station project, some stakeholders believe the ecological component remains in doubt. It can be said that the use of the station itself is one aspect of the project where sustainability is found: 1) The station provides safe drinking water to 150,000 customers in the growing regions of Bucks and Montgomery counties (Partners 2003; North Wales Water Authority, 2003); 2) The station meets its required withdrawal limits as determined to be sufficient to preserve the environmental integrity of the area; and 3) The Limerick Nuclear Power Plant is supplied by the station which produces electricity for over 1 million homes with 2,268 net megawatts (PECO, 2003).[18]

The pumping station, however, has been blamed for erosion of the banks of the Perkiomen and Neshaminy Creeks, causing changes to the natural habitats of living organisms. In addition, a 95 mgd intake entraps fish and other organisms. Finally, some believe that the use of potable

---

[18] Also, according to Gregory Cavallo, a DRBC hydrogeologist, the Point Pleasant project has been the primary developed surface water supply that has helped to reduce the need for ground water withdrawal in southeastern Pennsylvania in drought years.

water for the production of nuclear powered energy is unsustainable, and thermal pollution remains a concern.

The level of sustainability of a project outcome depends on the extent to which balanced **WE³** through conflict resolution has been achieved. Without opportunities for public participation in the conflict resolution process, the Point Pleasant outcome would have been considered less sustainable. A few sustainability implications, however, can be pointed out in the case of the Delaware River Basin with respect to conflict resolution:[19] 1) Active public participation started in the implementation stage rather than in the planning stage; 2) Initial approval was based on the EIA required by the National Environmental Policy Act which is basically an expert-driven approach, and environmental impacts were not fully evaluated; and 3) Litigation was an effective mechanism to resolve conflict, but it is time and resource consuming and is less public-driven.

The Susquehanna River Basin

The Susquehanna River is the largest river situated entirely in the U.S. and flows into the Atlantic Ocean. With a 27,510 square mile (71,251 sq. km) watershed that includes parts of New York, Pennsylvania, and Maryland, it accounts for approximately 50% of freshwater supply to the Chesapeake Bay (Figure 3). Such a large, multi-jurisdictional region requires a unique management strategy that addresses the many state, regional, and federal interests. On December 25, 1970 the Susquehanna River Basin Compact was signed into a law by the U.S. Congress and the state legislatures of New York, Pennsylvania, and Maryland. This document provides mechanisms to guide conservation, development, and administration of water resources in the river basin. It established the Susquehanna River Basin Commission (SRBC) as a federal-interstate commission agency to coordinate state and federal efforts.

The Commission develops and implements water resource plans, policies, projects, and facilities necessary to carry out the Compact mission (SRBC Compact, 1972). SRBC provides coordination, management, pathways for communication, and resolves controversies. It conducts water resource investigations, surveys, and studies, and initiates legal action when appropriate (SBRC website 2002). The objectives include: reducing

---

[19] DRBC's objectives for conflict mitigation in the Delaware River Basin are found in the Comprehensive Plan established in 1962. The criteria they use to endorse any project or proposal incorporates social, economic and environmental aspects as follows: The project must provide beneficial development of water resources in a given locality or region; it must be economically and physically feasible; It must conform with accepted public policy; and it must not adversely influence the development of the water resources of the basin (DRBC website 2003).

Fig. 3. Main tributaries in the Susquehanna River Basin
Source: Susquehanna River Basin Commission, 2003.

flood damage; providing for sustainable development and use of surface and groundwater; protecting and restoring fisheries, wetlands, and aquatic habitat; enhancing water quality; and ensuring future availability of flows to the Chesapeake Bay.

An Executive Director oversees a 25-member staff of technical, administrative, and clerical personnel. A commissioner who serves as spokesperson during periodic meetings represents each signatory member. At these meetings projects are discussed, regulations are adopted, and direct planning and management occurs. Each member has a single vote for situations where specific issues must be resolved by consensus. Commissioners can be presidential appointees, governors, or governor appointees (SRBC website 2003).

One place where multiple stakeholder interests collide is the Conowingo Pool. The Conowingo Pool is a 15 miles (25 km) long reservoir on the lower Susquehanna River formed by the Conowingo Hydroelectric Dam. Located 10 miles (16 km) north of the river's mouth in Maryland, the dam is owned by Exelon's subsidiary Susquehanna Electric Company. It is the largest of the four hydroelectric projects on the lower Susquehanna River, covering 8,650 acres (3,500 hectares) and reaching 90 feet (28 meters) at maximum depth. The reservoir straddles the Maryland-Pennsylvania border, with two-thirds falling on the Pennsylvania side in Lancaster and York counties. On the Maryland side, the pool borders both Harford and Cecil counties. The impoundment is a vital source of water for energy production, water supply, recreational opportunities, and habitat protection. Approved users of Conowingo Pool water include Peach Bottom Nuclear Power Plant, Muddy Run Pumped Storage Hydroelectric Project, Chester Water Authority, and the City of Baltimore Public Water Supply.

As part of a 1980 re-licensing program, the Federal Energy Regulatory Commission (FERC) mandated an operating condition requiring the dam to make periodic conservation releases. This guideline is intended to protect the fishery habitat and diadromous fish species that return to the Susquehanna each year to spawn. Additionally, this action maintains low salinity levels downstream where several municipalities have intake structures for public water supply (SRBC Conowingo Information Sheet, 2001).

The Conflict
The city of Baltimore first constructed an intake structure at Conowingo Pool in 1966 with capacity to withdraw and divert up to 137 mgd. Since that time, the City used Conowingo only intermittently during periods of low-flow and drought. In 1993, Baltimore announced its intention to sell up to 30 mgd of water to Harford County, Maryland. To offset the effects of this increased demand on the City water system, officials considered expanding use of Susquehanna River waters. In 1994 the SRBC notified the City it needed prior approval from the Commission before expanding their current intermittent use pattern (SRBC Settlement Agreement, 2001). It cited section 3.10 of the Susquehanna River Basin Compact where it

states that no project affecting water resources in the basin may be undertaken without submitting plans for approval to the SRBC.

Baltimore questioned the Commission's authority in this matter, citing the Maryland Reservation to the Susquehanna River Basin Compact in disagreement. This section of the Compact confirms the right of the mayor and City Council of Baltimore to construct and operate water facilities in a manner that benefits the City and its service area the most (Susquehanna River Basin Compact, 1972).

In July 1994, legal issues were set aside when both parties agreed to attempt to reach a solution that addressed the impacts of large water withdrawals from the river during low flows. Baltimore completed a study of its water systems ability to adjust to reduce river withdrawals during low flows in 1997. The parties determined there to be a need for a second study on the water quality impacts of withdrawing water during low flows. At this point, a settlement seemed imminent. However, the Commission and the City were unable to reach an agreement on a set of interim low flow requirements pending completion of the second study and execution of a final settlement. This sudden divide was possibly the result of Baltimore retaining a more aggressive counsel midway through the negotiation process. SRBC officials maintain the new lawyer effectively killed the settlement process by drastically changing the City's position on previously made points of agreement. As a result, no settlement was reached at this time (Cairo, 2003b).

The SRBC held a public hearing in April 1998 in York, PA to discuss the potential impact of future withdrawals or diversions by the city of Baltimore from Conowingo Pool. Approximately 40 different interest groups and stakeholders attended to provide insights and opinions. Among those in attendance were the city officials, citizens, fisherman, boaters, recreational seekers, and many other community groups with interest in the operation of Conowingo Reservoir. Approximately one month later, the Commission issued a final determination listing potential projects that, if undertaken by Baltimore, would be subject to SRBC review and approval. These specific actions included:

- Constructing a new water treatment facility to treat water withdrawn and diverted from the Susquehanna River that will be distributed within the City's water supply system or other areas;
- Installing additional, new, or upgraded pumps, motors, or other improvements to the City's Conowingo Pool intake and pumping facilities to increase the structures water withdrawal capacity;
- Modifying the Baltimore intake structure's historic method of operation. This includes: using Susquehanna River water at a constant rate instead of as a backup source, and withdrawing water when the

river's flows at the Marietta Gage are less than trigger flows set by the FERC; and

- Selling Susquehanna River water to Harford County (including implementing the 1993 Agreement Between Baltimore City and Harford County for Raw Water Supply) or to any other county or entity not presently served by the City's system (SRBC Press Release, 2000).

Baltimore, displeased with the regulatory nature of this determination, appealed to the United States District Court of Maryland in 1998 to declare the determination null and void. On March 30, 2000, Federal Judge William Nickerson ruled in favor of the SRBC, stating:

> There can be no serious dispute that, absent the applicability of some specific exception or overriding consideration, the Commission's authority encompasses the regulation of the City's withdrawal of water from the Basin. There is substantial evidence in that record that the potential projects identified, if undertaken by the City, would have the potential of causing adverse impacts to the basin's resources, particularly during critical drought conditions (SRBC Press Release, 2000).

Still unsatisfied with the situation, Baltimore appealed this ruling to the United States Court of Appeals for the 4th Circuit. Before a second ruling was handed down, the court agreed to temporarily suspend litigation in hopes of an out-of-court settlement being reached. In July 2001, a tentative agreement materialized when negotiations between the Commission and City were completed. The Commission agreed to accept additional comments and hold a public hearing on the agreement. On August 9, 2001, the agreement was unanimously approved by SRBC commissioners and officially put in action (SRBC Press Release, 2001).

The Outcomes

Out-of-court settlement negotiations between the SRBC and the City of Baltimore resulted in an agreement benefiting both parties. The City was authorized to divert and withdraw up to 250 mgd for use within its service area.[20] Any system improvement projects undertaken by the City will not require approval from the SRBC provided these activities do not increase withdrawal capacity above the 250 mgd limit. The term of the docket approval is August 2051, though the SRBC has the authority to impose other reasonable conditions during this time. During periods when river flows drop below QFREC levels[21] as measured by the USGS Marietta Gage, Baltimore is limited to a 30-day average of 64 mgd with a peak single day rate of 107 mgd. The 30-day average is based on 'pre-Compact'

---

[20] The Maryland counties of Baltimore, Carrol, Anne Arundel, Howard, and Harford are included in this service area.

[21] QFREC rates are flow levels established by the Federal Energy Regulatory Commissions related to the operation of Conowingo Dam.

use patterns, while the peak day rate is based on the amount Baltimore withdrew and diverted in 1966. If the Commission declares a drought emergency, the City is required to impose mandatory restrictions on its customers that are consistent with the Maryland Drought Monitoring and Response Plan. In this sense, the SRBC drought emergency powers remain intact with this settlement (SRBC Settlement Agreement, 2001).

Baltimore had to carry out a number of water conservation measures as a result of this agreement. Within 18 months of the effective date of the settlement, a complete review of the water conservation measures currently in effect throughout the service area had to be completed. Within 48 months of the agreement, the City had to implement water conservation measures consistent with SRBC standards or State of Maryland water conservation requirements. Baltimore is responsible for conducting monitoring programs so that records and documents are readily available if needed by the Commission. Meters capable of measuring the quantity of water diverted from the Susquehanna were installed and maintained to ensure an error of no more than 5%. Calibration and repair records should be maintained and made available to the Commission on request along with daily records of withdrawal quantities.[22] In periods of low flow or drought, these records may be requested more frequently.

This settlement agreement legally affirms SRBC's authority to regulate the City of Baltimore's withdrawals and diversions of water from the river. With approval of the agreement, Baltimore was provided with long-term certainty regarding the availability of water from the Susquehanna River. They could enter into a 1993 raw water supply agreement with Harford County without sacrificing overall productivity in their service area. The SRBC was recognized as a regulatory authority responsible for the effective environmental management of basin resources. The City cannot legally increase their use of Susquehanna River water without approval from the Commission.[23]

Sustainability Implications

Both positive and negative results are evident in the outcome of this conflict. Baltimore will considerably increase its use of river water, increasing the stress on the multi-use reservoir and possibly diminishing flows to the

---

[22] The SRBC will confer with the U.S. Geological Survey and the Safe Harbor Water Power Corporation to assure the Marietta stream gage is calibrated and maintained in accordance with applicable specifications.

[23] Also important to this settlement is a provision that orders the development of a Conowingo Pool Operating Plan. The SRBC convened a stakeholder group, including the City of Baltimore and other reservoir users, to participate in the development of an operating plan for the Conowingo Pool. Following completion of the plan, the Commission considered impacts of current diversions and uses of the reservoir, and the effect that may result from new or increased withdrawals, particularly during low-flow or drought periods (SRBC Settlement Agreement, 2001).

Chesapeake Bay. However, water conservation measures and withdrawal limitations (for drought emergency periods) mandated by the agreement are significant victories for those concerned with sustainability. Baltimore must comply with SRBC advisories and modify their resource use during low flow periods. Although more water will be consumed, preventative requirements for critical time periods will ensure availability and wise use of basin resources.[24] Metering and water conservation measures outlined in the agreement will considerably enhance the overall efficiency of the City supply system.

Significant to this case is the extent to which public participation and stakeholder involvement was included in the resolution process. Although the dispute eventually ended in litigation, the resolution process leading up to this point was participatory. The public hearing held in York, Pennsylvania in April 1998 was an important example of how SRBC involved the public in this resolution process. Having been unable to reach an agreement with City officials, the Commission held this hearing as a way to inform stakeholders about the issues surrounding the conflict. By acting on the opinions of river basin stakeholders, we believe SRBC's position in this conflict was strengthened. Both sides had apparently valid legal claims. Once the citizen-based determination was issued, however, the City's position became weakened. All aspects of **WE³** seem to be balanced in this case. Water conservation measures provide ecological harmony, while the participatory nature of the resolution process ensures social equity and accounts for all economic concerns.

The Nakdong River Basin

The management of water quality in the Nakdong River Basin is a crucial issue that affects the people and the industry in the south-east part of South Korea. The Nakdong River is the longest river in the country, flowing through two provinces (Gyeongsangbuk and Gyeongsangnam) and two large cities (Daegu and Busan) (Figure 4 illustrates the Nakdong and Han River Basins). The population in the river basin is about 13 million, with over 90% of that population receiving its drinking water from the river. Big cities and industrial complexes in its upper and midstream areas represent significant threats from pollution of water, especially for the people in Busan and Gyeongsangnam Provinces who draw water from the river downstream.[25]

---

[24] This provision protects water quality and ensures quantity of water flow through the Conowingo Dam to end of river communities like Port Deposit, Perryville, and Havre de Grace.

[25] The basic profile of the Nakdong River basin is as follows: Main stream length of 324 miles (522 km); Basin area of 12,464 square miles (32,280 km²); Average annual precipitation of 45 inches (1,137 mm); Population of 13,160,000; Livestock of 3,600,000; and Number of discharge sites of 12,058.

Fig. 4. River basins of South Korea
Source: Ministry of Environment. South Korea 2003a.

According to the water quality standard set by the Ministry of Environment, the Nakdong River corresponds to 'third-rate' water, which can be used for drinking water only after intensive purification treatment. Whereas, in comparison the Han and Geum Rivers correspond to 'second-rate' water that can be made portable by normal purification treatment (MOE, 2003b).[26]

---

[26] The BOD at Mulgeum, the location for drawing water for downstream use, is 3.5-5.0 mg/l which is much higher than the Han River and the Geum River whose BODs are 1.1-1.6 mg/l at their main drawing locations, Paldang and Daechung.

The central government created a comprehensive river basin management plan in 1992, investing US$2.5 billion (3 trillion won) for the enhancement of Nakdong water quality. The money was used for building basic environmental facilities like sewage treatment plants, wastewater treatment plants, and livestock manure treatment plants in the region. However, the government had difficulties in reconciling fundamental and political measures such as land use regulations due to economic development needs in upper and midstream regions of the basin. The government could only designate some areas near water intake points as "Areas for Limiting Wastewater Discharge Facilities."

The Conflict

The water quality of the Nakdong River had continued to deteriorate since the 1970s as urbanization and industrialization in the upper and midstream regions increased. In response, Busan City and Gyeongsangnam Province began to demand stronger regulations for the improvement of water quality. The phenol pollution accident that occurred in 1991, and the organic solvents pollution accident occurred in 1994 inflicted damage to public health of people in the basin. What made the concern more intense was the plan to construct the Wichon Industrial Complex near Daegu City.[27] Accordingly, the downstream demand for strict regulations became virulent. The Complex, built to boost the economy of Daegu City, brought about organized opposition from communities in the downstream area, especially Busan City.[28] A wide range of civil movement organizations, local governments, the press, politicians, and people in this area participated in the opposition campaign.[29]

The main demand of the downstream polity, especially the main city of Busan, was that regulation against pollution should be reinforced and the plan for the construction of new industrial complex canceled. Specifically, their demands included the reinforcement of environmental impact assessment systems, strengthening regulations on land use such as the designation of an additional source water protection area, restrictions on the new discharge facilities, and the designation of buffer zones.

Conflict increased as up- and midstream areas argued that the demands of the downstream to restrict the economic activities of the upstream severely inequitable, and that there was no 'trade-off' in return for such restrictions. To the further frustration of opponents, arguments were made that the construction of the industrial complex should be hastened, and to

---

[27] Daegu City is the largest city in the midstream area of the Nakdong River.

[28] Busan City is the second largest city in South Korea located at the mouth of the Nakdong River.

[29] The overall water quality issue and the Wichon industrial complex plan became the hottest issue in the 1990s in the basin.

facilitate this, the regulations on land use should be lessened. Thus, leaders were caught in a political vise.

It was quite difficult for the central government, political parties, or the National Assembly to resolve the conflict between the upstream and the downstream through existing institutional solutions or policies. Given the multi-stakeholder competing demands, it was difficult to come up with a solution to satisfy both upstream and downstream, also it was politically too risky to side with one coalition due to the resistance of the other.[30] Accordingly, the resolution of the problem was delayed for almost 10 years. The conflict exacerbated as the problem remained unsolved. In the last stage, even calmly hosting the actual discussion for resolving the conflict between both regions became difficult.

The Ministry of Environment (MOE) thought that this kind of daunting task required the mediation of the central government, and began to devise various measures to resolve the conflict. The measures included adoption of the total load management system (TLMS) to address concerns regarding pollution from upper reaches, and a water use charge collected from downstream residents intended for fulfilling upper and midstream needs not met due to regulation of water quality. These transactions were the core measures for harmonizing the interests of the upstream and downstream.

In February 1999, MOE launched a task force for formulating a comprehensive plan for Nakdong River water resource management. The task force was composed of MOE officials and experts in the field. The task force made its draft for the plan in July 1999 based on field research, many expert meetings and local hearings, and the collection of individual plans from related agencies and local governments. However, it was not easy to attain local consensus for the final plan. The local hearings on the draft plan were prevented from being held by angry resident people who fiercely opposed it.

What protesters opposed was the construction of more dams for supporting maintenance flows, and the failure to abolish the Wichon Industrial Complex plan. MOE reconsidered its plan for constructing dams and decided to set up a local level experts body that would search for alternative ways to secure the required quantity of water. The Minister of Environment sent a mailing to 22,000 residents appealing for their understanding and cooperation. MOE held over 40 meetings, hearings and seminars over the policy measures. All stakeholders, such as the central government officials (including MOE), local government officials, representatives of local communities, civic groups, specialists, and business

---

[30] Since over a quarter of the nation's population live in the Nakdong River basin area, any improper handling of the issue could bring serious political consequences.

representatives, participated in the meetings, hearings and seminars. Officials in MOE, including the Minister, worked extraordinarily hard to persuade local representatives in the meetings and seminars in local gathering places, which often continued as unofficial discussions that lasted deep into the night.

The Outcomes

Eventually, a consensus for improving water quality of the Nakdong River was settled through a tough negotiation process. A comprehensive plan was finalized under the agreement among up and downstream areas on December 30, 1999 (MOE, 2003a). To bolster the agreement, the government drafted "The Act Relating to Water Resource Management and Community Support for the Nakdong River" and presented it to the National Assembly in June 2000. However, it took time and effort for the Nakdong River Bill to pass the National Assembly due to differences in the positions of the congressmen who represented each area, as well as the revival of conflicts between up and downstream.[31]

Congressmen who were aware of the revived conflicts were not active at all in reviewing and passing the bill. The bill was essentially adrift in the National Assembly. In September 2001, although the regular session of the National Assembly began, there was no passage of the bill. Worries arose that the bill would actually be discarded if it did not pass in that regular session, due to the consecutive election schedule in the coming years. This inspired a few NGOs to take action to urge the National Assembly to pass the bill. They announced their position that they demanded the passage of the bill by the end of the year, and visited congressmen and the dominant party to express their demand. The press also criticized the National Assembly for failing to address the crucial issue due to defending their personal political interests.

The MOE seized the opportunity to save the bill. The Minister and staff visited the local communities that opposed the bill the most, and explained the contents of the bill and sought their support. Over 100 times the Minister and MOE officials, and to some extent environmental NGOs, had meetings and hearings with local people since June 1999. Thanks to these efforts, the bill was passed and put into effect in January 2002. This became a permanent institutional framework for ending conflicts between upstream and downstream stakeholders that has lasted almost 10 years and improved water quality of the Nakdong River (MOE, 2003a).

---

[31] In October 2000, Congressmen representing Busan City submitted a bill that demanded stricter regulations for upstream areas, regardless of the bill the government already submitted. In response to this, residents of local governments of upstream areas demanded the regulations they had conceded should be lessened and congressmen representing upstream areas became negative about the passage of the bill. The residents of upstream area demonstrated their objection to the bill in the National Assembly.

TLMS has been enforced in all areas of the Nakdong River, seeking to harmonize preservation and development by allowing regional development to be carried out in an environmentally-sensitive manner so that the desired water quality improvements are realized. Under the system, pollutant sources are managed so as to keep the total amount of pollutants in the watershed under a certain level (total allowance). In September 2002, MOE designated riparian buffer zones for protection of 88.3 square miles (228.8 km²) in the upstream areas of Youngchon, Unmun, Imha, and Milyang dams (MOE, 2003a). Within this 547 yard-wide zone (500 m) on both sides, any construction of restaurants, lodging facilities, saunas, multi-family housings, factories and barns has been banned.

Water use charges have been levied on, and collected from, downstream users to secure revenue for upstream community support and water quality improvement projects. However, users in water source management areas and other areas designated by presidential decree are exempt from having to pay the water use charge. To efficiently coordinate the imposition and collection of water charges, community support projects and other important policies concerning the watershed, a Watershed Management Committee (WMC) was established.[32] The Watershed Management Fund derives its revenue from the collected water use charge determined by WMC, donations from non-governmental parties, loans, and earnings from investing the fund. The Fund is used for purchasing riparian buffer zones and other lands and implementing community support projects. The Fund also contributes to the establishment, operation and maintenance of environmental infrastructure, and the operation of water pollution prevention facilities.

Sustainability Implications

The Nakdong case showed that it was very difficult for the central government, local governments, political parties, or the National Assembly to resolve conflicts between the upstream and downstream communities through institutional solutions or policies. A solution that would satisfy both regions was politically too risky to be with one side due to the resistance of the other side. Accordingly, the resolution of the problem was delayed for almost 10 years, and the conflict exacerbated as the problem remained unsolved. Despite several major confrontations, MOE, civic and religious groups, specialists and representatives from the regions were finally able to find a road to coexistence after holding numerous meetings and discussions. The breakthrough was finally possible because all stakeholders were committed to the common goal of reviving a dying river and public pressure became irresistible.

---

[32] WMC is composed of the Minister of MOE, mayors and governors of the Nakdong River Basin area, and the President of the Korean Water Resources Corporation.

The Nakdong River's water quality management projects stand as a sustainable outcome of how confrontations and conflicts can be resolved through dialogue and cooperation among stakeholders. Both TLMS and water use charges succeeded in fostering sustainability by harmonizing development needs with water quality preservation on a permanent basis. Ecological viability is accounted for, while economic issues are resolved. Both measures have also been successful in terms of equity, providing compensation for upstream sacrifice.

The Han River Basin

The population of the Han River Basin is approximately 24 million, nearly half the entire population of South Korea. The Han River represents the largest basin and includes three provinces (Gyeonggi, Gangwon and Chungcheongbuk) and two large cities (Seoul and Inchon). Most of the population is concentrated in the downstream area (Seoul and Gyeonggi Province), and only 9.1% live in the upstream areas). This makes for much easier water quality management compared to that of the Nakdong River. The primary sources of water intake from the Han River are Paldang Lake (2.03 billion gallons per day) and Jamsil Lake (1.67 billion gallons per day).[33]

The central government has designated the area near the Paldang Lake, which is 61 square miles (157 km²), as an 'Area for Water Source Protection', and prohibited many sources of pollution since 1975. In 1980 the government also designated seven cities and counties of the Gyeonggi Province above Paldang Lake as 'Areas for Environment Preservation', an area of 1,277 square miles (3,307 km²) to constrain the building of facilities that induce population inflow.[34] In addition, the government invested US$3.2 billion (3.81 trillion won) for the improvement of water quality in this region from 1993 to 1998. Despite such aggressive management strategies, water quality in the Han River Basin continued to be a significant problem throughout the decade.

The Conflict

Water quality of the Paldang Lake, which is the water source for 20 million people in the metropolitan area of Seoul, had become progressively worse

---

[33] The basic characteristics of the Han River are as follows: Main stream length of 299 miles (481.7 km); Basin area of 12,471 square miles (32,300 sq km²); Average annual precipitation of 51 inches (286 mm); Population of 23,490,000; and Number of discharge sites of 17,999.

[34] The government re-designated part of the seven cities and provinces in the upstream areas as "Area for Special Measures for Water Quality Management" (812 square miles or 2,102 km²) and has regulated the construction of restaurants, lodging facilities, and wastewater discharges since 1990.

in 1990s.[35] The deterioration of the water quality was partly because of the deregulation of land uses that took place in early 90s. Great pressure was placed on the government to come up with special measures to cope with the problem.   In May 1998, the central government decided that new special measures should be taken for the improvement of water quality of the Han River and began to prepare a special plan to improve the water quality of Paldang Lake to 'first-rate' water by 2005.[36] Communities in the upstream area regarded the measures as the introduction of new regulations and opposed them fiercely.

As described above, the government regulated the land use of the area above the Paldang Lake, redefining several special actions to promote water quality protection. Accordingly, communities in that area have held the view that they are victims of the 'power of Seoul'. They have believed that their development is being sacrificed for the water supply needs of downstream communities. They made it clear that they could not accept additional regulations. Their opposition movement became stronger, taking some organizational forms.   They even stopped by force the hearings prepared by the government for collecting the opinions of the local people.

The Outcomes

The Ministry of Environment, aware of the position of the upstream communities, prepared 'The Special Measures for the Water Quality Management of the Han River', based on a proposed 'win-win' spirit for upstream and downstream. The measures required upstream communities to use land in a manner that preserves water quality. Downstream areas are required to shoulder the financial burden that corresponds with restrictions placed on upstream users. By taking this reciprocal approach, the measure intended to promote a cooperative relationship between the upstream and the downstream.[37] The approach was so effective that it became the prototype of special plans for other river basins.

The government developed its Special Measures by collecting the opinions of specialists, residents, civic groups, and local governments. However, confirmation of the Special Measures was delayed due to the strong opposition of upstream residents. In August 1998, the residents of

---

[35] The government converted much of the agricultural land in the basin into quasi-agricultural land as one of the sweeping deregulation programs at that time. This led to the increase of polluting facilities in the upstream areas, leading to deterioration of the water quality. The BOD of the Paldang Lake was 1.0 mg/l in 1990 and rose to 2.0 mg/l in the spring of 1998.

[36] BOD 1.0 mg/l or less corresponds to —"first-rate" water, which can be used for drinking water after minor purification treatment such as filtration (MOE, 2003b).

[37] The measures were fundamental comprehensive plans that introduced effective preemptive measures, and the sharing of the burdens and costs between the upstream and the downstream made the arrangement sustainable to locals.

upstream areas that were under strong regulations prevented the hearing on the special measures from being held by force. The failure of the hearing elevated the national attention on the issue and prompted the participation of diverse stakeholders. The representatives of local residents, environmental NGOs, experts, and the press aggressively participated in the process. The Ministry of Environment recollected local opinions through resident polls and small-scale hearings and meetings.

Civic and environmental NGOs appreciated the integrity and advancement of the measures the government prepared and held a series of local meetings to reflect local opinions. As the dialogue between government officials and local representatives proceeded, it built relationships of mutual understanding and trust. Resident representatives and local governments in the upstream area tried to reduce regulations and expand compensatory and beneficiary measures, rather than object to the measures outright. As the issues in dispute were settled by continued dialogue between the government and upstream residents, the Water Management Policy Mediation Committee (chaired by the Prime Minister) formally adopted the Special Measures in November 1998 (Kim, 2000). With only a small NGO presence, governmental officials spent many late nights going from town to town – mixing recreational time with spreading their message to win local support.[38] To provide enduring legal support of the Measures, the Special Act of the Han River was enacted by the National Assembly in February 1999.

In order to prevent pollution in the Paldang Reservoir, restrictions were put in place. For example, land within 0.62 mile (1 km) of the main rivers and their tributaries, or 547 yards (500 m) in the case of land outside the Special Measures Zone for Water Quality Conservation, and for about 50 miles (80 km) upstream was designated as a Riparian Buffer Zone. Herein the location of pollution sources is severely restricted, and a special measure was adopted that forbids damaging publicly owned forests within 3.1 miles (5 km) of either the banks of tributaries or main rivers upstream of the Paldang Reservoir. In addition, the government has planned to purchase land in the upstream area to treat it like a buffer zone, creating a riparian forest that can mitigate pollution inflow from non-point sources.

Additionally, it was decided to gradually implement a total load management system (TLMS) to reduce pollution, while flexibly accommodating demand for regional development as far as science proved it was reasonable. The government planned to begin implementing a total load management system in the Han River basin starting in 2002 after research and the completion of legal details. TLMS would be applied only to those local communities that would want to adopt it voluntarily as a

---

[38] Such efforts would not reasonably be expected of civil servants in the U.S.

pilot program in its early stage. It would be expanded to other areas gradually as more local governments want to adopt it.[39]

If a local government accepts the burden of keeping the total load from the area under a certain level, it is allowed more flexibility regarding the matter of local development because the regulations on land use concerning special areas are exempted. This difference in the Measures between the Han River and the Nakdong River was the result of considering the fact that in the Han River Basin, there already existed strict regulations on land use such as the designation of Area for Water Source Protection, Area for Environment Preservation, Area for Special Measures for Water Quality Management of Paldang, etc. (MOE, 2003c).

A water use charge system was first adopted and implemented in the Han River Basin, and it became the model for other regions including the Nakdong River Basin. The characteristics of this system are nearly the same in both basins. The charge was negotiated at 80 won/ton (US$0.3/ gallon) originally. The total amount of water use charges levied in the Han River Basin is currently about US$220 million/year (260 billion won/ year). US$59 million (70 billion won) of this fund is spent on supporting residents in the upstream areas, and the rest is used on projects for improving water quality and the purchase of upstream land for conservation. As in the Nakdong River Basin, a Watershed Management Committee (WMC)[40] has been established to collect water use charges and coordinate community support projects and other important actions concerning the watershed.

Sustainability Implications
The Han River case is similar to that of the Nakdong River in that confrontations and conflicts were resolved through sustained dialogue and cooperation of all stakeholders. The main difference between the two basins is the implementation of TLMS. While the measures for the Nakdong River included a mandatory TLMS, the measures for the Han River involve a voluntary system, which gave the local government some incentives for local development. This difference was the result of the fact that in the Han River Basin, strict regulations on land use already existed. The designation of Riparian Buffer Zones, strong regulations regarding land use, and TLMS are regarded as having assured sustainability by harmonizing development and preservation of the Han River Basin.

---

[39] The manner in which the total load management system is implemented is one of the main differences between the Measures for the Han River and the Measures for the Nakdong River. While the official Measures for the Nakdong River included a mandatory total load management system, the Measures for the Han River adopted a voluntary system, which gave to the local government some incentives and flexibility for local development.

[40] The WMC of Han River is composed of the Minister of Environment, mayors and governors of the Han River Basin and the president of the Korea Water Resources Corporation.

Ecological viability was ensured, while economic coexistence and equity between upper and downstream users was institutionalized through water use charges and community support projects.

LESSONS FROM THE COMPARATIVE CASES AND CONCLUSIONS

Before deriving lessons from the case studies, comparisons of river basin management practices between the U.S. and South Korea are helpful. It should be noted that there are differences between the U.S. and South Korean cases. The Han and Nakdong Rivers serve 70% of the population in South Korea, whereas the Delaware and Susquehanna Rivers serve only 20% of the population of the U.S. Consequently, the South Korean government's role in its river basin managements had especially significant socio-political implications.

In the case of the U.S., conflict mediation occurs mainly through: 1) Regional governance at the federal-interstate scale, backed by law and significant resources; 2) Heavy reliance on the court system to clarify and strengthen federal-interstate compact law; and 3) Powerful NGOs and community groups at multiple scales for reconciling conflicting land and water uses, as well as acting as third parties between government and communities.[41] In the case of South Korea, the following is emphasized: 1) The central government's role in sorting out river basin management conflicts not addressed by law; 2) Less use of the courts compared to the U.S. cases; and 3) Heavy reliance upon compensation schemes to build community consensus.

Although the core principles underlying the U.S. and South Korea approaches vary due to differences in government and civil society, as well as culture and geography, important similarities exist nevertheless . First, there is an emphasis on giving local communities an avenue for expressing their concerns. Second, a form of conflict articulation and debate is offered, rather than a simple 'command and control' structure, albeit with heavy NGO influence in the U.S. and more emphasis on civil service involvement in the case of South Korea. And finally, a commitment to equitably meet the needs of both upstream and downstream users guides those in the management and mitigation process.

---

[41] The U.S. government has in recent times often supported a consensus-driven model for issues that, for instance, involves working with private property owners and water users. This controversial approach is opposed to turning to regulation and litigation for leverage in dealing with conflicts. Notably, in rural areas a desire to devolve control to the local scale is often enhanced, as in remote areas enforcement can be problematic, and voluntary action is even more favorable.

The results of our four case studies generally support our assertion that conflict resolution enhances opportunities for balanced **WE³**, leading to a higher level of sustainability in river basin management. The U.S. cases (Delaware and Susquehanna River Basins) show that litigation has been the major mechanism to resolve conflict, with aid from NGOs. Whereas in the South Korean cases (Nakdong and Han River Basins), an administrative role (played by the Ministry of Environment) has been the dominant mechanism in conflict resolution. In both the U.S. and South Korean cases, community-based transactions in the process of conflict resolution brought about litigation or administrative initiatives. Without these community-based transactions, environmental, socio-political and economical goals of river basin management (**WE³**) would have been considered less sustainable.

Additional valuable lessons can be drawn from the case studies. Major conflicts, as shown in these cases, can take more than 10 years (20 years in the case of the DRBC) to be resolved in the courts. To minimize transaction costs and to be more effective, conflict resolution mechanisms should be institutionalized to allow for community-based transactions in the planning stage of project. Without transparent, democratic and participatory transactions, such mechanisms as EIA, litigation, or administrative initiatives can result in less sustainable outcomes.

Additionally, the balanced consideration of both supply-side and demand-side options is important as shown in the Delaware and Susquehanna River Basin cases. Diversions and allocations from the rivers are certain to increase in years to come. Demand-side conservation strategies can work to offset potential negative effects of such changes. Instead of participating in lengthy, expensive battles over water rights and usage patterns, the SRBC case shows how comprehensive agreements that promote conservation measures and mandate drought period restrictions can be a more effective strategy and help for a "soft-path" approach to making resources available.

Finally, as illustrated in the Nakdong and Han River Basin cases, equity-driven programs are critical in the resolution of conflicts. Without the benefit of law or significant regulatory authority, MOE officials worked directly with stakeholders to ensure that an equitable solution was reached. In sum, community-based transactions, guided by the need to balance economic, environmental and social interests, can be vital to the resolution of major upstream-downstream conflicts. It appears that through these transactions, across widely differing policy, geographical, and cultural settings, the goal of sustainable river basin management can be secured.

REFERENCES

Ariel, D. 2004. Exploring Transboundary Water Conflict and Cooperation. Water Resources Research. W05S01: 1-3.

Beach, H.L., Hamner, J., Hewitt, J.J., Kaufman, E., Kurki, A., Oppenheimer, J.A., and Wolf, A.T. 2000. *Transboundary Freshwater Dispute Resolution.* The United Nations University, Tokyo, Japan.

Cairo, R. 2003a. Re: SRBC vs. City of Baltimore. Personal communication, email to General Council for Susquehanna River Basin Commission. April 27, 2003.

Cairo, R. 2003b. Re: Baltimore Case and Settlement. Email to General Council for Susquehanna River Basin Commission. 16 Jan. 2003.

Carluccio, T. 1987. New Jersey Opinion: A Battle to Protect the Delaware. The New York Times. 13 Sept. 1987. 11NJ 30.

Delaware River Basin Commission. 2005. Internet: www.state.nj.us/drbc.

—. Accessed 26 Oct. 2002. Home Page. Internet: www.state. nj.us/drbc.

Delaware Water Emergency Group, et al. v. Gerald M. Hansler, et al. and Neshaminy Water Resources Authority and Philadelphia Electric Company. 1981. 536 F. Supp. 26.

Delaware River Basin Collaborative Environmental Monitoring and Research Initiative. Accessed 4 2003 05. Internet: www.fs.fed.us/ne/global/research/drb/basin.html.

Espey, M. and Towfique, B. 2004. International bilateral water treaty formation. Water Resources Research. W05S05: 1-6.

Featherstone, S.P. 1999. An evaluation of federal-interstate compacts as an institutional model for intergovernmental coordination and management: Water resources for interstate river basins in the United States. Doctoral Dissertation, Philadelphia, PA: Temple University.

Featherstone, J.P. 2001. Interstate Organizations for Water Resource Management. Annual Meeting of the American Political Science Association. Atlanta, Georgia. September 1, 2001.

Flack, J. Ernest and David A. Summers. 1971. "Computer-Aided Conflict Resolution in Water Resource Planning: An Illustration." Water Resources Research. 7 (6): 1410-1414.

Fraser, N. 1993. "Rethinking the Public Sphere: A Contribution to the Critique of Actually Existing Democracy," in Craig Calhoun, ed. Habermas and the Public Sphere. Cambridge, MA: MIT Press. pp. 109-142.

Fraser, N. 1999. An evaluation of federal-interstate compacts as an institutional model for intergovernmental coordination and management: water resources for interstate river basins in the United States. Doctoral Dissertation, Philadelphia, PA: Temple University.

Gleick, P. 2000. Accessed 23 2003 03. The World's Water Information on the World's Freshwater Resources, Water Conflict Chronology. Internet: http://www.worldwater. org/conflict.htm.

Just, R.E. and Netanyahu, S. 2004. Implications of 'victim pays' infeasibilities for Interconnected games with an illustration for aquifer sharing under unequal access costs. Water Resources Research. W05S02: 1-11.

Foster K.E. and Wright. N.G. 1980. Jojoba: An alternative to the conflict between agricultural and Municipal ground-Water requirements in the Tucson Area, Arizona. Ground Water 18 (1), 31-36.

Kim, S.H. 2000. A study on the process of the legalization of watershed management policies: Focusing on the Han River cases. Master's thesis, Seoul National University, Seoul, Korea.

Lazar, K. and King, L. 1994. Bucks plans to sell point pleasant pump. Philadelphia Inquirer. 16 June 1994. B7.

Leach, W.D. and Pelkey. N.W. 2001. Making Watershed Partnerships Work: A Review of Empirical Literature. Journal of Water Resources Planning and Management. November/December 2001: 378-385.

Lepawsky, A. 1950. Water Resources and American Federalism. The American Political Science Review (44) 3, 631:649.

McCormick, Z.L. 1994. Interstate Water Allocation Compacts in the Western United States—Some Suggestions. AWRA Water Resources Bulletin (30) 3, 385:395.

McGinnis, Michael Vincent, John Wooley and John Gamman. 1999. "Bioregional Conflict Resolution: Rebuilding Community in Watershed Planning and Organizing." Environmental Management 24 (1): 1-12.

McPhee, J. 1989. The Control of Nature. The Noon Day Press, New York, USA.

Ministry of Environment. 2003a. River for all achieved by participation and cooperation: Success story of making Nakdong River Special Measures. Seoul, Korea.

_____. 2003b. Water quality and regulation standard, Available from World Wide Web http://www.me.go.kr>accessed on June 22nd, 2003.

_____. 2002. Environmental White book 2002. Seoul, Korea.

_____. 2002. Green Korea 2002. Seoul, Korea.

_____. 2001. Environmental White book 2001. Seoul, Korea.

_____. 2000. Environmental White book 2000. Seoul, Korea.

_____. 1999. Environmental White book 1999. Seoul, Korea.

_____. 1999. The Special Comprehensive Measures for the Water Quality Improvement of the Nakdong River. Seoul, Korea.

_____. 1998. The Special Comprehensive Measures for the Water Quality Improvement of the Han River. Seoul, Korea.

Mostert, E. 2003. Conflict and Cooperation in the Management of International Freshwater Resources: A Global Review. Delft University of Technology, Delft, the Netherlands. A contribution from UNESCO's International Hydrological Programme to the World Water Assessment Programme. It was prepared within the framework of the joint UNESCO.Green Cross International project entitled From Potential Conflict to Cooperation Potential (PCCP): Water for Peace. Available at: http://www.hidropolitik. hacettepe.edu.tr/gfreshwater.pdf.

Murphy, I.L. and Sabadell. J.E. 1986. International river basins: A policy model for conflict resolution. Resources Policy, 12 (1), 133:144.

Neshaminy Water Resources Authority v. Del-Aware Unlimited, Inc., et al. (1984) 332 Pa. Super. 461.

North Wales Water Authority. Accessed 20 Jan. 2003. Internet: www.nwwater.com.

Partners in PA – Partnership for Safe Water Update. 2 (1). www.dep.state.pa.us/dep/ deputate/watermgt/wsm/WSM_DWM/Complian/PA-AWWA-2_(1).pdf. Accessed 20 Jan. 2003.

Philadelphia Electric Company. Accessed 20 Jan. 2003. Internet: www.peco.com.

Schogol, Marc. Accessed 6 2003 04. Pumped-up ideals in Bucks battle. The Philadelphia Inquirer. Internet: www.philly.com/mld/inquirer/news/local/5567077.htm.

Sherk, G. William. 2004. Interstate Water Compacts: A Bibliography. Available at: https:// repository.unm.edu/bitstream/1928/372/1/bibliography+pt+1.pdf.

_____. 1994. Resolving Interstate Water Conflicts in the Easter United States: The Re-Emergence of the Federal-Interstate Compact. AWRA Water Resource Bulletin 30 (3), 397:407.

Smith Jr., W. James. 2003. The Clearinghouse Approach to Enhancing Informed Public Participation in Watershed Management Utilizing GIS and Internet Technology. Water International 27 (4), 558:567.

Smutko, L.S., Klimek, S.H., Perrin, C.A. and Danielson. L.E. 2002. Involving Watershed Stakeholders: An Issue Attribute Approach to Determine Willingness and Need. Journal of the American Water Resource Association 38 (4), 995:1006.

Sneddon, C. 2002. Water Conflicts and River Basins: The Contradictions of Comanagement and Scale in Northeast Thailand. Society and Natural Resources 15: 725-741.

Steinberg, P.E. and Clark. G.E. 1999. Troubled water? Acquiescence, conflict and the politics of place in watershed management. Political Geography 18: 477-508.

Stevens, W.K. 1987. Bucks County's Long Water War: Decision Time. The New York Times. 12 June. A16.

Susquehanna River Basin Commission. 2003. Internet: http://www.srbc.net.
_____. Accessed 25 October 2002. Internet: http://www.srbc. net.
_____. 2001. SRBC Approves Settlement Agreement With Baltimore for Withdrawals From the Susquehanna River. Press Release: August 10, 2001. Available: http://www.srbc. net/press.htm.
_____. 2001. Information Sheet: Conowingo Pool. Harrisburg, USA.
_____. 2001. Settlement Agreement between the Susquehanna River Basin Commission and the Mayor and City Council of Baltimore. Harrisburg, PA, USA.
_____. 2000. Court Rejects Baltimore's Challenge to SRBC's Regulatory Authority. Press Release: April 4, 2000. Available: http://www.srbc. net/press.htm.
_____. 1972. Susquehnna River Basin Compact. Harrisburg, PA, USA.
Thomas, C. 1997. Bureaucratic landscapes: Interagency cooperation and the preservation of biodiversity. Ph.D. dissertation. University of California, Berkeley, Berkeley, CA, USA.
Wang, Y.D., Smith Jr., James. W. and Bynne J. 2005. Water Conservation-Oriented Rates: Strategies to Extend Supply, Promote Equity, & Meet Minimum Flow Levels. American Water Works Association, Denver, CO, USA.
Yoder, Diane E. 1999. A Contingency Framework for Environmental Decision-Making: Linking Decisions, Problems, and Processes. Policy Studies Review 16(3/4): 11-35.
Yoffe, S., Fiske, G., Giordano, M., Giordano, M., Larson, K., Stahl, K. and Wolf. A.T. 2004. Geography of international water conflict and cooperation: Data sets and applications. Water Resources Research W05S04: 1-12.
Zimmermann, F.L. and Wendell, M. 1952. New Experience with Interstate Compacts. The Western Political Quarterly 5 (2), 258:273.

North America

12

International Joint Commission: A Model of Cooperation in the Great Lakes Region

Velma I. Grover[1] and Haseen Khan[2]

[1]Natural Resources Consultant, Adjunct Professor, York University, Toronto, Canada
[2]P. Eng., Manager, Water Resources Management Division,
Department of Environment and Conservation, P.O. Box-8700,
Confederation Building, West Block, 4th Floor, St. John's NL A1B 4J6, Canada

INTRODUCTION

Canada and the United States are in an envious and unique position of sharing the world's most valuable and strategic Great Lakes system. The

Fig. 1. Five Great Lakes

Great Lakes (a group of five interconnected lakes, as shown in Figure 1 - water in these lakes flows from Lake Superior into Lake Michigan and Huron, then into Lake Erie, Lake Ontario, along the St. Lawrence River and finally into the Atlantic ocean), a treasure that constitutes about 20% of the entire world's accessible freshwater, are lumped together between the two countries and indeed they often provide the natural boundary between the two countries. Although the lakes can be cut in half on political maps, but they act as one entity in reality, and thus it is impossible for one country to make use of their share of the resource without affecting the other's interest.

There have been several treaties[1] to ensure proper sharing of the waters by the two countries. One of such treaties resulted in the formation of the International Joint Commission (IJC) to deal with such issues. This chapter discusses how IJC and its committees have helped solve the problems between two countries peacefully. IJC is an institution which can be used as a role model for other countries sharing transboundary water and associated problems.

INTERNATIONAL JOINT COMMISSION

Both Canada and United States have thrived on the richness of Great Lakes, but have not protected[2] them adequately to ensure the rights and interest of future generations. The past and current water use practices in Great Lakes and development activities in shoreline and surrounding

---

[1] Ever since this treaty of 1909, a number of other treaties/conventions have been signed. Following is a list of few of such treaties:

Treaties
- Boundary Water Treaty   1909
- Niagara River Treaty   1950
- Columbia River Treaty and Protocol   1961
- Skagit River Treaty   1984

Conventions:
- Lake of Woods Convention and Protocol   1925
- Rainy Lake Convention   1938
- Convention on Great Lakes, Fisheries (Great Lakes Fisheries Commission)   1955

Agreements:
- St. Lawrence Seaway Agreement   1959
- Great Lakes Water Quality Agreements   1972 & 1978
- Water Supply and Flood Control in Souris River Basin   1989

[2] The proper management of the Great Lakes water becomes all the more important because water is of glacial origin from the last ice age. Only 1% of the Great Lakes volume is renewed on an annual basis by rain and other sources. In other words, using more than 1% of the volume in a year will reduce the lake levels beyond natural renewal. In many ways this is a non-renewable resource that needs protection from any water extractions beyond its renewal rate. During the last century or so, inhabitants of the Great Lakes region have witnessed a high level of industrial and commercial development which has contributed to the economic prosperity of the area. The area is home to a number of industries such as automobile manufacturing, shipbuilding, and other industrial activities. Besides, the area is popular for tourism and also for commercial fishing.

areas have resulted in multifaceted issues and challenges[3] ranging from ecosystem integrity to governance.

Concern about sharing of the Great Lakes started way back in the 19th century – long before the IJC came into existence, there were other Joint Commissions like the International Boundary Commission created under the provisions of the Treaty of 1889[4] or the Commission of 1892 to conserve the fish. The United States and Great Britain (acting on behalf of Canada) agreed to create a joint commission of two experts to investigate the fisheries in the contiguous waters of the two nations on December 6, 1892 and to recommend regulation to conserve fisheries. A relatively small commercial fishing business on the Great Lakes in the 1830s had escalated in response to ever-broadening regional and national markets of the 1850s into burgeoning markets involving 12,600 workers, a 131 million-pound annual catch valued at US $3,849,000, and capital investment of US $6 million of the 1890s. This was the first time that the two governments recognized the Great Lakes as a natural geographic unit requiring uniform policies protecting resources and water pollution.

One of the challenges presented to this commission of 1892 was the political realities which included legal precedents, nationalism, and divided authority over tasks. The Commission also functioned in an atmosphere

---

[3] Ecologically, the Great Lakes are the heart of the region, impacting the surrounding nature in an important way. The Lakes are home to a wide variety of flora and fauna. Historically, the Great Lakes have had a large number and variety of fish species, which have generally been a good indicator of the condition of the Lakes. However, in recent times, species native to the Great Lakes have been declining for several reasons. Invasive species, namely the sea lamprey and the zebra mussel, have made their way into the Lakes and are not only causing ecological damage but economical damage as well. The sea lamprey has decimated the trout fishery in the past, and the zebra mussel not only destroys native species but also causes clogs in water intakes which cause over a billion dollars in damages annually. The Great Lakes ecosystem faces many serious challenges; including the fact that on an average one new aquatic invasive species arrives every eight months, adding to over 160 already causing serious ecological and economic damages.

Invasive species are not the only cause for concern in the Great Lakes watershed. Past and ongoing development has affected Great Lakes habitats, and threatens plants and animals that rely on them for survival. Many coastal areas also suffer from massive sewer overflows that contaminate water and close beaches. Continued pollution from non-point and point sources, and many others contribute to water quality and related problems. Despite the large amounts of data and information on the Great Lakes that have been collected over the years, not enough has been transformed into knowledge about the key indicators of the health of the ecosystem. In summary, the Great Lakes are subjected to a number of stressors both natural and anthropogenic such as: urban sprawls, farmland development, aging sewer systems, industrial pollution, invasive species, water diversion, global warming and climate change, airborne toxic chemicals, emerging toxins, pharmaceuticals, endocrine disruptors, water export, dead zones, etc.

[4] Minutes of Meeting of the Joint Commission, June 16, 1911, The American Journal of International Law, 5 (3): 832 – 833.

charged with nationalistic frictions that had been building up since the American Revolution of 1776.

The experiment of 1892 highlights another main theme in the Canadian-United States history, the political differences between the two nations – differences in their fishery policies. The Americans subscribed to maximum production and minimum regulation, whereas, the Canadians adopted a more conservative view, long-term sustained yield, using regulation based on the English model of the colonial period.[5]

Great Lakes fishermen, American or Canadian, generally disliked the idea of the regulation. The idea of a joint commission staffed by experts with power to modify regulations of two national governments enforcing those regulations was disliked and the fishermen wanted free open-entry fishery. While this first attempt of the American and Canadian governments to conserve the Great Lake fisheries failed, its two major policy thrusts – joint regulation of fishing and the control of water pollution became and remained an agenda for the future.[6]

With different types of conflicts like fishing, pollution control[7] during 19th century, some very farsighted people both in Canada as well as in the United States thought that water problems could haunt the two nations and such a conflict could be avoided and/or solved by signing an agreement, which enshrined some principles to help in resolving the conflicts. Such an agreement, Boundary Waters Treaty,[8] was signed on January 11 1909 by the Right Honorable James Bryce, O.M., His Majesty's Ambassador Extraordinary and Plenipotentiary at Washington, on behalf of Great Britain (and Canada) and by Elihu Root, the then Secretary of State of the United States on behalf of the President of the United States thereby establishing a permanent institution which was dubbed as the

---

[5] Margaret Beattie Bogue, To Save the Fish: Canada, the United States, The Great Lakes, and the Joint Commission of 1892, The Journal of American History, 79 (4): 1429–1454.

[6] Treaty between the United States and Great Britain Concerning the Fisheries in Water Contiguous to the United States and The Dominion of Canada. The American Journal of International Law, 2 (3): Supplement: Official Document (July 1908), 322 – 325.

[7] It might be of interest to note that the stress in treaties up to 1908 was on conservation of the fish and prevention of pollution. The stress after 1909 is more on maintaining water levels and also prevention of pollution in such a manner that no harm is done to the interests of the other party until/unless the party being harmed is appropriately compensated. This way it is ensured that future developments are not hindered and no one is harmed without appropriate compensation – a typical aim of economic regulation. Pollution still remains one of the major concerns and has become of greater significance over period of time. Between 1950 and 1968, IJC was made responsible for monitoring the pollution of the Great Lakes connecting channels – Rainy River, the St. Croix River, and the Red River – through its appointed boards.

[8] http://www.sfu.ca/cstudies/science/water/pdf/Water-Ch21.pdf

International Joint Commission (IJC)[9], with equal representation of both the United States and Canada, to which questions would be referred arising from action(s) by either government upon the interests of the other for examination and would report its findings and would advise the governments to avoid action(s) inflicting unnecessary injury upon its neighbor.[10]

The IJC, undoubtedly, is the Boundary Waters Treaty's most important institutional innovation. It is an independent body of six commissioners, three from Canada, appointed by the Governor in Council, and three from the United States, appointed by the President with advice from the Senate (Article VII). Of the six commissioners, there are two co-chairs (each national division chooses its own chairman), one representing Canada and one representing United States. Each national division appoints its own secretary, who is the custodian of all records of the Commission for its own delegation. The Commission, may employ engineers and clerical assistance as required. The expenses to be paid in equal share by both governments. The salaries and personal expenses of Commissioners and secretaries are borne by their respective governments (Article XII). The functions performed by the Commission have expanded and include—consultation and consensus building; providing a forum for public participation; engagement of local governments; joint fact-finding; objectivity; and flexibility.[11]

Thus the Commission is a binational and not a bilateral institution, which is an important feature of the Commission and can be used for other models. There is parity/equality between the U.S.[12] and Canadian members of the Commission. Members act as a single body seeking solutions to common problems with similar interests and not as bi-partisans seeking national advantages under instructions from their governments[13]

---

[9] By Article VII of the treaty Great Britain and the United States agreed to establish a permanent International Joint Commission. The International Joint Commission Between the United States and Canada, The American Journal of International Law, 1912, 6(1).

[10] Canada and the United States: An International Commission, Pacific Affairs, 1929, 2(1).

[11] The IJC and the 21st Century: Response of the IJC to a Request by the Government of Canada and the United States for Proposals on How to Best Assist them to meet the Environmental Challenges of the 21st Century.

[12] In 1939 President Roosevelt introduced the practice of appointing two top-level civil servants as part-time commissioners, while reserving the third position for a political appointee who was expected to devote most of his time working for the Commission. This practice has been carried out till recently. Since, 1950 the Canadian cabinet has filled the chairmanship with political appointee and the other two members are either employees or former employees of the provincial governments. Willoughby, William R.

[13] The IJC and the 21st century.

The general purpose of the treaty is stated in the preamble as follows:

> "To prevent disputes regarding the use of boundary waters, and to settle all questions which are now pending between the United States and the Dominion of Canada, involving the rights, obligations or interests of either in the relation to the other or to the inhabitants of the other, along their common frontier, and to make provision for the adjustment and settlement of all such questions as may hereafter arise".

The Preliminary Article sets the geographical jurisdiction of the Commission. "For the purposes of this Treaty boundary waters from main shore to main shore of the lakes and rivers connecting waterways, or the portions thereof, along which the international boundary...."

The Commission has four key functions: approval, investigative, administrative and arbitral function.

I. Approval

Article (III) of the Boundary Waters Treaty authorizes the IJC to approve or disapprove applications for the use, obstruction or diversion of boundary waters on either side of the border that would affect the natural level or flow on the other side. IJC is also responsible for the approval and management of structures built for hydroelectric power generation and navigation. The commission has dealt with over 30 such projects involving dams, diversions, obstruction and remedial works.

The Commission acts as a court of law and renders judgement on matters specified in Article III (obstructions and diversions) and/or Article IV (the construction or maintenance on their respective sides of the boundary of any remedial or protective works or any dams or other obstructions in waters flowing from boundary waters or in waters at a lower level than the boundary in rivers flowing across the boundary, the effect of which is to raise the natural levels of water on the other side of the boundary unless the construction or maintenance thereof is approved by the aforementioned International Joint Commission. In 1938 the two countries signed a convention for the regulation of the level of the Rainy Lake watershed which empowered the IJC to adopt control measures it deemed necessary with respect to dams or works on the boundary waters of the watershed.

It is further agreed that...water flowing across the boundary shall not be polluted on either side causing injury to health or property on the other side.

II. Investigative/Quasi-judicial Powers

Article VIII empowers the Commission to adjudicate over matters defined under Articles III and IV. Article IX empowers the Commission to examine and report on the laws and facts at the request of either government. It thus acts as a quasi-judicial body to consider applications for approval to build and operate certain works in boundary waters and in rivers that flow across the boundary – to the extent of dispensing with The Hague Tribunal[14] as far as their bilateral disputes were concerned.[15] However their findings are not binding either as a decision or as an arbitral award. Since, the recommendations are non-binding – these are also known as 'reference' functions.

Had the Commission been limited to these important categories it would not have been able to make significant contribution to international peace and understanding. Article X invests the tribunal with greater moral rather than legal powers, because no reservations or qualification of any kind is contemplated between the contracting parties. It is worth while to have a look at the wordings of the article

> "Any questions or matters of difference arising between the high contracting parties involving the rights, obligations, or interests of the United States or of the Dominion of Canada either in relation to each other or to their respective inhabitants, may be referred for decision to the International Joint Commission by the consent of two parties, it being understood that on the part of the United States any such action will be by and with the consent of the Senate, and on the part of His Majesty's Government with the consent of the Governor General in Council. In each case so referred, the said commission is authorized to examine into and report upon the facts and circumstances of the particular questions and matters referred together with such conclusions and recommendations as may be appropriate, subject, however, to any restrictions or exceptions which may be composed with respect thereto by the terms of reference".

In a way, it may be compared to the Treaty of Ghent[16], which was signed between the USA and Britain on December 24, 1814 to mark the end of war of 1812–1814, and opened the way to demark the exact international boundary between Canada and the United States. It was followed by an agreement to dismantle all forts and warships along that boundary. The 20th century began with the treaty of 1909 with a commission to which any disputes may be referred with regard to and adjudication all possible questions of disagreement (regarding obstruction, diversions of boundary waters affecting the natural levels or flow of such waters on either side of the line, dams, navigation and other similar issues) between Canada and

---

[14] no wonder that one of the founders dubbed the Commission as a "little Hague". Maxwell Cohen.

[15] Willoughby

[16] http://members.tripod.com/~war1812/treaty.html.

the United States, their provinces and states and their respective citizens.[17] It was this commission which decided what dams would be built and to what height. What levels of water will be maintained to ensure navigation between two countries.

IJC does not deal with just transboundary water issues, but other transboundary issues also as provided by Article IX which states that the Commission can be called to investigate and report on "questions of differences arising along the common frontier" like air pollution – basically recommending principles or developing resources with the aim of preventing and resolving transboundary conflicts. One of the typical examples is the Canadian Copper Smelter Trail case[18] in which it was the damaging effect of fumes from the Canadian side of the boundary line upon the owners of the land adjoining State of Washington. Even though the smelting interests were willing to buy the land thus affected, but other factors like state land ownership laws had to be taken into consideration. It has also the ability and the mandate to adapt the terms of approval to changing circumstances. The governments asked the Commission to commence work under a 1975 reference on air quality in the Windsor-Detroit and Port-Huron-Sarnia area in 1988.

To be more proactive, the hearings are held near the scene of the proposed work, enabling the concerned private persons to state their case with a minimum of expense and inconvenience and to present relevant oral and documentary evidence. It arrives at the decision by considering both the Canadian and the American Statutes and never feels bound by them. The most significant fact is that the Commission has never hesitated to make authoritative determination regarding its own jurisdiction. Thereby it has the reputation of handling applications with a commendable measure of neutrality and impartiality. This is another point which can be adapted in other models – to be near the communities which are involved in the conflict/issue.

---

[17] The International Joint Commission Between the United States and Canada, The American Journal of International Law, 1912, 6(1): 194–195.

[18] Following expression by the U.S. government and at the request of the two governments, the Commission investigated and recommended remedial measures to reduce emissions from the smelter at Trail in British Columbia. The proposed solution was the payment of compensation to cover damages suffered by the residents of the United States in 1920s. Thus the Commission was not only able to establish the precedent-setting principle in international law that activities in one country must not be allowed to cause environmental damage in another, but was able to act as an impartial referee. By offering binational scientific and technical advice, it helped avert a serious conflict. IJC report.

For further details about the case, please, refer to "The Trail Smelter Dispute" by John E. Read in the Canadian Yearbook of International Law, 1963, pp. 213 – 229 by C.B. Bourne, University of British Columbia Press, Vancouver.

Not everything handled by the Commission was kosher – one of the exceptions was the controversial application by St. Lawrence River Power Company in 1918 to construct a submerged weir in the South Channel of St. Lawrence River near Massena, New York. The Canadians immediately objected based on Article VII of the Webster-Ashburton Treaty of 1842. which specifically stated that the channels on both sides of the Long Sault Island shall be equally free and open to ships. The Canadians felt that the proposed construction would completely block navigation. The Americans, on the other hand, felt that the Webster-Ashburton Treaty was superseded by the Treaty of 1909. Some of the Canadian Commissioner joined their American brethren to approve the construction, much to the annoyance of the Canadian government.[19] In this case, no division along the national lines took place like it did in the Milk and St. Mary River case stated above.

Another typical example where the division in the Commission took place strictly along national lines was that of the Milk and St. Mary Rivers – a controversy which lasted for more than two decades – the source of the difficulty was the divergence of interpretation of several words in the sentences of Article VI by the Americans and the Canadian spokespersons. The Americans wanted the IJC to reopen the issue and the Canadian response was polite but firm refusal. IJC finally decided to listen to the American arguments in support of reopening the issue. When the matter was finally put to vote after many years and sessions, the outcome was division along strictly national lines. One of the rare occasions when the Americans voted for the American point of view and the Canadians voted for theirs.[20]

The Columbia basin controversy is a good example of how the powers of federal government override the state water rights, if used by a strong federal politician as Chairperson of the Commission with strong ties in Washington. Federal regulation of power production and the federal Endangered Species Act have greatly influenced the institutions that relate to water resources management in the basin.[21]

In keeping with the changing times, the first Great Lakes Water Quality Agreement was signed in 1972 which was amended in 1978 and further revised in 1983 and 1987. Article VIII of this Agreement, the Twelfth Biennial Report on Great Lakes Water Quality was produced in 2004.

---

[19] Willoughby, pp. 31 – 33.

[20] Willoughby

[21] There was controversy between the State of Washington and the Province of British Columbia. Washington wanted to raise the height of their dam to increase the amount of electricity generated. British Columbia was concerned about the effects (flooding). British Columbia was ready to provide extra electricity required and this solution was not acceptable to Washington.

These reports disseminate latest information keeping both the governments as well as the public informed about the latest development.[22]

The Commission also conducts investigation and makes recommendations on issues referred to it by the Canadian and United States governments. As a part of this role, in the late 1970s, IJC issued a landmark reference study on pollution from land-use activities. Later in 2000, the IJC completed another reference study on the controversial issue of consumption, diversion and removal of water from the Great Lakes.[23] IJC also oversees the implementation of the Great Lakes Water Quality Agreements[24] signed by Canada and United States in 1972 and ratified in 1978 and 1987. The Commission has conducted over 50 investigations concerning flow regulations, water and air pollution, water apportionment, invasive species, sediment contamination, etc. The IJC has yet to use its arbitral function for the resolution of disputes between Canada and the United States.

Until the 1950s the Commission was involved primarily in orders of approvals and investigations relating to pollution and international cooperation. In recent decades, the Commission's investigations have focused on issues relating to environmental problems (pollution of the Great Lakes and water levels), water diversion, water export, global warming and climate change, etc.

The Commission has played a very important role in maintaining the confidence of the two federal governments. It has adopted a fair and balanced approach in providing sound scientific and technical advice, and

---

[22] International Joint Commission, Twelfth Biennial Report on Great Lakes Water Quality, September 2004.

[23] In the United States, 36 out of the 48 continental states have declared that they are predicting water shortages within the next ten years, and there have been a number of attempts to divert/withdraw Great Lakes water for its use in other regions. Diversions of water from the Great Lakes would surely do wonders in relieving the soon to come stress, but it would result in serious and irreversible ecological and hydrological imbalance with long lasting impacts on Great Lakes and the region. Regardless, the pressure to divert Great Lakes water will continue to grow, and there will have to be difficult choices made by American and Canadian law makers and courts regarding the future use of the Great Lakes system. To meet these challenges steps were taken way back in the nineteenth century.

[24] Implementation of these agreements requires a greater degree of accountability, benchmarks for measuring progress and an aggressive implementation schedule that reflects the urgency of basin ecosystem restoration and protection efforts to deal with new non-chemical stressors, new chemicals and new effects caused by atmospheric deposition of chemical stressors. Science Advisory Board (SAB) recommends that the Parties conduct a comprehensive review of the operation and effectiveness of the Agreements by seeking public input. Also recommends the development of ecosystem forecasting capability, establishing a binational Integrated Great Lakes Observing System establishing an "International Field Year for Great Lakes Research". 2001- 2003 Priorities Report, Emerging Great Lakes Issues in the 21st Century. pp. 105–111.

avoiding disputes and conflicts. It has also proven itself capable of responding to emerging issues and challenges.

III. Administrative/Executive Duties

Article VI of the Treaty makes provision for direct administrative control by the Commission. The administrative function of the IJC generally deals with orders of approval for major structures in the basin in order to maintain a balance between conflicting interest of water withdrawals, navigation, hydroelectric generation and environmental concerns. The major orders issued by the IJC include the 1914 Lake Superior Control Board, 1952 St. Lawrence River Board and 1953 International Niagara Board of Control.

The Commission appoints boards[25] to compile information needed to grant approvals and to ensure compliance with the terms and conditions of approvals. It also forms task forces to conduct the technical work required during investigation.

---

[25] Since its inception, the IJC has built an enviable reputation as an institutional mechanism for cooperative problem solving on a wide range of water- and boundary-related problems. The Commission sees the creation of international watershed boards as a refinement that can assist the parties greatly in addressing new challenges. The Commission has created the following Water Quality Boards:

Water Quality Boards:

1	Accredited Officers for the St. Mary and Milk Rivers	1914
2.	International Lake Superior Board of Control	1914
3.	International St. Croix River Board of Control	1918
4.	International Lake of the Woods Board of Control	1925
5.	International Lake Champlain Board of Control	1937 (inactive)
6.	International Kootenai Lake Board of Control	1938
7.	International Columbia River Board of Control	1941
8.	International Rainy Board of Control	1941
9.	International Osoyoos Lake Board of Control	1946
10.	International Niagara Board of Control	1950
11.	International St. Lawrence River Board of Control	1952
12.	International Souris River Board of Control	1959

Pollution Boards:

1.	International Advisory Board on Pollution Control St. Croix River	1962
2.	International Rainy River Water Pollution Board	1966
3.	International Red River Pollution Board	1969
4.	Great Lakes Water Quality Board	1972

Advisory Board:

1.	Great Lakes Science Advisory Board	1978
2.	International Air Quality Advisory Board	1978

Investigative Boards:

1.	International Souris-Red River Engineering Board	1948
2.	International Technical Information (Network) Board	1979 (inactive)

Except for limited initiative powers conferred by the Great Lakes Quality Agreement, the Commission has no independent power of inquiry except those initiated by either of the two partners. It has become an accepted principle that neither country will reject the request for investigation by the other, until/unless it has compelling reasons to do so. For example, the United States agreed to the Canadian request for investigation of the regulation of the use and flow of the Souris River in the 1940s – though without much enthusiasm. Similarly, the Canadians agreed with equal unenthusiasm to second the investigation of the proposed Passamaquoddy tidal power project in 1960s.

IV. Arbitral

By virtue of Article X the Commission is also a permanent court of arbitration between two countries. The article states:

> Any questions or matters of differences arising between the High Contracting Parties involving the rights, obligations, or interests of the United States or of the Dominion of Canada, either in relation to each or to their respective inhabitants, may be referred to the International Joint Commission by the consent of the two Parties.

Even though, the Commission was given an arbitral function which it has never been asked to exercise to date, and hence remains untested in the real world. Often disputes are submitted for arbitration only after it has proved impossible to settle them by diplomacy, and when national feelings have been aroused. It is probably a boon that it has never exercized arbitral function, because in that case, it would be imposing its will on the sovereign rights of the partners. Besides, arbiters are required to have

---

3. Flathead River International Study Board	1985 (inactive)
4. Red River Flood Study Board (to be established under	1997 (IJC Reference)

The following International Boards report to the Canadian and US governments:

Control Boards:
1. International Lake Memphremagog Levels Boards	1920

Pollution Boards:
1. Canada-United States Committee on Quality in St. John River	1972
2. Poplar River Bilateral Monitoring Committee	1980
3. Souris Basin Bilateral Water Quality Monitoring Group	1991

Treaty Boards:
1. International Niagara Committee	1950
2. Columbia River Permanent Engineering Board	1964

Study Boards:
1. Canada-United States Garrison Consultative and Technical Committee	1981
2. Niagara River Toxics Committee	1981

some legal training. Even though some of the Commissioners did have a legal background, but not all of them do not have the required legal qualifications. It could be that the Commission has operated so efficiently as it was not necessary to use its arbitral powers. Washington refused to refer the Trail Smelter case to the IJC, because of lack of confidence in the Commission. It is/was felt that complex waterways system problems are ill-suited for resolution through judicial proceedings. Or may be none of the Partners wanted to involve the Commission in acrimonious controversies.

CONCLUSION

The IJC has been in existence for 100 years (even though it was to remain in force for a fixed period of five years from the date of ratification and thereafter until terminated by a twelve-month written notice by either Government to the other.[26] But no one has found it necessary to cancel the Treaty (and serve notice.)[27] IJC has conducted well over 100 different studies and projects, for both the governments. Personnel of the Commission share information and data and they are not beholden to their agencies or their governments and act as part of the IJC team. When the engineers from the former Soviet Republic or from any part of Africa or South America hear about the working of the IJC, they are very impressed. They simply cannot imagine how data and information could be shared between two countries in a way which does not seem to be possible in their home countries.[28] One of the reasons could be that Canada and US share a more cordial relationship than most of the other countries facing transboundary issues. Another reason can be that both the countries are prosperous and have enough funds to support such an organization. Such laurels are not only bestowed upon the IJC in more recent times, but have been done ever since its very inception. Lord Curzon spoke of the IJC as a possible model for the adjustment of differences at the Dardanelles at the Lausanne Conference in 1923. According to the French Minister, Aristide Briand, the IJC model of conciliation and arbitration between Canada and the United States could be used to regulate the Franco-German frontier.[29]

---

[26] Editorial Comment, Boundary Waters Between the United States and Canada, The American Journal of International Law, 1910, 4(3).

[27] One of the reasons of this success might be the cordial relations between these two countries and can best be expressed in words of Marcel Cadieux, "Canada's relations with the United States have been good, complex, fluctuating, but on the whole effectively handled."

[28] http://www.sfu.ca/cstudies/science/water/pdf/Water-Ch21.pdf

[29] Canada and the United States: An International Commission, Pacific Affairs, 1929, 2(1).

The treaty has served a valuable function in protecting the interests of both Canada and the United States. Its principle of equality is of particular importance to Canada. Without this treaty, both countries might have proceeded with unilateral developments, resulting in a strained relationship between the two neighbors and the compromise of the long-term integrity of boundary and transboundary watercourses. While the treaty has served as a valuable instrument for the governance of boundary and transboundary waters, it is considered to be vague in relation to modern social and economic needs (fisheries and recreation), and pollution control. The other countries adopting this model can take these issues into consideration from the very beginning.

The Commission would continue to play an important role in dealing with emerging issues in continental waters on a pro-active basis. In view of the important role of the IJC and its past accomplishments, it is incumbent upon both federal governments to strengthen the role and authority of the Commission, and to use its expertise to deal with boundary and transboundary water governance related matters.

The International Joint Commission (IJC) by virtue of its regulatory, administrative, investigative and arbitral mandate, as provided under Boundary Waters Treaty, has played and would continue to play an important role in protecting the long-term integrity and sustainability of Great Lakes. The IJC's proposal to the governments of the United States and Canada to develop a system of biological indicators to monitor and assess the health of the Great Lakes, is a positive step for their long-term integrity.

Independence of joint institutions allows it to play an effective mediating role in assisting member governments to resolve common problems or to monitor the implementation of their agreed regime. That is why it is recommended that the other countries adopting the similar model should stress on the independence of such a commission – that is it should not be affiliated to any National agency and should be jointly funded by the participating nations.

Being a permanent body, it meets regularly and has frequently induced a feeling of *esprit de corps* among its members which is scarcely possible in ad hoc tribunals. For decisions or conclusions, the Commissioners act independently like judges in a municipal government. The procedures are rather simple which enables any private person whose rights are affected to be heard by the counsel.

One of the greatest strengths of the IJC has been to allow its institutions grow and adapt as member countries develop closer relations and better understanding of each other's interests. Both Canada as well as the United States relied increasingly on the IJC for solution of the transboundary environmental problems despite the fact that its basic instrument (the

Boundary Waters Treaty of 1909) neither foresaw many of the problems described below nor expressly conferred competence upon the IJC to deal with them. The organic[30] institutional developments in increased Opportunities for cooperative action and growing confidence of member states in the institutions played an important role.

The IJC has developed expertise in addressing complex ecosystem management issues and has wide experience in handling a full range of water issues which are usually bilateral in nature could be used in a multilateral environment. The Great Lake Quality Agreement which can be used by other major transboundary basins by establishing watershed boards of the type described above.[31]

The Commission is optimistic about the future of the Canada-U.S. transboundary relationship despite the challenges the two nations will face in the 21st century. Challenges such as

> Population growth and urbanization
> Climate change
> Economic expansion, energy demands, and waste generation
> Technological development, and
> Environmental awareness

The United States and Canada have demonstrated the ability to engineer new institutions and mechanisms to ensure that the interests of their citizens in the boundary area, as well as their common environment and their natural resources are properly managed and protected. The very flexibility of the Boundary Waters Treaty and the Commission itself has enabled the IJC to respond to changing times.

---

[30] Since 1970, the government, citizens, and industry have enhanced their efforts to restore and protect the Great Lakes. These collaborative efforts have made some progress in protecting the long-term integrity of the Great Lakes, but have not been very effective in preventing their further degradation. However, asymmetry of economic development and population size distribution between the two countries further complicates conflict over water supply and water pollution. Therefore, a much more concerted and pro-active approach is essential for the prevention of future problems and restoration and protection of this valuable resource. Specific strategies, framework or agreements need to be developed and implemented to deal with water use appropriation, climate change and global warming, long range transport of pollutants, introduction of new aquatic invasive species, protection of near shore waters and coastal areas against point and non-point pollution sources, implementation of measures to clean-up the identified areas of concern, reduce or eliminate the entry of toxic pollutants, develop sound information base and ecosystem health reporting indicators, and to ensure long-term sustainability of the Great Lakes.

In 1985 the International Joint Commission (IJC) identified 43 areas of concerns in Great Lakes with 31 areas in need of immediate remedial action. These are considered as most severely degraded areas with multiple problems such as: contaminated sediments, degraded fish and wildlife habitat, and impaired beaches. It has been recommended that remedial action plan be developed for the restoration of these areas.

[31] The IJC and the 21st Century.

Public participation systems needs to be made more robust along the Nile-TAC or CEC lines as provided under Articles 14 and 15 of NAAEC. There is no formal line of public access other than the request for non-confidential information as described and prescribed under Article XII.

The IJC has gone to great lengths to balance its findings and recommendations to such a degree that there are no absolute winners or losers.

The Commission has the capability to expand its activities significantly in areas like upper atmospheric air pollution, point source air pollution, in coastal and marine matters, outer-continental and other similar matters. This expansion could well extend to issues which, while bilateral, have clear multilateral ramifications.[32] Basically, the whole area of environment where states and provinces have jurisdiction and parochial interests and responsibilities, an institution like the Commission may enable national and regional authorities to resolve their difference and arrive at practical solutions like the Garrison Diversion project.[33]

---

[32] John E, Carroll.
[33] Marcel Cadieux.

South Pacific

13

Water in the Pacific Islands: Case Studies from Fiji and Kiribati

Eberhard Weber

The University of the South Pacific
Faculty of Islands and Oceans, Head of the School of Geography
Suva, Fiji Islands

INTRODUCTION

Water can be a scarce resource in the Pacific Islands. This statement seems to be rather contradictory considering that the Pacific Ocean is by far the single largest body of water on the globe, and most of the Pacific Islands are humid tropical islands, where precipitation is plentiful (Carpenter et al., 2002). Drought and water scarcity therefore is not easily recognized as a problem (Terry and Raj, 2002). Despite this a growing number of islands in the Pacific Ocean are reporting water scarcity, many of them even facing severe water problems. White et al., therefore, highlight that population centres in small islands of the Pacific have water supply problems that are amongst the most critical in the world (White et al.,1999, 2004).

Quite often islands in the Pacific Ocean are very small and without meaningful catchment areas. The harvesting and storage of freshwater is a constraint with a number of factors such as small land areas, atoll geology, pressures of human settlements, conflicts over traditional resource rights, capacity limitations, frequent droughts and inundation by the sea during storms (White et al., 2004). At the same time the Pacific Islands are experiencing the increasing demand for water from a fast growing population, the expanding tourism sector and sometimes also industries. In addition many countries in the Pacific Island region are threatened by a continued over-exploitation and pollution of limited surface and groundwater resources and the environmental degradation of coastal areas (including coral reefs) (Baisyet 1994).

Water scarcity cannot be seen isolated from the other threats that the small islands are facing. Natural hazards like cyclones and flash floods have caused severe damage to the island's water supply since long. Global climate change and the anticipated rise in sea-levels are especially threatening the low-lying atoll islands. These new risks, which will have an impact not only on freshwater resources on these islands, but also on the ability of island populations to cope with their changing environments. The danger that a growing number of islands will become uninhabitable and their residents environmental refugees (Dow et al., 2005) is thus rather real and urgent.

There are many issues regarding water in the Pacific Islands. As it is not possible to cover all—two major aspects therefore will be in the foreground in this chapter. First to describe the background under which water-related issues are dealt with in the Pacific. This means the physical, environmental, economic as well as the political and institutional sides of water supply and distribution. Such a perspective should serve the purpose of discussing whether the islands in the Pacific are different in their water-related problems from other regions in the world.

This general overview will then be developed by a number of case studies. These case studies will highlight the various topics involved in water- related issues in the Pacific. The first case study looks at the 1997-98 in Fiji. The vulnerability of Fiji's economy and the people to cope with such a lack of rainfall will be the centre of discussion. After that we will have a look at the situation in Suva, the capital of the Fiji Islands, and the biggest urban agglomeration in the Pacific Island region. Annual rainfall of more than 3,100 mm makes Suva one of the 'wettest' capitals on the globe, but nevertheless water scarcity has become a major problem for the 200,000 odd citizens. The episodes of 'urban droughts' are becoming more frequent in recent years and it seems that in the near future not much hope for an improvement can be expected. Another important issue on water will be taken up in the third case study on Fiji: conflicting property rights create economic, political and social instability in Fiji. The parallel existence of traditional and modern institutions create fields of uncertainty that also make water-related issues even more complex than they are already. Finally, the last case study is on South Tarawa, the capital of Kiribati. South Tarawa is one of the most densely populated areas within the Pacific Island region. Today almost half of Kiribati's population is concentrated on South Tarawa, and the population is rising at a very fast pace. Migration from the outer islands of Kiribati might reduce water-related problems there, but in South Tarawa the limits for a sustainable water supply have been reached since long.

WATER IN THE PACIFIC ISLANDS – AN OVERVIEW

The Pacific Ocean is the biggest in the world. It is 16,700 km wide at the equator and more than 19,000 km at its widest point from Singapore to Panama. The huge body of water, more than 150 million km², covers an area larger than the world's entire land mass combined.

The Pacific Islands region occupies a vast 30 million km² of the Pacific Ocean, which is an area more than three times larger than the United States of America or China. The region has a very small land mass. Geographically it extends from Pitcairn in the east to Papua New Guinea in the west. It has 7,500 islands of which less than 1000 are inhabited. The 22 countries and territories of the Pacific Islands region consist of approximately 550,000 km² of land with 7.5 million inhabitants. If Papua New Guinea is excluded, the figures drop to 87,587 km² of land and 2.7 million people.

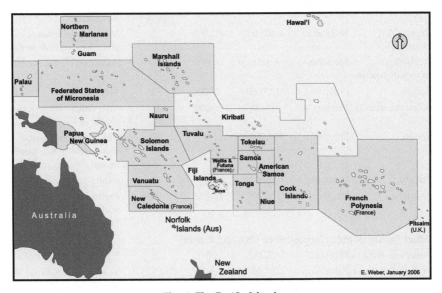

Fig. 1. The Pacific Islands

The countries in the South Pacific have large coast-to-land ratios. Their economies mainly depend on coastal resources, and large proportions of their populations are concentrated in coastal areas. The only exception in the region is Papua New Guinea, which has a more dispersed population and economic base (Bleakley, 1995).

The Pacific Islands region is geographically extremely diverse. The largest state, Papua New Guinea, has a land area of 462,000 km², while the smallest islands states such as Nauru, Pitcairn, Tokelau and Tuvalu are smaller than 30 km² each. Some countries and territories, like Nauru and

Table 1. Summary data for Pacific Island countries and territories, Source: Falkland (2002)

Country or territory	Sub-region	Approx. population (in 2000)	Total land area (km²)	Number of islands or atolls	Island type according to geology
Pacific Island countries					
Cook Islands	Polynesia	16,000	240	15	Volcanic, volcanic & limestone, atoll
Federated States of Micronesia	Micronesia	114,000	702	607	Volcanic, atoll, mixed
Fiji	Melanesia	785,000	18,300	300 (Approx.)	Volcanic, limestone, atoll, mixed
Kiribati	Micronesia	85,000	810	33	32 atolls or coral islands, 1 limestone island
Nauru	Micronesia	11,000	21	1	Limestone
Niue	Polynesia	1,700	260	1	Limestone
Palau	Micronesia	22,000	487	200 (approx.)	Volcanic, some with limestone
Papua New Guinea	Melanesia	4,400,000	462,000	?	Volcanic, limestone, coral islands and atolls
Republic of Marshall Islands	Micronesia	60,000	181	29	Atolls and coral islands
Samoa	Polynesia	175,000	2,930	9	Volcanic
Solomon Islands	Melanesia	417,000	28,000	347	Volcanic, limestone, atolls
Tonga	Polynesia	99,000	747	171	Volcanic, limestone, limestone & sand, mixed
Tuvalu	Polynesia	11,000	26	9	Atoll
Vanuatu	Melanesia	182,000	12,190	80	Predominantly volcanic with coastal sands and limestone
Other Pacific islands (Territories of USA and France)					
American Samoa	Polynesia	67,000	199	7	5 volcanic and 2 atolls
French Polynesia	Polynesia	254,000	3,660	130	Volcanic, volcanic & limestone, atolls
Guam (USA)	Micronesia	158,000	549	1	Volcanic (south) and limestone (north)
New Caledonia (France)	Melanesia	205,000	18,600	7	Volcanic, limestone
Island countries in other regions					
East Timor	SE Asia	800,000	24,000	1 main island	Volcanic
Maldives	Indian Ocean	270,100	300	26 atolls	Approx. 1,900 islands

Niue, consist of one single small island while others such as French Polynesia and the Federated States of Micronesia have more than a hundred islands each (Table 1), which are in some cases spread out over enormous distances of several thousands of kilometers like Kiribati. This state has a land territory of just 810 km² scattered over an Exclusive Economic Zone of more than 3.5 million km². The country stretches almost 5,000 kilometers from East to West and more than 2000 kilometers from North to South. To fly from South Tarawa, the country's capital, to Kiritimati Island, the biggest island, one has to cross over in two other countries, Fiji and the USA (Hawaii).

Small islands are often classified according to their topography, which is mainly a result of their geological structure (Fig. 2). We roughly differentiate between 'high' and 'low' islands, and somehow in the middle 'raised' islands. Geographical factors play a major role in development problems of the islands in the Pacific. Many of them belong to the group of the least developed countries, but differences in resource endowment and living standards of the population are vast. High islands have fertile soils for agriculture and generally good water resources, both surface water as well as groundwater, however there are usually much bigger problems in low islands.

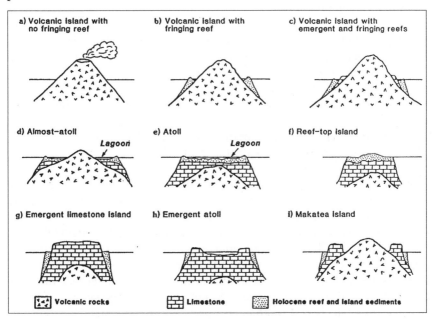

Fig. 2. Main types of mid-oceanic islands in the Pacific
Source: Scott et al. (2003)

The Melanesian countries are relatively big, mountainous and of volcanic origin. Rich soils that provide a good base for agriculture, exploitable mineral resources and plentiful marine resources are found here. Many of the Polynesian and Micronesian island nations however are much smaller and disadvantaged by a number of physical features. Kiribati, the Marshall Islands, Tokelau and Tuvalu consist mainly of low-lying atolls, many less than 100 km² in size and not more than five meters above sea level.

Atolls are often described as one of the earth's harshest environments. In most cases they are extremely isolated like oases amidst an oceanic desert. Tiny long stretches of land, only a few hundred metres broad, and hundreds, if not thousands of kilometers away from larger landmasses create unique ecosystems, but also unique conditions for human existence. Here people do not have the safety margin available in continental and large insular regions. Considerable effort is needed to exploit poor atoll soils. Atoll production systems include a limited selection of tree crops, root crops, fisheries and cottage industry (Rapaport, 1990). Some atoll states such as the Maldives Islands in the Indian Ocean have managed to set up commercial fishing and are also rather successful in the tourism sector.

Smaller volcanic islands such as the Cook Islands, parts of the Federated States of Micronesia, Tonga and Samoa have some fertile land, but their small size and the lack of natural resources are enormous obstacles to economic development.

High islands of volcanic origin usually have good potential for the development of surface water resources as well as groundwater resources. They often have perennial rivers (e.g. many islands in Fiji, Papua New Guinea, Samoa, Solomon Islands and Vanuatu). Volcanic islands frequently have springs, both in elevated and coastal areas that are used as important sources for water supply schemes, especially on the rural community level.

Depending on their location many of the high islands also receive a lot of precipitation, most as orographic rain. The availability of water is therefore not the major problem, but the storage and distribution of water alongwith the very high capital costs for the water supply of fast growing urban centres are areas of concern. Low coral islands as well as (raised) limestone islands have little surface water resources and largely depend on groundwater resources that are often complemented by the collection of rainwater (and unconventional sources of freshwater such as desalination). Raised limestone islands generally have little or no surface water owing to the high permeability of the rock. On smaller islands and small catchments of larger islands, stream flows may become very low or cease during extended droughts. Surface water on low islands, if present, is likely to be in the form of shallow, brackish lakes unless the rainfall is very high when it may be fresh. Nauru, for example, is a limestone island which has an interior brackish lake near sea level (Falkland, 2002). Freshwater on atolls is the most limiting factor for human settlements. The island soil and underground is usually so porous that water seeps down to a lens of freshwater saturating rocks and sand almost instantaneously.

Rapaport (1990) highlights that many Pacific atolls once supported larger populations than they do today. Early accounts indicate a "miserable existence for the inhabitants of many atolls visited" (Wiens, 1962). Starvation, emigration, and war were very much the reality on many atolls (Alkire, 1978, Pollock, 1970).

Freshwater resources of small island states are often classified as either 'conventional' or 'non-conventional'. Falkland (2002) differentiates between "naturally occurring water resources" that require a relatively low level of technology to develop them and "water resources involving a higher level of technology" (Falkland, 2002). Naturally occurring or conventional resources include rainwater collected and stored, groundwater and surface water. Non-conventional resources include the use of seawater or brackish groundwater, desalination, water importation by ships or pipelines, treated wastewater, and substitution of water (such as the use of coconuts during droughts).

The collection of rainwater is rather common on the Pacific Islands. Sometimes entire collection systems are developed, especially where other sources of freshwater are limited and where sufficient precipitation can be expected during longer periods of a year. Rainwater collection systems are found on the roofs of individual houses, administrative buildings or even especially paved runways. On some very small low-lying countries, such as Tuvalu, the northern atolls of the Cook Islands, and some of the raised coral islands of Tonga, rainwater collection on roofs of community buildings is the sole source of fresh water.

For small, low-lying islands groundwater is often the most reliable and important water resource. Groundwater occurs either as perched or as basal aquifers. Perched aquifers develop above an impermeable layer, or when groundwater is retained in compartments by a series of vertical volcanic dikes (Falkland and Custodio, 1991). They are similar to the aquifers found on large islands or continents. Basal aquifers occur on high as well as low islands in the form of coastal aquifers or rainwater that percolates through an island and floats on the denser salt or brackish water in what is called a Ghyben-Herzberg lens (Whittaker, 1998, see Figure 3).

The size of such a freshwater lens is more or less proportional to the width and surface area of an island. It is also influenced by factors such as rainfall levels, the permeability of the rock beneath the island, and salt mixing due to storm- or tide-induced pressure (Roy and Connell, 1991). In some cases such lenses may be as thick as 20 metres providing secure and long-lasting water supply. On raised coral atolls, such as Nauru and many of the islands of Tonga the freshwater lens may be no more than 10–20 centimetres thick, and is thus very vulnerable to over-exploitation (Falkland and Custodio, 1991).

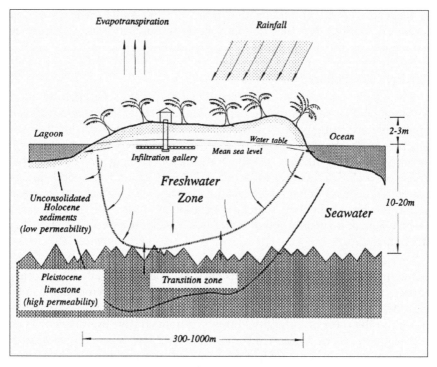

Fig. 3. Freshwater lens of a coral atoll
Source: White et al. (2002)

Surface water in bigger quantities and potable qualities is mainly restricted to high islands. It occurs here in the form of ephemeral and perennial streams, springs, lakes, and swamps (Falkland, 1999). On coral atolls and limestone islands surface water is rarely found because of the high infiltration capacity of the soils and rocks (Falkland and Custodio, 1991). The usual small size and altitude of these islands also restrict the size of potential catchment areas. In those rare cases where surface water is found on low islands, it is likely to be in the form of shallow, often brackish lakes that easily get polluted especially in places with a high population density.

A number of high islands in the Pacific use mainly surface water as their major source of freshwater supply because gravity-fed water systems are more cost-effective than the development and maintenance of systems based on groundwater sources. For example, surface water provides more than 95% of the water requirements in French Polynesia. Surface water also contributes to the freshwater supply in Samoa, Fiji, Nauru, Palau, and on the high islands of the Cook Islands (Falkland and Custodio, 1991).

Almost 98% of global water resources is saltwater. It therefore seems logical when countries in the middle of the ocean start to use saltwater to generate freshwater. There are several technologies at hand (distillation, reverse osmosis and electrodialysis) and some countries in the Pacific region started to use this unconventional source for their regular water supply. On Nauru about 60% of the island's water supply is from desalinisation. The heat from the power station is used for saltwater distillation. Experience with desalination on South Tarawa, Kiribati, has however shown the problems and limitations of desalinisation. A number of reverse osmosis units have been installed (Metutera, 2002), but some had been non-functional for long periods as it took months to get spare parts. The plants installed on Tarawa supply only a small proportion of the total water supply requirements. It would be rather risky to depend entirely on a technology that obviously cannot be well maintained. In addition water produced in desalinisation plants is much more expensive than 'conventional' sources of freshwater due to the high energy costs and other operating expenses. Desalinisation is therefore an option where the high costs can be recovered easily, or where due to the lack of other freshwater sources no alternatives to this rather expensive solution are available. In a number of tourist islands reverse osmosis is used to supply water to hotels and beach resorts (e.g. Mana Island, Fiji and Akitua island, Aitutaki atoll, Cook Islands).

Another non-conventional source of freshwater is the transportation of water from one island to another through ships or pipelines. Before Nauru started to produce freshwater through distillation, ships supplied water to the people of this tiny island nation. Also some of the small islands of Fiji and Tonga regularly receive water from nearby islands by barge or boat. During severe droughts or natural disasters small islands of Fiji, Kiribati, and the Marshall Islands have relied on coconuts for drinking water. Finally, non-potable sources, including sea water, brackish groundwater and wastewater, are used to flush toilets and fire fighting on a number of Pacific Islands such as Kiribati and the Marshall Islands (Falkland, 1999).

In the Pacific region there are a growing number of people, who do not have as much water as is required. A survey by the Asian Development Bank (ADB) found that only 50–75% of the residents of Samoa and only 44% of the residents in Kiribati had access to safe water (Burns, 2002). In the Federated States of Micronesia only 30% of the population has access to safe water, and in the Marshall Islands not more than 50%. (UNEP, 1999). In Papua New Guinea just 10% of the rural population has access to safe drinking water (Burns, 2002). Rapid growth in population , and an increasing demand from the tourism sector and industry are placing a lot of strain on the limited water resources of many Pacific Island Developing Countries (PIDCs) (Falkland and Custodio, 1991). In the major centres that are growing at alarming rates the existing supply and distribution systems are no longer capable of satisfying demand. Very often capital is not available to make necessary expansions or even to safeguard that the existing system is well maintained. Water leakages have let large quantities of this already scarce resource go waste. In extreme cases up to 70% of water is lost through leaks in the system. (SOPAC, 2001).

Burns (2002) highlights that many islands in the Pacific rely on a single source of water. This makes them rather vulnerable to all sorts of risks. Here vulnerability is different from the risk of water scarcity elsewhere. Sometimes the next freshwater source is thousands of kilometers away; at times there are no frequent transport links to many of the islands, so that it can take many days, if not weeks to provide supply. The ability to sustain a particular level of freshwater supply thus decides whether an island is permanently suitable for human settlement or whether people have to move elsewhere. In the years to come many islands might get deserted by people because water scarcity makes them uninhabitable. Environmentally induced migration will aggravate the concentration of population on major islands and increase environmental problems even further.

Problems with water are caused not only by population growth and migration to the major cities in the region. Cyclones with heavy winds and torrential rain are causing floods in many areas, destroying crops in the fields and people's homes. Flooding also often leads to water pollution, making water unsafe for human consumption. This occurs especially where no adequate sewage disposal infrastructure exists (UNEP, 1999).

Last but not the least, El Niño/Southern Oscillation (ENSO) episodes have reduced the amount of rainfall considerably in many parts of the western Pacific. Droughts caused by El Niño were reported in 1978, 1983, 1987, 1992, 1997-98, 2001 and 2003. Some stations recorded a decline in precipitation by as much as 87% in the western Pacific while resulting in unusually high rainfall in the central Pacific (Burns, 2002; Terry and Raj, 2002).

Droughts caused by El Niño have a severe impact on both high as well as low islands. They put a lot of strain on agricultural production, and have also depleted rainfall collection supplies and the freshwater lenses and perched aquifers on many atoll islands in the Pacific. For example, in 1998, 40 atolls of Micronesia ran out of water during an ENSO event, resulting in the declaration of a national emergency (Field, 1998a, b). In the same year, rainwater tanks in substantial parts of Kiribati dried up and shallow groundwater reserves became brackish (World Bank, 2000). The main island of the Marshall Islands only had access to drinking water for seven hours every fourteen days, and rationing occurred on all islands in the North Pacific (East-West Center 2001). In 1998 Fiji had one of the severest droughts in history. In the seven months from September 1997 to

March 1998, rainfall recorded was 60% lower than the average – the lowest ever recorded in the country since 1942 (Bolataki, 1998; Lightfood, 1999).

Water and Vulnerability of Small Island Developing States

Although small island states are not at all homogeneous they share many common features that lead to an increase of their vulnerability. They are small in size and in most cases surrounded by larges expanses of ocean; they usually have limited resources, are prone to natural disasters and extreme events. Geographically they are often rather isolated and their economies are extremely open and vulnerable to external shocks. These and other characteristics limit the capacity of small island states to mitigate risks and stress and to adapt to changing natural and cultural environments (IPCC, 2001a).

Freshwater resources and their management have many aspects not only in the Pacific. Severe drought conditions in Fiji, Kiribati, Samoa and many other countries in the Pacific Island region highlighted the urgency to develop more efficient ways of water use, and better and safer ways of water supply and distribution. Restrictions are not only due to natural reasons, but very often because financial commitment Pacific Island nations can contribute is much less than it would be necessary. Capital costs are very high especially for urban water supply systems.

Small Island Developing States (SIDS) in the Pacific attract the highest per capita aid in the developing world. A justification of this could be that due to their very special situation, SIDS are more vulnerable to natural hazards, and to economic recession than 'normal' developing countries. The vulnerability of the SIDS is also often mentioned, as regards to water-related problems.

Many of the Pacific Islands are amongst the Small Island Developing States. SIDS are small islands and low-lying coastal countries that have a small population, lack of resources, are remote, and are especially vulnerable to natural disasters. They largely depend on international trade. They suffer from lack of economies of scale, high transportation and communication costs, and costly public administration and infrastructure.

Fifty one states and territories work together in the Alliance of Small Island States (AOSIS). In April 1994, at their first Global Conference on Sustainable Development of SIDS in Barbados, they adopted the Barbados Programme of Action that at various levels would promote sustainable development of SIDS. Since then the United Nations confirmed the special situation of the SIDS at various meetings and conferences.The special status of SIDS was also confirmed at the World Summit on Sustainable Development in 2002.

Undoubtedly life on a tiny island, in the middle of a huge ocean, with a small population and hardly any resources would be difficult. Cyclones or hurricanes frequently destroy crops and bring floods. However the per capita income of many of these SIDS is higher than in many other developing countries. Countries that have their own problems that might be different from those of the SIDS, but not necessarily less severe and easier to solve.

Pacific Islands are not homogenous at all. There are vast differences in the physical, economic, historical and cultural backgrounds of island nations, and also among the individual islands themselves. Drought on a low-lying atoll island that largely depends on its groundwater is very different from drought on a big, mountainous high island where the water supply is mainly dependent on rivers. Both types of islands are very vulnerable to internal processes such as a high population growth and also to external pressure of no rainfall due to El Niño events. But both types of islands are vulnerable in a different way, and their vulnerability requires different coping strategies and adaptations.

One could get the impression that due to the (small) scale of cities and a low population on the Pacific Islands problems related to water are better and easier manageable than elsewhere. The cities of the Pacific Islands are not huge settlements with millions of inhabitants such as many cities in Asia. In the Pacific the cities are tiny. The solutions should be within the financial means of the municipalities and the Governments of these islands. However we should not forget that the budget outlay of all these nations is also very small. Investments in water supply systems are costly, and thus the size of a population is not really a good indicator on the affordability of modern technology especially in case where the majority of the people is rather poor and not able to pay much for their water needs.

In a recent report on Urbanization in the Pacific Islands the World Bank remarked:

> "The problems associated with delivering satisfactory water supply in Pacific island towns are primarily political and institutional rather than technical. They reflect inappropriate policies, undue government interference, and the lack of appropriate incentives for consumers to reduce demand to sustainable levels, all of which undermine the ability to operate and maintain water supply systems properly" (World Bank, 2002).

In the case of Suva, the capital of the Fiji Islands, a lack in rainfall can hardly be seen as a major reason for water scarcity. Also an unequal distribution of rainfall during the year does not explain the situation. As in many other places in the Pacific Islands, Suva has high rainfall throughout the year, without a real distinct dry season. Nevertheless there are tremendous problems with water and its supply.

The three case studies from Fiji and one from South Tarawa, the capital of Kiribati, will give insights in the very complex causes of water-related problems. The case studies show that scarcity created by nature is only marginally the reason, why human societies face problems regarding water. The case studies also clearly demonstrate that even in individual countries like Fiji and Kiribati, regional variations play a very important role as far as supply of water is concerned.

CASE STUDY 1: EL NIÑO AND AGRICULTURAL DROUGHT IN FIJI

The Fiji Islands are located between 12°-21°S latitude and 176°E-178°W longitude. Fiji consists of more than 300 islands of which about 1/3 are inhabited. With a land mass of 18,272 km² Fiji is the third largest state in the region next to Papua New Guinea and the Solomon Islands. The Exclusive Economic Zone (EEZ) of the country covers 1.3 million km². The two biggest islands, Viti Levu and Vanua Levu, have the majority of the total population of about 900,000, with about 50% living in urban centres such as Suva (177,000), Lautoka (45,000), Labasa (25,000) and Nadi (33,000). The two largest islands account for 87% of the land area and 90% of the population.

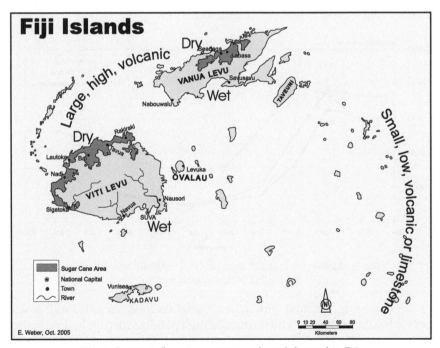

Fig. 4. Factors influencing water supply and demand in Fiji

The larger islands such as Viti Levu, Vanua Levu, Taveuni, Kadavu and the islands of the Lomaiviti group are rather mountainous and of volcanic origin. They are rising more or less abruptly from the shore to impressive heights. The southeast or windward sides of the islands record the highest rainfall of up to 5,000 mm per year. The western and northern parts of the major islands are in the rain shadow of the volcanic mountain ranges. They are therefore much drier and frequently threatened by droughts (Terry and Raj, 2002).

The climate in Fiji is dominated by the southeast trade winds. Exposure and topography control the distribution of rainfall on the islands. Average annual precipitation over the Fiji Group ranges from 1500 mm on smaller islands to over 4000 mm on the larger islands. Topographic effects mean however that much of this falls within the windward side of the islands.

The wet season from November to April is also the season of tropical cyclones. In the western parts of the bigger islands up to 80% of the annual total rainfall falls during this period. The western and northern parts of the major islands receive only 60-70% of the rainfall recorded in the eastern parts (Fig. 5). Here drought conditions are more likely to occur, especially during El-Niño episodes. These drier parts of Viti Levu and Vanua Levu are the centre of Fiji's sugar cane production. Drought therefore frequently affects the livelihood of a huge number of people and also cause a lot of harm to Fiji's export earnings (Lightfood, 1999).

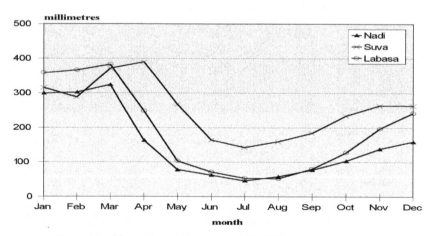

Fig. 5. Monthly total rainfall averages (1961–1998) for selected places
Source: Fiji Meteorological Service

Decline in agricultural production threatens food security, and poses severe health problems, whilst errant rainfall patterns disrupt hydroelectric power generation on Viti Levu. These are some of the more visible impacts of water shortages over these periods.

Table 2. Rainfall during Nov-April and May–Oct. in major centres of Fiji

	Nadi		Suva		Labasa	
	mm	%	mm	%	mm	%
Nov-April	1385	76.6	1991	63.4	1794	78.7
May-Oct	424	23.4	1150	36.6	486	21.3
Total	1809	100.0	3141	100.0	2280	100.0

Source: Fiji Meteorological Service

Drought in the western and northern parts of Fiji's major islands were recorded in 1983 and 1987. Both were connected to El Niño events. The latter one caused severe water shortages all over Fiji, including the usually wet areas of Suva/Nausori. The shortages were so severe that water supply needed to be restricted in most urban centres. Agriculture, especially Fiji's sugar industry, was badly affected.

The drought of 1987 was considered the worst drought in more than 100 years – until ten years later when an even more severe one afflicted the country. The 1997/98 drought affected most of the Pacific Island countries. The impact on water supply and agriculture created a number of economic, social and health problems. A direct consequence of the drought was Fiji's economic recession of 1998, when GDP was 8% lower than in 1996 (Lightfoot, 1999). In the western and northern provinces of the two main islands, Viti Levu and Vanua Levu, agricultural activities and production suffered a lot. These provinces are the centre of Fiji's sugar cane cultivation, and until recently the most important economic sector of the country.

There are more than 22,000 sugar cane farmers in Fiji. The vast majority of them have holdings that are not larger than 3–4 hectares. They rarely produce more than 200 tons of sugar cane a year. They earn around F$3,000 not enough to be above the poverty line. Another 20,000 low paid harvesting workers earn their livelihood directly through sugar cane production. A few thousands work as lorry drivers to bring the sugar cane to the mills or are employed in the sugar mills. Altogether some 50,000 people find employment in the sugar sector. This is about a quarter of Fiji's economically active population. Drought and income loss in sugar cane thus have severe consequences for the wellbeing of hundreds of thousands of people. In addition to that sugar cane is also one of the major export sectors that provide Fiji with foreign exchange. In a good year more than F$300 million is earned through the export of sugar mainly to the European Union.

During the drought of 1997-98 Fiji's sugar cane production declined by more than half. Instead of about four million tons not even two million tons of sugar cane were harvested. Officials from the Fiji Sugar Corporation estimated that the country lost about US$ 50 million in export earnings. For the first time in history, Fiji had been forced to import its domestic sugar requirement. About 35,000 tons of sugar was bought from Thailand,

Guatemala and Australia. The country was also forced to cancel export commitment of more than 100,000 tons of sugar to Japan, Malaysia and other Pacific Island countries (Radio Australia, July 13, 1998). The price of shares of the Fiji Sugar Corporation dropped to a record low of US$ 0.18 (Radio Australia, June 5, 1998).

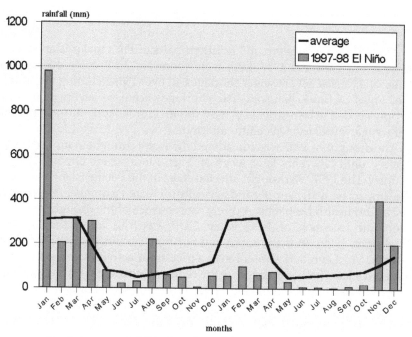

Fig. 6. Average rainfall in Lautoka and during the 1997-98 drought
Source: Fiji Meteorological Service

In some districts small farmers lost their entire sugar cane crop. In the Ba province, almost all of 43,000 people needed government assistance. In all drought – affected provinces about 80,000 people – 10% of Fiji's population – depended on government support to avert starvation and destitution (Daily Post, June 5, 1998). A year later sugar cane production had fully recovered and a production of 4 million tons was realized (Fiji Times, March 22, 1999). This time Fiji also benefited from a severe drought in Mauritius. As this island nation in the Indian Ocean was not able to supply its sugar quota to the European Union, Fiji was asked to make up Mauritius' deficit (Radio Australia, August 12, 1999).

As sugar cane production declined the demand for cane cutters, mill workers and other workers dependent on sugar production dropped. In total there was 50% less work available for sugar cane cutters. The burden of this reduction fell mainly on casual workers (Lightfood, 1999).

The drought of 1997-98 demonstrated how vulnerable Fiji's economy and society is to water scarcity. The losses in the sugar industry were surely the most profound and led to severe consequences. In 2004-05 dry conditions again seriously affected the cane farming belt in Viti Levu. The Save the Children Fund Fiji estimated that 5,000 to 8,000 children dropped out of school during the year. Parents had no money to pay for school fees, lunches, bus fares and school books (Fiji Times, February 2, 2005). Water scarcity however is not only connected to insufficient rainfall. The case study of Suva shows that people can suffer from a drought even when plenty of freshwater is around them.

CASE STUDY 2: PROBLEMS OF WATER SUPPLY IN SUVA

The urban agglomeration that comprises Fiji's capital Suva and two smaller independent towns, Lami to the west of Suva peninsula and Nausori to the east at the Rewa River is at the south-eastern side of the main island of Fiji, Viti Levu. Between Suva and Nausori a number of larger settlements have come up during recent decades. Quite often the area therefore is called the Suva-Nausori-Corridor. Today more than 270,000 people live in this greater Suva urban area, almost a third of Fiji's population. One water supply system serves the entire area, but Suva, Nausori and Lami have their own sewerage systems. The water for the present system is mainly supplied by the Waimanu and Tamavua rivers and the Savura creek.

The central piece of the greater Suva water supply system is the clear water reservoir at Tamavua. It is located at an elevation of 124 m, and water from here is fed into the distribution network and distributed by gravity. The Tamavua plant is supplied by three gravity sources located in the headwaters of the Tamavua river catchment, and two pumped sources on Savura Creek and the upper Waimanu River. In addition water from the Waila water treatment plant is pumped to the Wainibuku Reservoir at 81 m and to the Raralevu Reservoir at 55 m. Also from these two reservoirs water is fed by gravity into the distribution network. The Waila water treatment plant is supplied by water pumped from the lower Waimanu River.

The water supply in the Lami-Suva-Nausori area is often disrupted, and the quality of the water is often rather poor as streams and the coastal waters are more and more polluted. The existing water supply system has reached the limit of its capacity, and has already placed restrictions on housing development.

In the year 2002, many of Suva's citizens often went to work without a shower. Or they got up an hour earlier to visit a friend or relative who lived in a part of the town where no water disruption occurred. In the same year, more rain fell on Fiji's capital than in an average year: 3,627 mm in Suva, and even 1,000 mm more on the township of Lami, eight kilometres away from Suva, and a notorious area as far as water scarcity is concerned. Whoever has lived in Suva for a while wonders how water scarcity could occur here, as Suva is one of the 'wettest' capitals on earth. People living in the west of Viti Levu always felt sorry for the residents of Suva: as so much rain is not easy to tolerate.

During the early 1990s water disruptions occurred during 'drought' years. By the mid-1990s disruptions became frequent during the drier periods of each year. Disruptions now occur during all periods of the year. Provision of water by tank trucks, with storage in plastic tanks, which is expensive, is becoming increasingly common.

Water scarcity is a real problem for a larger part of Suva's population. It happens and those who are not affected by it would not even recognize the suffering of people living just in the next suburb: It occurs on days when there is a heavy downpour; during the rainy season, and it happens not because there is no rain. Sometimes areas in Suva, Lami or the small town of Nausori people must arrange their lives without water for days, weeks and sometimes even months (Keith-Reid, 2003). They get used to waking up in the middle of the night to see if the pressure is strong enough for some water in their taps. At the roadside they wait with buckets and cans for the lorries and trucks of the Public Works Department – quite often in vain. Employers send their workers home as the low water pressure does not reach higher levels in multiple-floor buildings. Schools have to close and the University of the South Pacific has sent thousands of students home as there was no water available on campus.

There are many reasons for the water scarcity in Suva, and a lack in rainfall is not one of them.

According to the Suva-Nausori Water and Sewerage Masterplan more than 50% of the water provided through Suva's distribution system gets lost before it reaches consumers. Leakages, illegal connections and other errors cost the water supply authorities a lot.

The population of greater Suva will increase from 248,000 in 1999 to 371,000 by the year 2019 (Fiji Times, December 27, 2001). The present distribution system is already too small to cater for the existing population, not to speak of the annual population increase of 2.1% over the next 20 years. In addition to shortages in water supply not even 40% of Suva's population is connected to the sewerage reticulation system. More than 60% use septic tanks and pit latrines, which perform poorly in Suva's low permeability soils. These large numbers of unconnected households, overflows from the sewerage system as well as industrial discharges to drains, creeks and the bay are causing environmental damage and pose a potential risk to public health (Fiji Times, December 27, 2001).

People living in one of the many squatter settlements in and around Suva are not connected to the water supply and sewerage system. It is estimated that by the year 2006 the Suva-Nausori corridor will be home to 90,000 squatters. 50 to 60% of these squatters live below poverty line and do have not enough money for even basic needs (Fiji Sun, August 18, 2004). Between 1996 and 2003 there had been a 73% increase in squatter settlements in Suva (Fiji Times, Feb. 9, 2005).

But not only the very poor suffer from water scarcity. In December 2002 one of Suva's most popular hotels had its own experience: within less than a day the hotel lost all its 105 guests. From full occupancy to empty, and to make matters worse most of the guests refused to pay their bills. The hotel had not received a single drop of water for more than 20 hours (Fiji Times, December 14, 2002). Tradewinds Hotel was not the only prominent victim of the water crisis. At Lautoka hospital major surgical operations were cancelled as the hospital had no water. In many parts of Viti Levu people had to improvise their water supply, in most cases without support from government authorities who just did not have the resources to deal with a crisis of such a dimension. In the first week of December 2002 a key pumping station in Suva broke down three times, one reservoir was empty, and a second one close to empty. Two lines between the reservoir burst, and aging pipes were leaking all over Suva. To ease the problem the Public Works Department employed 12 water trucks to bring water to residents in Suva's suburbs. Most of them however newer saw such a truck.

There are many reasons for Suva's ailing water distribution system. They range from chronic governmental neglect over years, insufficient budget allocations for the Public Works Department, lack of skilled and experienced engineers and administrators, and chronic corruption at various levels within the system. However the major reason is that over decades little had been done to maintain the water distribution system and adjust it to the fast growing population. Now as the collapse of the system is highly visible the costs to get out of the mess are just too high.

Suva is Fiji's capital. The government is located here and also the headquarters of most of the government agency. Suva beyond any doubt gets the highest government allocations as far as infrastructure development is concerned. The problems are even larger in other places in Fiji.

In June 2002 a state of emergency was declared in Kadavu, Fiji's fourth largest island about 80 kilometres south of Suva. Six hundred people live in the government station Vunisea, Kadavu's 'capital'. It consists of a government hospital, the only one in Kadavu, a government primary and secondary school, a police station and a post office and a few shops. Some

280 students from the government schools as well as 10 of the 12 patients from the hospital were sent home because of lack of water. Only two bedridden patients were allowed to stay in the hospital. The water for these patients had to be brought from a creek half a kilometre away. The Public Works Minister explained the collapse of the water supply system because of old and faulty pipes. "We need to replace the pipes for up to five kilometres. We need to find the money for that" (Fiji Times, June 11, 2002). What elsewhere is not really a big problem – To replace five kilometre water pipes—can become a major problem on an island, where water pipes are not available nor the workers who could do the work. It also can become a big problem when Governments are more concerned with the major islands, and just do not care much about the infrastructure of the islands in the periphery.

History and Background of the Present Water Supply System

In its present form Suva's urban water supply and sewerage system was developed in the 1970s and 1980s. Since then it could not keep pace with the ever increasing demand and has suffered from deferred maintenance and upgrading.

The need for a reliable water supply system became evident with the establishment of Suva as the capital of Fiji in 1882. The first piped supply was installed in 1890 from an intake in the Tamavua hills. With the growth of the city an appraisal of the existing supply became necessary and a number of improvements were recommended in 1911. Population and economic activities increased and thus water demand continued to become bigger and was met by piecemeal improvements in treatment and storage.

In 1961 the Tamavua water treatment plant was established, which needed urgent upgradation by 1970. Ten years later a new treatment plant was required – with support from the Australian Government – the Waila treatment plant was constructed in 1982. This, however, should have been the last major investment in Suva's water supply system for long.

Officials from the Public Works Department and representatives of the Government frequently blame each other for the neglect. The Director of Water and Sewerage (DWS) in the Public Works Department has the overall responsibility for Fiji's water supply and sewerage treatment and disposal. The PWD falls under the Ministry of Works and Energy and operates and maintains 13 regional, city or town water supply systems that produce and distribute about 170,000 cubic metres water daily serving some 610,000 people, or more than 80% of Fiji's population. Most of the freshwater distributed comes from surface water. At the moment the exploitation of groundwater does not play a big role in urban supply systems.

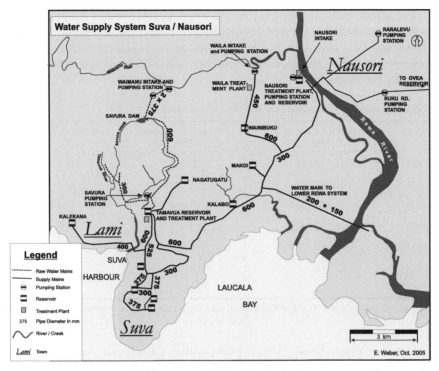

Fig. 7. Suva – Nausori Water Supply

To secure water supply to the residents of Suva, the PWD often complains of low budgetary support for the maintenance and improvement of its water and sanitation system. According to the PWD engineers it is technically not much of a problem to ensure that the residents of greater Suva have secure water supply 24 hours a day. The problems with delivering satisfactory water supply rather in financial and institutional bottlenecks. Like many other departments in Fiji with a more technical orientation the PWD has a shortage of personnel with appropriate managerial, financial, and technical qualifications and experience. The drafting of new projects, their technical and financial feasibility thus requires in most of the cases experts from overseas.

The Government on the other side often blames PWD for the insufficient system, but also highlights that it is not able to bear the high investment and operational costs of the existing system. It needs to be seen whether the water supply improves in the greater Suva area after a U$150 million project by the Asian Development Bank will be completed by 2007. The PWD is also in charge of water billing and the collection of water rates. Revenues earned are turned over to the Government's consolidated fund and the PWD in turn is financed through government budgetary allocations.

There is little incentive for the PWD to improve its financial efficiency. The revenues collected at present are insufficient to cover even operating costs. The situation is even worse as a huge proportion of water revenues are not collected reducing the financial capability of the Government to tackle the various problems (Table 3).

Table 3. Collection and arrears of water revenues in Suva

Year	Revenue collected US$	Arrears of revenue US$	Outstanding (%)
1998	11,047,150	9,252,285	45.6
1999	10,054,348	11,127,944	52.5
2000	11,470,720	16,126,473	58.4
2001	12,872,907	18,231,934	58.6
2002	n.a.	21,078,072	n.a.
2003	n.a.	23,412,841	n.a.
2004	n.a.	24,152,022	n.a.

Source: Report of the Auditor General, Audit report on the Infrastructure Sector, Suva, various years

Another major cause is corruption within the department. Having no water is just more than an annoyance. Having no water over an extended period of time makes life in a particular place miserable, if not impossible. 'Solving the problem' often costs money, even though the services ought to be free. In October 2003 there were reports that truck drivers from private companies distributing water to areas where supply had ceased asked a fee for their valuable load. Acting on behalf of the PWD charging money for this water service was illegal. Earlier in 2003 an investigation was started regarding allegations that private water truck operators had given PWD officials bribes for cutting off the water supply to entire neighborhoods to stimulate the water-trucking business. Watermains had intentionally been shut off so that water trucking was required (PINA Nius, October 9, 2003). For a few companies Suva's 'water crisis' turned out to become a spring of money that never runs dry. In an investigation report on irregularities that occurred during Suva's water crisis in 2002, the Auditor General noted that a range of issues hinted at corruption (Auditor General 2003). One company, not observing tender procedures, was awarded contracts for supply of hoist trucks and water tankers avoiding existing contractors. The PWD hired trucks and other equipment at much higher rates than that which the cheapest contractors had asked for. Payments were made for services provided by vehicles that were under repair/scrapped at the time for which the invoice had been issued. Payments were made for a day that does not exist in the calendar (September 31 (!!!)). Water tankers were hired at exorbitant rates; hiring rates paid for several vehicles were equivalent to the purchase price of new vehicles. The report concludes: "From the various defects, anomalies and irregularities noted during our investigation on vehicle/plant hires, it appears highly likely that official corruption exists at Suva Water Supply & Kinoya Sewerage Treatment Plant" (Auditor General 2003).

It is very difficult to establish how much the Government lost through such fraudulent practices. The Opposition in Parliament gave figures that the PWD scam costs tax payers more than the $30 million lost during the Agriculture scam (Fiji Times, April 12, 2003). All in all it is estimated that Fiji annually loses more money through the abuse of public funds than the amount of foreign aid given to the country (PINA Nius, Nov. 27, 2003).

Wastewater and Sewerage

In 1986 a little more than 61% of Suva's population had a flush toilet in which wastes were either retained in septic tanks or discharged into the local sewerage system. Septic tanks are rather problematic to operate in the Suva area as much of the city's area is located on soap stone which does not allow septic tank effluents to percolate properly. In addition to this the high annual rainfall of more than 3,000 mm on an average results in frequent saturation of the soil which tends to prevent oxygen penetration. As a result the natural treatment in Suva's septic tanks is rather slow and inefficient. Widespread seepage of sewerage waste into Suva's numerous creeks occurs rather frequently and some of the creeks, such as Nubukalou Creek, have been described as "sewers rather than creeks" (ESCAP, 1999).

What happens in Suva is symptomatic to many Pacific Island capitals where planning and development activities in the water sector cannot keep pace with rapid urban growth. Islands with a tiny land mass are especially affected by this, as sewage waste often accumulates very close to human settlements causing many waterborne diseases. But not only freshwater resources are a concern, marine pollution is a reason for concern in places such as Suva and the lagoon of Tarawa atoll in Kiribati. The results of a study undertaken by the University of the South Pacific indicate that the general water quality status of Suva harbor is above acceptable levels. The faecal coliform counts in the water showed frequent occurrences of unacceptably high values at several sites, far above the levels of the World Health Organization (WHO) give as limits for recreational waters (ESCAP, 1999).

Average faecal coliform concentrations greatly exceed internationally acceptable standards in most, if not all, of Suva's creeks. Of particular concern is Nubukalou Creek which drains a major area of the city that is without sewerage. The National State of the Environment Report states that "with faecal coliform levels thousands of times above an acceptable level it should be regarded as a sewer. The continued sale of fish along the creek bank, with the consequent use of its water for washing them, is a serious health hazard" (Watling et al., 1992).

In Suva the principal sewage treatment plants of Kinoya and Raiwaqa are frequently not able to function efficiently. Effluents that are discharged to surrounding waters are very often not completely treated. Through the Vatuwaqa River they finally end in the Laucala Bay. A similar scenario exists in Lami and Nasova where sewage effluents are discharged into the Wailada Creek and the Leveti Creek respectively.

Industrial areas around Suva, such as Lami, Walu Bay, Vatuwaqa and Laucala Beach Estate are another source for water contamination. The water pollution from these areas significantly reduces water quality in the near-shore waters around Suva and its neighboring settlements, especially Lami. There are no effective regulations to control the profusion of water pollution sources in those industrial estates, and the streams and creeks that drain those areas are probably the most polluted in the country (Gangaiya et al., 2001).

CASE STUDY 3: LAND TENURE AND WATER

Land tenure is the way in which people obtain, use and distribute rights to land. "There is no place in the world where anyone owns all rights to any piece of land. People own rights in and over land which may or may not be freely disposable. These rights are always subject to rights of other persons, entities and institutions in and over the same land" (Namai, 1987).

Land is a very sensitive issue in Fiji. Land in the Fijian language is called vanua, which is more than a resource. Vanua gives identity to the people of Fiji as their lives are closely connected to their land. Land presents the major source of security, both in a material, social and even psychological sense. Through the land an individual is tied to a social group. Closely related family groups live together in villages, cultivating well-defined land areas originally acquired by conquest or occupation of empty land. Several such family groups, claiming descent from a common ancestor, are linked in a larger social units – the *mataqali*. A number of mataqalis are grouped into yavusa of varying rank and function. Several yavusa form a vanua and several vanuas make up a province. There are 14 provinces in all: Ba, Ra, Serua, Nadroga, Namosi, Bua, Macuata, Lau, Cakaudrove, Naitasiri, Lomaiviti, Tailevu, Rewa and Kadavu. The provinces form three confederacies: Kubuna, Tovata and Burebasaga.

The land belongs to all of the mataqali members, not to private people. They can use it like private land. The land is given to them by their chief (Turaga ni mataqali). It is the individual who benefits from his or her efforts when cultivating the land, but it is not theirs.

Each distinctive social unit and subunit (except the *tokatoka* subdivision) is headed by a chief (turaga). Consequently, chiefs are placed at different levels on the hierarchy, but the *mataqali* subdivisions are the basic landowning social units. They own about 90% of the total land area in Fiji. However when payment for land leases are made the chiefs of all the other, higher social units get a hefty share, despite the fact that they are not the owners of the leased land.

Although native land is owned by the mataqali, the leases are administered by the Native Lands Trust Board (NLTB) established in 1940 under the Native Land Trust Act. The NLTB even has the power to lease the land that is not required for occupation by the members of a mataqali without the consent of the mataqali. Some of the excess land has historically been used for growing sugar cane and other crops, mainly by descendants of indentured Indians. More recently coastal land has been used for tourism schemes.

The formula for sharing the rent from the leases is 15% for the NLTB, 5% for the chief of vanua, 10% for the chief of the yavusa and 15% for the mataqali chief. The remaining 55% is shared amongst the members of the mataqali (which can number in hundreds, Table 4).

Table 4. Distribution of rent amongst different levels of traditional land-owners

Unit	Chief/Headman	No. of families	Share from lease
Native Lands Trust Board (NLTB)			15 %
Vanua	Turaga-Ni-Taukai	1	5%
Yavusa (tribe)	Turaga-Ni-Qali	1	10%
Mataqali (clan)	Turaga-Ni-Mataqali	1	15%
Members of Mataqali		Upto several 100s	55%
Totatoka (family group)		0%	

On September 25, 2005 more than 300 people, mainly villagers from the central highlands of Viti Levu, filled the Supreme Court room in Suva to follow the F$52.3 million compensation case between Monasavu landowners and the Fiji Electricity Authority (FEA). Monasavu dam is Fiji's biggest hydro-project. The 80 MW power station about 60 kilometres northwest of Suva supplies electricity to most parts of Viti Levu. 80–90% of FEA's customers on Viti Levu receive power generated at this site. Before Monasavu dam was completed in 1983 Fiji depended mainly on diesel-driven generators for its electricity supply. In 1982 Fiji generated 924 TJ (Terajoule) of its electricity supply through diesel generators. By 1986 this had come down to 67 TJ. Hydro-power increased from nothing in 1982 to 1067 TJ in 1986 (Prasad 1998).

Even today FEA and its customers are very aware of the importance of Monasavu dam. In drought years, when the water table of the dam is too low to allow an outflow to run the generators, FEA has to switch back to expensive diesel generators. High prices for fossil energy makes this an expensive endeavor. Fiji has a huge hydro-power potential. High volcanic mountains on all of the bigger islands, an average annual rainfall of more than 4000 mm, and extensive water catchment areas in the very sparsely populated interior parts of the major islands make ideal conditions for the use of water for electricity generation. Not many countries in the Pacific Island states are fortunate enough to have such extensive potentials for hydro-power. It is estimated that Fiji's total potential of hydro-power is over 1 GW, more than 10 times the amount used at the moment.

Despite all these positive news Monasavu hydro-dam and other projects there had been often in the limelight in recent years. Huge problems can emerge when various institutional arrangements over land control and management conflict with each other. It is not the conflicting water rights that have created all the problems. The land on which the Monasavu dam was built, the artificial lake that has submerged a lot of land as well as the catchment to provide water to the Monasavu reservoir are under dispute till today between the traditional landowners and the FEA.

When the Monasavu dam was constructed in the late 1970s an agreement between the landowners and government representatives had been reached that the landowners should be paid F$400 (US$238) per acre. FEA thus holds the view that it bought the land on which the dam and the reservoir is built for about F$1 million. Later however the landowners demanded a compensation for the 22.500 acres of catchment area. They argued that they were restricted in the use of the land as logging is prohibited to avoid a siltation of the lake. In June 1998 violent clashes near the power station brought the landowner's demands into the national press. They demanded an additional F$35 million compensation from the Fiji Government, and threatened to close down the power station, if their demands were not met (Daily Post, June 30, 1998). The Fiji Government then sent 100 armed soldiers and riot police to the site (Daily Post, July 2, 1998), but also started negotiations with the landowners about their claims (PACNEWS, July 8, 1998). Early October the Fiji Government offered the 19 land-owning groups a compensation of F$10.3 million for their land and another F$4.3 million for the timber standing on the land as compensation for the loss of productive use of the land (Fiji Times, Oct. 7, 1998). Some of the landowners rejected this offer and started legal action against the Government. In

November the Fiji High Court issued an injunction to a group of Monasavu landowners preventing the Native Lands Trust Board (NLTB) from paying F$14.6 million compensation to the landowners until the matter is sorted out by the court (PACNEWS, Nov. 10, 1998).

It took almost two years before the landowners of Monasavu again appeared on the front-pages of Fiji's newspapers. On May 19, 2000 Fiji experienced its third coup d'état after the two military coups of 1987. While coup leader George Speight and his followers kept the elected Fiji Government hostage in the Parliament building for almost two months Monasavu landowners again seized the power station, took FEA workers hostage and stopped the flow of water from the dam (PINA Nius Online, July 7, 2000). Although the military could get control over the dam by August 10, most places in Viti Levu experienced frequent power blackouts until the last week of August (Radio Australia, August 11 and 23, 2000). After the military intervention to regain the dam the Fiji Army established a permanent presence at the Monasavu dam. Despite that the Fiji Government wanted to have the dispute settled. In August 2000 there had been a meeting between FEA officials and the landowners, where the FEA offered a compensation of F$52.8 million. In June 2001 the same amount was reported in the national press. The Cabinet agreed to allow the FEA to sign an agreement that sets the compensation at F$52.8 million to be paid out over a 99-year period, provided that the landowners discontinue all legal proceedings against the FEA and to refrain from disrupting FEA work (Fiji Times, June 13, 2001). Despite all the positive announcements no payment has been made until today.

Monasavu surely is the most spectacular case where land rights of traditional landowners are conflicting with modern development efforts. However it is by far not the only case. In recent years a large number of such cases came up, and very often water- related issues were involved. In a conflict comparable to the one in Monasavu landowners of the Wainiqeu Mini-Hydro Scheme outside Labasa, the biggest town on Vanua Levu, wanted the Fiji Electricity Authority to pay F$ 7,302,880 (US$ 3,373,931) as compensation for the hydro-plants's water catchment area. Also landowners of the Navau water catchment area on Vanua Levu were demanding F$ 2,851,680 (US$ 1,317,476) from the Government (Daily Post, Sept. 8, 2000). In August 1999 another landowner group near Labasa had threatened to block the road to the Benau reservoir and water treatment plant, if the Government would not pay a compensation of F$150,000 for the land the reservoir is build on (Fiji Times, September 1, 1999). In Suva the landowners of the Wainibuku reservoir are demanding US$150,000 as compensation for the PWD's use of the land (Radio New Zealand International, June 5, 2003). Also the water supply in Lautoka, Fiji's second biggest city, is at the mercy of landowners. On April 15, 1999 landowners closed the pipeline that brings water from the Varaqe Dam in the highlands near Lautoka to the Suru Reservoir, which supplies water to Lautoka hospital, the main city area and outer areas of the south (Daily Post, April 16, 1999). In early 2003, Qerelevu Hindu School had to close down as landowners demanded money for the water supply of the school. For the past 30 years water for the school had been drawn from the community water supply system of Toge village, but then the landowners from the village stopped water supply demanding a goodwill payment. The school's headteacher is quoted by Fiji Sun (Feb. 3): "Now, without any written order, the landowners are demanding that we pay F$5,000 in goodwill and F$1,000 per household to get water. After we informed them that it was impossible for us to pay, as most of the people here can not afford it, they disconnected the water supply. It's almost three weeks now".

Not only water supply of many places are threatened by high demands of traditional landowners. In November 2000 landowners on which the government owned Ratu Kadavulevu Secondary School is built demanded F$0.5 million as goodwill payment to renew the lease for the land. They threatened to close the school if the amount is not paid within a week. The Government finally decided to pay F$196.000. For many years the Nadrau landowning unit of Saunaka village near Nadi International Airport were demanding compensation for the land Fiji's biggest international airport is built on. In 1999, when the Airport was privatized, they demanded F$48 million for the land. The conflict dragged on for a while and in July 2001 the landowners threatened to close down the airport if the Government wouldn't pay F$7 million compensation for the 434 acres of land. The landowners argued that during World War II, when the airport was built as an airfield used by the United States Air Force, they had given their land for public use, but now it is used for commercial purposes and they want to have a share in it (Daily Post, July 24, 2001). In August 2001 the government finally agreed to pay F$1.1 million to the landowners.

In March 2001 it was decided to relocate the Sigatoka Hospital. To renew the lease for the land the hospital had been built on the landowners demanded goodwill payment of F$700,000. While the Government and the Native Lands Trust Board (NLTB) had repeatedly pointed out that Fiji's land laws make no provision for such goodwill payments or any other payments to facilitate the renewal of a lease, the NLTB apparently had offered F$200,000 as a goodwill payment (Fiji Sun, April 1, 2004). The cases are endless. Land is used more as an economic and political weapon. However there is justification in some of the demands of traditional landowners. Often they become victims of modernization, of a modern administrative structure that just did not acknowledge their rights. As in the case of Fiji's mahogany plantations, the world biggest and worth several hundred millions of dollars. When the lease was negotiated in the

1950s and 1960s the landowners just did not know how valuable the trees were. No wonder that now – when the trees are ready for harvesting – they have come up with their demands.

However there are also so many cases, when the landowners misuse their powers. Such cases can be very small and only locally relevant like the demands of landowners near Lautoka, who had erected road blocks and were charging people one dollar each time they crossed the barrier. The road in question serves about 80 families, a primary school, a Hindu temple and a tourist attraction (Radio New Zealand International, May 7, 2003). A similar case occurred in the rural hinterland of Sigatoka, Fiji's salad bowl. A roadblock saw several hundred ethnic Indian farmers trapped and being forced to pay US$20 each trip to reach markets or hospitals. Finally the government agreed to pay US$14,000 to the landowners to lift the roadblock (Radio New Zealand International, Feb. 25, 2003).

Other cases however are large enough to threaten Fiji's economic and political well-being, such as the expiry of land leases in Fiji's major export sector, the sugar economy. Land leases in the sugar sector are regulated by the 1966 established Agricultural Landlord and Tenant Act (ALTA). According to ALTA the lease expires after 30 years and the land goes back to the landowners if the lease is not renewed. More than 6,000 leases expired between 1997 and 2004 and a large number of them were not renewed. In each individual case this meant the loss of livelihood for the family of a sugarcane farmer. It also meant the loss of the home as the houses were built on leased land too. Between 2005 and 2028 another 7100 sugarcane leases will expire. In such cases farmers could convince the landowners to renew their leases after a goodwill payment of several thousand dollars.

The reasons why the Fijian landowners are very reluctant to renew the leases are complex. Sometimes they want to use the land for themselves, e.g. to profit from the high subsidies the European Union gave sugar producers in Fiji for decades. Often the landowners are no longer happy with the rent payment that earns them between F$45 and F$480 per hectare depending on the quality of land. The rent is not lower than what would be paid in other countries, between 3 and 11% of the value of the gross production, but the number of those who want to benefit from the payment can be huge. A share of 15% straightaway goes to the NLTB, another 30% to the different chiefs and the rest is distributed amongst the members of the mataqali. The question here is how big the landowning group is: there are cases where a member gets just F$2 from the rent, in other cases this can be more than F$4,000. 'Goodwill payments' are therefore welcomed, when the negotiations about the renewal of the lease are due. What would be considered as blackmailing elsewhere has become common practice in Fiji: to ensure that the lease is renewed a few thousand dollars change owners, tax-free of course.

In many cases however there are obvious political motivations behind the non-renewal of land leases. In 1975 Sakiasi Butadroka, leader of the Fijian Nationalist Party brought in a motion in the Fiji Parliament to deport all people of Indian descent back to India. Families who lived in Fiji for three generations, people who never had been to India, without any land or other form of property there. Although the motion was declined it expresses fairly accurately the ethnic conflict in Fiji. On the one side there are the indigenous Fijians. Many of them still live in rural areas in a semi-subsistence agricultural economy. For them land rents are a way to improve their cash income. On the other hand are the descendants of Indian plantation workers. Many of them are still in the agricultural sector, cultivating sugarcane on small holdings of not more than 4-5 hectares which they have leased from indigenous landowners.

However it would be too simplistic to see the land and water conflicts in Fiji mainly as the result of the ethnic conflicts of the country. There are various conflict lines: the conflict between traditional leaders and the modern state. Traditional leaders who want to get their share from the development efforts of the state. They are powerful as traditionally they can mobilize support from their kin and clan members. Also economically the chiefs can count on their support as they get some of the lease money or compensation. However in the end the chiefs are the ones who get the major share, who get huge amounts. Politically they can find support with their demands in Fiji's parliament and from Fiji's government, but only as long as they hold the political power.

Such conflicts are not unique to Fiji. One can find them in many parts of the Pacific Islands. Since many years the water and power supply of Port Moresby, the capital of Papua New Guinea has been under constant threat as landowners demand compensation for the land of the Rouna hydro-electrical scheme and the Sirinumu water supply project (The Post-Courier, Aug. 24, 2005). Earlier water and power supply to Port Moresby had been frequently interrupted (The Post-Courier, Nov. 19, 2004). Similar conflicts occur in the tourist town of Madan (The Post-Courier, April 2, 2003), and the two towns in PNG's highlands, Goroka and Mt. Hagen (The Post-Courier, Jan. 21, 2001).

In recent years similar conflicts occurred in Honiara, the capital of the Solomon Islands. In June 2000 militants blew up a pumping station with dynamite. As a result, 90% of the homes in Honiara were left without water supply for more than a year (Radio Australia, March 24, 2002). In November 2001 the landowners of the Kongulai water supply shut down the supply of water to most parts of Honiara. While negotiations with the government continued over months power and water were frequently interrupted.

Insecure water supply in larger islands of Melanesia is the result of conflicts in institutional arrangements such as property rights, scarcity of financial resource and a lack of Good Governance and accountability in water administration. Insecure water supply often is also the result of civil unrest or even civil war. There are situations when naturally caused water scarcity like ENSO-induced droughts can create difficulties for particular sections of the societies in Melanesia, but proper resource management and disaster mitigation efforts surely would be able to minimize these problems as we can assume that enough water could be made available, if a proper water resources management would exist.

The last case study looks at an environment that is much more vulnerable to natural hazards than rather big, mountainous islands are. Coral atolls are amongst the harshest environments on earth: flat ribbons of sand, little land for an expanding population, with scarce freshwater supply; supporting a limited range of vegetation making agriculture difficult and restricted to a few crops, extreme geographic fragmentation and isolation making transport and communications costly and difficult. South Tarawa, the capital of Kiribati, is such an atoll island. Not only water-related problems make one wonder how people can survive in such an adverse environment. But they do, despite the fact that over time conditions do not improve, but deteriorate.

CASE STUDY 4: WATER SUPPLY IN SOUTH TARAWA, KIRIBATI

The Republic of Kiribati consists of 33 islands spread over three island groups, the Gilbert, Phoenix and Line Islands. 32 of the islands are coral islands or coral atolls, while one, Banaba, is a raised island. All the three groups are often subject to severe droughts (Rapaport, 1990).

With a land area of about 800 km^2 Kiribati belongs to the world's smallest states. However is stretches over almost 5,000 km from East to West and more than 2,000 km from North to South. The Exclusive Economic Zone is more than 3.5 million km^2, giving Kiribati's territory an expansion comparable to the land area of the USA. Most of the islands are not more than 2 km wide and less than 5 m above sea level. There are only four flights a week in and out of the country. Distance and access to markets are challenges of a magnitude faced by few countries. This is even true internally, where it is difficult to maintain a communication and transport network that covers the entire territory of the country. Kiribati has the largest atoll population in the Pacific with about 85,000 people in the last census of 2000. The annual growth rate between 1995-2000 had been 1.7% with high urban growth rates (UN 2002).

The resource base is very narrow. The climate and poor soil offer little potential for agricultural or industrial development. The public sector dominates all spheres of economic activity. Fishing licensing fees are the major source of foreign exchange and government revenue. However compared to the annul value of Kiribati's tuna resources that are caught by foreign fishing vessels,the US$20 million that the Kiribati's Government is getting for issuing fishing licenses is rather low (Islands Business Magazine, Feb. 11, 2005). Import duties and remittances from I-Kiribati employed in foreign shipping fleets provide significant additional government revenue and foreign exchange, respectively. The population is concentrated in the Gilbert Islands Group, which includes Tarawa, the capital. Soon a half of the population will live in South Tarawa, while the Phoenix Group is virtually uninhabited. If present trends continue, population will double within 20 years presenting even greater challenges to overcome environmental and health problems, particularly in Tarawa.

The scattered nature of the islands, isolation from each other and the region, poor soil and harsh climate, pose a big development challenge to the government of Kiribati and its people. Kiribati ranks 11th of 14 Pacific island countries and 129th in the world in the UNDP's Human Development Index. In terms of infant mortality and child morbidity, per capita GDP, and access to water and sanitation, Kiribati is among the lowest in the region (UN 2002).

Amongst the most restricting factors for development is the scarcity of freshwater. Rainfall is very unevenly distributed within and between the years. Droughts that can last ten months or more are common in Kiribati (Metutera, 2002). With an increase in population Kiribati's very fragile freshwater lens has become extremely vulnerable to depletion, intrusion of seawater, contamination with sewerage and other pollutants and is causing severe health problems. This is especially true for South Tarawa, Kiribati's capital, a place with the highest population density in the Pacific Islands.

South Tarawa – A Thirsty Capital

Tarawa atoll in the Gilbert Group consists of more than 20 islands, of which 8 are inhabited. The western part from Bonriki, where Kiribati's international airport is located to Betio with the port and the copra mill, Tarawa's only industry worth mentioning, is called South Tarawa. This stretch of Tarawa is about 35 km long. Various causeways connect the islands of South Tarawa from Betio to Bairiki. The longest is the Betio-Bairiki Causeway with 3.4 km. It was built in the middle of the 1980s by a Japanese company.

Most of the land is less than 3 metres above sea level, with an average width of only 430-450 metres. About half of Kiribati's population of 93,100 people live on South Tarawa's land area of 12,56 km². Migration from outer islands of Kiribati resulted in an annual population growth rate of 5.2%. Should this rate continue South Tarawa's population will double in just 13 years reaching 73,400 by 2013 (Haberkorn, 2004).

The 2000 census counted a population of 84,494 for the whole of Kiribati, an average growth rate of 1.7% per year from 1995-2000 with an urban growth of 5.2% and rural decline of 0.6%. South Tarawa had 36,717 people or 44% of the national population compared to 37% only five years earlier (UN 2002). When Tarawa became the capital of the Gilbert and Ellice Islands Colony in 1947 South Tarawa had a population of 1,643, representing 6% of Kiribati's population (Lea/Connel, 1995). Environmental problems (water quality; waste; sanitation; lagoon pollution) are created by South Tarawa's congestion. Crowded and unsanitary conditions contribute to a high incidence of diarrhoeal diseases with more than 700–800 reported cases per month (Hunt, 1996) and a high death rate especially among young children.

Although the exact number is not known a big share of South Tarawa's population live as squatters, especially those who in the last decades migrated to Tarawa from the outer islands. A concentration of squatter settlements is in the area of Betio, Bairiki and Bikenibeu. In Betio it is estimated that more than one-third of all households are squatters. Squatting occurs mainly on Government leased or owned lands such as along causeways, foreshore areas and areas adjoining rubbish dumping sites. On Government leased lands, 'informal' housing arrangements based on kinship obligations can be made with landowners, despite the fact that the land has been leased to the Government. Most squatters lack access to land, water supply and sanitation facilities. Based on population estimates, the number of housing plots required to accommodate the proposed 2010 population is in excess of 2,000 units (Urban Management Plan, 1995).

The extremely dense, congested and unregulated housing arrangements have contributed to many health, social and environment problems. The 2000 census records an average household size of 6.7 persons for the whole of Kiribati. However in South Tarawa 8.1 persons live on an average in a household. More than 30% of all households have more than 10 persons. The negative impact on Tarawa's water supply is surely the most serious one. Most of the water is pumped from a subterranean water lens in the northern and eastern parts of Tarawa. Some of the lenses are already so polluted that they are no longer safe to use (ADB 2004) and others are significantly overstretched by growing demand, illegal connections, spread of settlements onto land above the water lens, and the widespread use of pit and water seal toilets. A traditional I-Kiribati practice is to use the sea as a toilet. In many of the squatters' houses there are no proper toilet facilities. On islands that are less populated this poses little risk to public health but on over-crowded Tarawa, where 53% of households regularly use the beach, it has created serious health problems (UN 2002).

Water Supply in South Tarawa

There are three different sources for potable water available to the population of South Tarawa: rainwater, ground/well water and reticulated water. In addition to this there had also been a number of trials with desalination which technically turned out to be too risky to entirely depend on. Two of the three desalination plants broke down soon after installation and it took months to get spare parts from overseas to repair them (Metutera, 2002).

Rainwater is used in Kiribati as a supplementary water source since long. The collection of rainwater is a more efficient way to produce freshwater than groundwater extraction. To rely entirely on rainwater is rather risky. Unequal distribution of precipitation can lead to months-long drought and storage facilities are not large enough to help bridge such events. However rainwater collection can reduce the pressure on other water resources, especially the one on the freshwater lenses.

The reticulated water is extracted from the freshwater lenses at Buata and Bonriki (north to the airport) at a maximum rate of 1,250 cubic metres per day. In 1989 the sustainable yield for the two pumping zones had been estimated at 950 cubic metre a day. In 1992 a review was carried out and after that the sustainable yields estimates were 1000 and 300 cubic meters a day for Bonriki and Buata respectively. The latest calculations (2002) hint at a sustainable yield 1700 m^3/day (Bonriki: 1350 m^3/day, Buota: 350 m^3/day) (Metutera, 2002). There are more than 3,500 connections that provide 1-2 hours of water each in the morning, at lunch time and in the evenings. (Urban Management Plan 1995). As reticulated water is not available around the clock there is still a very heavy reliance on traditional wells and the collection of rainwater. According to the 1990 census for 49% of all households in South Tarawa wells were the major sources of water supply, while 35% used mainly rainwater (Urban Management Plan 1995). A problem with intermittent water supply is that an equitable distribution of water cannot be achieved this way. Those living close to the elevated tank/reservoir will always receive more water than those living downstream who receive little water or no water at all.

South Tarawa's water problems grow as fast as the population does. In 2000 water demand for a population of 36,227 persons was 1159 m^3/day. Even if Kiribati's government finds means to slow down the population

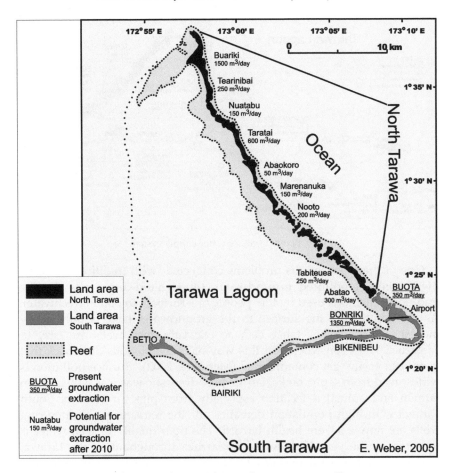

Fig. 8. Present and potential groundwater sources on Tarawa

growth on South Tarawa, more than 2100 m³/day will be required by 2020 to ensure 40 litres water per head and day. If population increases at the present rate there will be almost 100,000 persons living on South Tarawa in 2020, 7,600 persons per km², consuming 3151 m³ water per day, almost three times of the demand in 2000.

It is estimated that the present freshwater lenses can meet South Tarawa's water demand until 2010. Beyond that new water sources have to be found or developed. North Tarawa, which at present is not used to procure water for the capital might be a solution (Fig. 8). An estimated sustainable yield of 3450 m³/day could be achieved there. Extracting water from lenses in North Tarawa however might become expensive and time consuming as the landowners in North Tarawa are hesitant to lease or sell their land to the Government (Metutera, 2002).

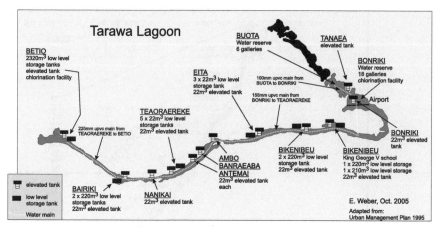

Fig. 9. South Tarawa's water supply system.

One of the most serious problems connected with traditional wells is the danger of polluted groundwater. Open hand-dug wells are the traditional methods used by the people of I-Kiribati to obtain freshwater. As the depth from the surface to the groundwater table is only a few metres and the soil is fairly easy to excavate open wells are easy to construct. Wells constructed in this way have the obvious drawback that water can easily get contaminated (Metutera, 2002). Surface pollution is widespread from septic tanks, latrine pits, domestic waste as well as open human and animal defecation especially from pigs. In areas of South Tarawa with high population densities all the remaining old open dug wells are now a severe health hazard. The poor quality of well water is reflected in the very high rates of diarrhea throughout South Tarawa. Polluted water was the reason why the former water lenses in Betio and Teaoraereke were closed for the reticulated water supply.

The pumping zones for South Tarawa's water supply are away from highly populated areas. In addition to this a number of technical and institutional measures have been started to protect the freshwater lenses that contribute to South Tarawa's reticulated water supply. In order to protect the freshwater lenses in the areas that provide reticulated water for South Tarawa a Water Reserves Committee has established a 50 meter 'setback zone' around each zone as water reserves. The dwellers of informal settlements within these zones were asked to resettle voluntarily, which turned out to be rather difficult in some cases, as alternative vacant land is very scarce in Tarawa. In addition new migrants arriving from outer islands are putting extra pressure on available open land areas and the newcomers are eyeing the cleared reserves in the setback zones with increasing interest (ADB 2004). Also cultural issues contributed to the problems of the project. Like many atoll societies the I-Kiribati society is very egalitarian. The major social units are extended families, where values like cooperation, solidarity and reciprocity play very important roles (Frisbie, 1921). The resettlement of squatter communities within the water reserve zones turned out to be difficult also because I-Kiribati culture expects people to be modest and humble, and not placing oneself above others. Even forcing squatters to leave protected areas thus is against cultural norms of the people of Kiribati (ADB 2004; White et al., 1999).

To allow the freshwater lenses to recharge faster the South Tarawa Water Supply System has used infiltration galleries since the late 1960s (for technical details, see Metutera, 2002; White, 1996; White et al., 1999). The galleries consist of some form of horizontally laid permeable conduit to allow water to infiltrate from the surrounding saturated zone. An open area such as Tarawa's airport acts as a catchment that can rapidly recharge the aquifer after heavy rains.

Altogether there are 24 infiltration galleries strengthening South Tarawa's water supply, six on Buota and eighteen on Bonriki. The length of each of the galleries using perforated pipes is about 300 m. Thus the total length of the galleries is 5,100 m for Bonriki and 1,800 m for Buota.

Water quality in South Tarawa is closely connected to the quality of sewerage treatment/disposal. A reticulated sewerage system had already been introduced in South Tarawa between 1978 and 1982 in response to a major cholera epidemic in 1977. The system covers only areas of Betio, Bairiki and Bikenibeu, the most populated parts of South Tarawa, about 30% of the size and 60-70% of the population. "In reality however, the coverage in these urban centers is mainly limited to the low density permanent houses, Government building and communal toilet blocks" (Metutera, 2002). The households connected to the system use salt water for toilet flushing, which discharge into three separate ocean outfalls over the reef edge.

The water supply system like the sewerage system is under increasing strain and already operating above capacity. Lack of financial resources, spare parts and expertise has deteriorated the system. Pump stations are frequently out of functioning, sewer mains are leaking with salt water and sewage, the communal toilet blocks are in a state of disrepair having been vandalized or just let to run down with salt water pipe leakages, structures collapsing and an estimated 70% of cisterns broken (Urban Management Plan 1995).

Global Climate Change and Water Resources

Like many atoll islands Kiribati is also highly vulnerable to global climate change. The 2001 synthesis report of the Intergovernmental Panel on Climate Change concludes that global warming is underway. It is very likely to increase during this century at rates unprecedented in the past 10,000 years. For small islands, the IPCC warns of deteriorating coral reefs, mangroves, and sea-grass beds; major species loss; worsening water balance in atoll nations such as Kiribati; and declines in vital reef fisheries. For the Pacific islands as a whole, the World Bank warns of reductions in agricultural output, declines in groundwater quantity and quality, substantial health impacts (increased diarrhea, dengue fever and fish poisoning), extensive capital damage due to storm surges, and lost fish production. Countries like Kiribati and the neighboring Tuvalu are predicted to suffer the greatest impact of climate change including disappearance in the worst case scenario. Although most media attention has focused on sea level rise, the expected impacts – particularly for atolls – are likely to be reduced agricultural output (due to changing rainfall patterns and increased temperatures), a decline in groundwater quantity and quality (sea level rise and possibly drought) (IPCC 2001). The World Bank concludes that: "Managing change will be particularly critical in the area of climate change, a subject of immense and immediate impact on Pacific Island countries. Choosing a development path that decreases the islands' vulnerability to climate events and maintains the quality of the social and physical environment will not only be central to the future well being of the Pacific Island people, but will also be a key factor in the countries' ability to attract foreign investment in an increasingly competitive global economy" (World Bank, 2000).

All atolls in Kiribati will be severely affected by global climate change. As infrastructure is most developed in Tarawa it is expected that the material losses will be highest there. By 2050 Tarawa could experience annual damages of about US$8–US$16 million. "In years of strong storm surge, Tarawa could face capital losses of up to US$430 million in land and infrastructure assets destroyed by inundation. Relocation of communities might be the need if the loss of land and freshwater supplies becomes critical. Climate change is thus likely to place a substantial burden on the people and economy of Kiribati. The projected losses could be catastrophic for a country with a 1998 GDP of only US$47 million" (World Bank 2000).

Global climate change affects Tarawa through variations in sea-level, rainfall, evapo-transpiration and through extreme weather events. In a worst case scenario with a sea-level rise by 0.4 meters, a decline of rainfall by 10% and a reduction of the width of the atoll the thickness of the groundwater lens could decline by as much as 38% (World Bank 2000).

Global climate change is surely the biggest threat to many of the Pacific Island states, not only to the small, low atoll islands. The case studies above however show that many severe problems around a safe and secure water supply exist throughout the Pacific Islands. Problems severe enough to threaten the existence of settlements in particular places, and to endanger the livelihoods of big proportions of Pacific Island populations. Especially in fast growing urban areas in the Pacific Islands the gap between demand and supply of water is increasing rapidly. However one needs to be aware that water supply is only one of many issues that affect the well-being of the urban population. The major problems in urban centers in the Pacific include a serious shortage of land and conflicts with traditional land tenure, falling standards of infrastructure, an increase in the number of squatter settlement and informal housing, poverty, vulnerability and environmental degradation (UNEP 1999). It is very likely that these problems not only continue to exist, but that they are becoming bigger even when huge investments are made to improve living conditions. The future especially for urban areas in the Pacific Island thus does not look too optimistic.

REFERENCES

Alkire, W.H. 1978. Coral Islanders, AHM Press, Arlington Heights, South Pacific.

Asian Development Bank. 2004. Bringing Water to the Poor. Selected ADB Case Studies, Manila, the Phillipines.

Auditor General of Fiji. 2003. Auditor General's Report to Parliament – Special Investigation, Parliamentary Paper No 11 of 2003, Suva

Baisyet, P.M. 1994. Overview of Watershed Management in the South Pacific Island Countries. Proceedings of the Pacific Water Sector Planning Workshop, UNESCO/SOPAC/UNDDSMS Honiara, Solomon Islands, pp. 173-179.

Bleakley, C. 1995. Marine Region 14 SOUTH PACIFIC, A Report to the World Bank Environment Department, in Greame Kelleher, Chris Bleakley, and Sue Wells (eds), A global representative system of marine protected areas. Volume 4: South Pacific, Northeast Pacific, Northwest Pacific, Southeast Pacific and Australia/New Zealand. Canberra, ACT, Australia: Great Barrier Reef Marine Park Authority; Washington, D.C.: World Bank; Gland, Switzerland: World Conservation Union. (http://www.deh.gov.au/coasts/mpa/nrsmpa/global/volume4/chapter14.html)

Bolataki, M. 1998. Worst drought hits Fiji, Daily Post (Fiji), 17/4/98.

Burns, W.C.G. 2002. Pacific Island Developing Country Water Resources and Climate Change, *In* The World's Water 2002-2003, Gleick, P.H. (ed). The Biennial Report on Freshwater Resources, Island Press, Washington, D.C., USA, pp. 113–131.

Carpenter, C., Stubbs J. and Overmars M. (eds) 2002. Proceedings of the Pacific Regional Consultation on Water in Small Island Countries, Sigatoka, Fiji Islands, 29. July–3. August, Asian Development Bank and South Pacific Applied Geoscience Council.

Dow, K., Carr E.R., Douma, A., Han G. and Hallding K. 2005. Linking Water Scarcity to Population Movements: From Global Models to Local Experiences, Poverty and Vulnerability Programme, Stockholm Environment Institute.

Falkland, A. 1999. Impacts of climate change on water resources of Pacific Islands, PACCLIM Workshop, Modeling the Effects of Climate Change and Sea Level Rise in Pacific Island Countries, Auckland, New Zealand.

Falkland, A. and Custodio, E. 1991. Hydrology and water resources of small islands: A practical guide. UNESCO, Paris, France.

Falkland, T. 2002. From Vision to action – Towards sustainable water management in the Pacific, Theme 1 Overview Report Water Resources Management, Pacific Regional Consultation on Water in Small Island Countries, Sigatoka, Fiji, 29 July–3 August 2002.

Field, M. 1998a. Major global climate belt moves: affects Pacific Countries, Pacific Islands Report (http://archives.pireport.org/archive/1998/august/08%2D10%2D05.htm)

Field, M. 1998b. U.S. El Nino drought aid begins arriving in FSM and Marshalls, Pacific Islands Report (http://archives.pireport.org/archive/1998/april/04%2D10%2D04.htm)

Frisbie, R.D. 1921. The Book of Puka-Puka, New York, USA.

Gangaiya, P., Tabudravu, J., South, R. and Sotheeswaran, S. 2001. Heavy metal contamination of the Lami coastal environment, Fiji, South Pacific Journal of. Natural Science, 19: 24-29.

Haberkorn, G. 2004. Population and Environment: The Pacific Challenge and the MDGs in the Pacific, ESAO Region Roundtable on ICPDat 10, Kuala Lumpur, 20-23 July 2004.

Hill, D. 2002. Water in Small Island Countries, Consultation Meeting on Sustainable Water Resources Management in the Pacific, Theme 5: Institutional Arrangements, Regional Consultation Workshop on Water in Small Island Countries, Sigatoka, Fiji, 29 July-3 August

IPCC, 2001a. Climate Change 2001: Impacts, Adaptation, and Vulnerability, Small Island States, Cambridge University Press, 17: 845-875.

IPCC, 2001b. Climate Change 2001: Synthesis Report for Policymakers, Cambridge University Press.

Lea, J.P. and Connel, J.H. 1995. Managing Urban Environmental Sanitation Services in Selected Pacific Island Countries: Annotated Bibliography of Regional Literature and Data Sources, Washington, DC, USA.

Lightfood, C. 1999. Regional El Niño Social and Economic Drought Impact Assessment and Mitigation Study, prepared for Disaster Management Unit and South Pacific Geiscience Commission, Suva, Fiji.

Metutera, T. 2002. Water Management in Kiribati with Special Emphasis on Groundwater Development Using Infiltration Galleries, presented at: Pacific Regional Consultation Meeting on Water in Small Island Countries, 29 July-3 August 2002, Sigatoka, Fiji.

Namai, B. 1987. The Evolution of Kiribati Tenures, *In* Land Tenure in the Atolls, R.G. Crocombe (ed.), Institute of Pacific Studies, University of the South Pacific, Suva, Fiji.

Pollock, N. 1970. Breadfruit and breadwinning on Namu Atoll, Marshall Islands. Honolulu, University of Hawai'i, Ph.D. Dissertation.

Prasad, Surendra. 1998. Renewable energy resources and utilisation in Fiji: an Overview. Paper presented to the Seminar on Renewable Energy Sources for Rural Areas, Nadi, Fiji Islands, July 20-25, 1998.

Rapaport, M. 1990. Population Pressure on Coral Atolls: Trends and Approaching Limits, *In*: Atoll Research Bulletin, No. 340, September, Washington, D.C., USA.

Roy, P. and Connell, J. 1991. Climate change and the future of atoll states, Journal of Coastal Research, 7, (4) 1057–1064.

Scott, D., Overmars, M., Falkland, T. and Carpenter, C. 2003. Pacific Dialogue on Water and Climate, Synthesis Report, SOPAC Miscellaneous Report N. 491, Suva, Fiji.

Terry, J. and Raj, R. 2002. The 1997-98 El Niño and Drought in the Fiji Islands, *In* International Hydrological Programme (ed), Hydrology and water management in the humid tropics, Proceedings of the Second International Colloquium 22–26 March 1999 Panama, Republic of Panama, IHP-V. Technical Documents in Hydrology. No. 52, UNESCO, Paris

United Nations. 2002. Common Country Assessment, Kiribati, Suva, Fiji.

United Nations Environment Programme, 1999. Pacific Islands Environment Outlook, London, UK.

Watling, D. and Chape, S. 1999. Environment: Fiji, The National State of the Environment Report. Prepared and published by IUCN.

While, Ian. 1996. Fresh groundwater lens recharge, Bonriki, Kiribati, UNESCO, International Hydrological Programme, IHP-V Project 6-1, Preliminary Report, Paris, France.

White, I., Falkland, T. and Scott, D. 1999. Droughts in small coral islands: Case study, South Tarawa, Kiribati, IHP Humid Tropics Programme, Technical Documents in Hydrology, No. 26, UNESCO, Paris, France.

White, I., Falkland, T., Perez, P., Dray A. and Overmars, M. 2004. Sustainable Development of Water Resources in Small Island Nations, Proceedings of the 2nd Asia Pacific Association of Hydrology and Water Resources, July 5-8.

White, I., Falkland, T., Crennan, L., Jones, P., Metutera, T., Etuati, B. and Metai, E. 1999. Groundwater recharge in low coral islands Bonriki, South Tarawa, Republic of Kiribati: Issues, traditions and conflicts in groundwater use and management, UNESCO, International Hydrological Programme, IHP-V, Technical Documents in Hydrology, No. 25, Paris, France.

Whittaker, R.J. 1998. Island Biogeography, Oxford University Press, New York, USA.

Wiens, H. 1962. Atoll ecology and environment. Yale University Press, New Haven, USA.

Index

WATER CONFLICT CHRONOLOGY[1]

Dr. Peter H. Gleick

Pacific Institute for Studies in Development, Environment, and Security

Available online at www.worldwater.org/conflictchronology.pdf —Updated October 12, 2006

Date	Parties Involved	Basis of Conflict	Violent Conflict or In the Context of Violence?	Description	Sources
3000 BC	Ea, Noah	Religious account	Yes	Ancient Sumerian legend recounts the deeds of the deity Ea, who punished humanity for its sins by inflicting the Earth with a six-day storm. The Sumerian myth parallels the Biblical account of Noah and the deluge, although some details differ.	Hatami and Gleick 1994
2500 BC	Lagash, Umma	Military tool	Yes	Lagash-Umma Border Dispute-The dispute over the "Gu'edena" (edge of paradise) region begins. Urlama, King of Lagash from 2450 to 2400 B.C., diverts water from this region to boundary canals, drying up boundary ditches to deprive Umma of water. His son Il cuts off the water supply to Girsu, a city in Umma.	Hatami and Gleick 1994
1790 BC	Hammurabi	Development disputes	No	Code of Hammurabi for the State of Sumer - Hammurabi lists several laws pertaining to irrigation that address negligence of irrigation systems and water theft.	Hatami and Gleick 1994
1720-1684 BC	Abi-Eshuh, Iluma-Ilum	Military tool	Yes	Abi-Eshuh v. Iluma-Ilum- A grandson of Hammurabi, Abish or Abi-Eshuh, dams the Tigris to prevent the retreat of rebels led by Iluma-Ilum, who declared the independence of Babylon. This failed attempt marks the decline of the Sumerians who had reached their apex under Hammurabi.	Hatami and Gleick 1994

Date	Parties Involved	Basis of Conflict	Violent Conflict or in the Context of Violence?	Description	Sources
circa 1300 BC	Sisra, Barak, God	Religious account, Military Tool	Yes	This is an Old Testament account of the defeat of Sisera and his "nine hundred chariots of iron" by the unmounted army of Barak on the fabled Plains of Esdraelon. God sends heavy rainfall in the mountains, and the Kishon River overflows the plain and immobilizes or destroys Sisera's technologically superior forces ("...the earth trembled, and the heavens dropped, and the clouds also dropped water," Judges 5:4; "...The river of Kishon swept them away, that ancient river, the river Kishon," Judges 5:21).	New Scofield Reference Bible, KJV; Judges 4:7-15 and Judges 5:4-22.
1200 BC	Moses, Egypt	Military tool, Religious account	Yes	Parting of the Red Sea- When Moses and the retreating Jews find themselves trapped between the Pharoah's army and the Red Sea, Moses miraculously parts the waters of the Red Sea, allowing his followers to escape. The waters close behind them and cut off the Egyptians.	Hatami and Gleick 1994
720-705 BC	Assyria, Armenia	Military tool	Yes	After a successful campaign against the Halidians of Armenia, Sargon II of Assyria destroys their intricate irrigation network and floods their land.	Hatami and Gleick 1994
705-682 BC	Sennacherib, Babylon	Military weapon /target	Yes	In quelling rebellious Assyrians in 695 B.C., Sennacherib razes Babylon and diverts one of the principal irrigation canals so that its waters wash over the ruins.	Hatami and Gleick 1994
6th Century BC	Assyria	Military target; Military tool	Yes	Assyrians poison the wells of their enemies with rye ergot.	Eitzen, E.M. and E.T. Takafuji. 1997

Date	Parties Involved	Basis of Conflict	Violent Conflict or in the Context of Violence?	Description	Sources
Unknown	Sennacherib, Jerusalem	Military tool	Yes	As recounted in Chronicles 32.3, Hezekiah digs into a well outside the walls of Jerusalem and uses a conduit to bring in water. Preparing for a possible siege by Sennacherib, he cuts off water supplies outside of the city walls, and Jerusalem survives the attack.	Hatami and Gleick 1994
681–699 BC	Assyria, Tyre	Military tool, Religious account	Yes	Esarhaddon, an Assyrian, refers to an earlier period when gods, angered by insolent mortals, created destructive floods. According to inscriptions recorded during his reign, Esarhaddon besieges Tyre, cutting off food and water.	Hatami and Gleick 1994
669–626 BC	Assyria, Arabia, Elam	Military tool, Military target	Yes	Assurbanipal's inscriptions also refer to a siege against Tyre, although scholars attribute it to Esarhaddon. In campaigns against both Arabia and Elam in 645 B.C., Assurbanipal, son of Esarhaddon, dries up wells to deprive Elamite troops. He also guards wells from Arabian fugitives in an earlier Arabian war. On his return from victorious battle against Elam, Assurbanipal floods the city of Sapibel, and ally of Elam. According to inscriptions, he dams the Ulai River with the bodies of dead Elamite soldiers and deprives dead Elamite kings of their food and water offerings.	Hatami and Gleick 1994
612 BC	Egypt, Persia, Babylon, Assyria	Military tool	Yes	A coalition of Egyptian, Median (Persian), and Babylonian forces attacks and destroys Ninevah, the capital of Assyria. Nebuchadnezzar's father, Nebopolassar, leads the Babylonians. The converging armies divert the Khosr River to create a flood, which allows them to elevate their siege engines on rafts.	Hatami and Gleick 1994
605–562 BC	Babylon	Military tool	No	Nebuchadnezzar builds immense walls around Babylon, using the Euphrates and canals as defensive moats surrounding the inner castle.	Hatami and Gleick 1994

Date	Parties Involved	Basis of Conflict	Violent Conflict or in the Context of Violence?	Description	Sources
590-600 BC	Cirrha, Delphi	Military tool	Yes	Athenian legislator Solon reportedly had roots of helleborus thrown into a small river or aqueduct leading from the Pleistrus River to Cirrha during a siege of this city. The enemy forces became violently ill and were defeated as a result. Some accounts have Solon building a dam across the Plesitus River cutting off the city's water supply. Such practices were widespread.	Absolute Astronomy 2006
558-528 BC	Babylon	Military tool	Yes	On his way from Sardis to defeat Nabonidus at Babylon, Cyrus faces a powerful tributary of the Tigris, probably the Diyalah. According to Herodotus' account, the river drowns his royal white horse and presents a formidable obstacle to his march. Cyrus, angered by the "insolence" of the river, halts his army and orders them to cut 360 canals to divert the river's flow. Other historians argue that Cyrus needed the water to maintain his troops on their southward journey, while another asserts that the construction was an attempt to win the confidence of the locals.	Hatami and Gleick 1994
539 BC	Babylon	Military tool	Yes	According to Herodotus, Cyrus invades Babylon by diverting the Euphrates above the city and marching troops along the dry riverbed. This popular account describes a midnight attack that coincided with a Babylonian feast.	Hatami and Gleick 1994
430 BC	Athens	Military tool	Yes	During the second year of the Peloponnesian War in 430 BC when plague broke out in Athens, the Spartans were accused of poisoning the cisterns of the Piraeus, the source of most of Athens' water.	Strategy Page 2006.
355-323 BC	Babylon	Military tool	Yes	Returning from the razing of Persepolis, Alexander proceeds to India. After the Indian campaigns, he heads back to Babylon via the Persian Gulf and the Tigris, where he tears down defensive weirs that the Persians had constructed along the river. Arrian describes Alexander's disdain for the Persians' attempt to block navigation, which he saw as "unbecoming to men who are victorious in battle."	Hatami and Gleick 1994

Date	Parties Involved	Basis of Conflict	Violent Conflict or in the Context of Violence?	Description	Sources
210-209 BC	Rome and Cathage	Military tool	Yes	In 210 BC, Scipio crossed the Ebro to attack New Carthage. During a short siege, Scipio led a breaching column through a supposedly impregnable lagoon located on the landward side of the city; a strong northerly wind combined with the natural ebb of the tide left the lagoon shallow enough for the Roman infantry to wade through. New Carthage was soon taken.	Fonner 1996, Gowan 2004
537	Goths and Rome	Military tool and military target	Yes	In the 6th century AD, as the Roman Empire began to decline, the Goths besieged Rome and cut almost all of the aqueducts leading into the city. In 537 AD this siege was successful. The only aqueduct that continued to function was that of the Aqua Virgo, which ran entirely underground.	Rome Guide 2004, InfoRoma 2004.
1187	Saladin and the Middle East	Military tool	Yes	Saladin was able to defeat the Crusaders at the Horns of Hattin in 1187 by denying them access to water. In some reports, Saladin had sanded up all the wells along the way and had destroyed the villages of the Maronite Christians who would have supplied the Christian army with water.	Lockwood 2006, Priscoli 1998
1503	Florence and Pisa warring states.	Military tool	No: Plan only	Leonardo da Vinci and Machievelli plan to divert Arno River away from Pisa during conflict between Pisa and Florence.	Honan 1996
1573-74	Holland and Spain	Military tool	Yes	In 1573 at the beginning of the eighty years war against Spain, the Dutch flooded the land to break the siege of Spanish troops on the town Alkmaar. The same defense was used to protect Lieden in 1574. This strategy became known as the Dutch Water Line and was used frequently for defense in later years.	Dutch Water Line 2002
1642	China; Ming Dynasty	Military tool	Yes	The Huang He's dikes breached for military purposes. In 1642, "toward the end of the Ming dynasty (1368-1644), General Gao Mingheng used the tactic near Kaifeng in an attempt to suppress a peasant uprising."	Hillel 1991

Date	Parties Involved	Basis of Conflict	Violent Conflict or in the Context of Violence?	Description	Sources
1672	French, Dutch	Military tool	Yes	Louis XIV starts the third of the Dutch Wars in 1672, in which the French overran the Netherlands. In defense, the Dutch opened their dikes and flooded the country, creating a watery barrier that was virtually impenetrable.	Columbia 2000
1748	United States	Development dispute; terrorism	Yes	Ferry house on Brooklyn shore of East River burns down. New Yorkers accuse Brooklynites of having set the fire as revenge for unfair East River water rights.	Museum of the City of New York (MCNY n.d.)
1777	United States	Military tool	Yes	British and Hessians attacked the water system of New York. "...the enemy wantonly destroyed the New York water works" during the War for Independence.	Thatcher 1827
1841	Canada	Development dispute, terrorism	Yes	A reservoir in Ops Township, Upper Canada (now Ontario) was destroyed by neighbors who considered it a hazard to health.	Forkey 1998
1844	United States	Development dispute, terrorism	Yes	A reservoir in Mercer County, Ohio was destroyed by a mob that considered it a hazard to health.	Scheiber 1969
1850s	United States	Development dispute; terrorism	Yes	Attack on a New Hampshire dam that impounded water for factories downstream by local residents unhappy over its effect on water levels.	Steinberg 1990
1853–1861	United States	Development dispute, terrorism	Yes	Repeated destruction of the banks and reservoirs of the Wabash and Erie Canal in southern Indiana by mobs regarding it as a health hazard.	Fatout 1972, Fickle 1983
1860–1865	United States	Military tool; Military target	Yes	W. T. Sherman's memoirs contain an account of Confederate soldiers poisoning ponds by dumping the carcasses of dead animals into them. Other accounts suggest this tactic was used by both sides.	Eitzen and Takafuji 1997

Date	Parties Involved	Basis of Conflict	Violent Conflict or in the Context of Violence?	Description	Sources
1870s	China	Development dispute	No	Local construction and government removal (twice) of an unauthorized dam in Hubei, China.	Rowe 1988
1870s to 1881	United States	Development dispute	Yes	Recurrent friction and eventual violent conflict over water rights in the vicinity of Tularosa, New Mexico involving villagers, ranchers, and farmers.	Rasch 1968
1887	United States	Development dispute, Terrorism	Yes	Dynamiting of a canal reservoir in Paulding County, Ohio by a mob regarding it as a health hazard. State Militia called out to restore order.	Walters 1948
1890	Canada	Development dispute, terrorism	Yes	Partly successful attempt to destroy a lock on the Welland Canal in Ontario, Canada either by Fenians protesting English Policy in Ireland or by agents of Buffalo, NY grain handlers unhappy at the diversion of trade through the canal.	Styran and Taylor 2001
1908-09	United States	Development dispute	Yes	Violence, including a murder, directed against agents of a land company that claimed title to Reelfoot Lake in northwestern Tennessee who attempted to levy charges for fish taken and threatened to drain the lake for agriculture.	Vanderwood 1969
1863	United States Civil War	Military tool	Yes	General U.S. Grant, during the Civil War campaign against Vicksburg, cut levees in the battle against the Confederates.	Grant 1885, Barry 1997
1898	Egypt; France; Britain	Military and political tool	Military maneuvers	Military conflict nearly ensues between Britain and France in 1898 when a French expedition attempted to gain control of the headwaters of the White Nile. While the parties ultimately negotiates a settlement of the dispute, the incident has been characterized as having "dramatized Egypt's vulnerable dependence on the Nile, and fixed the attitude of Egyptian policy-makers ever since."	Moorhead 1960

Date	Parties Involved	Basis of Conflict	Violent Conflict or in the Context of Violence?	Description	Sources
1907–1913	Owens Valley, Los Angeles, California	Terrorism, Development dispute	Yes	The Los Angeles Valley aqueduct/pipeline suffers repeated bombings in an effort to prevent diversions of water from the Owens Valley to Los Angeles.	Reisner 1986, 1993
1915	German Southwest Africa	Military tool	Yes	Union of South African troops capture Windhoek, capital of German Southwest Africa. (May.) Retreating German troops poison wells – "a violation of the Hague convention."	Daniel 1995
1935	California, Arizona	Development dispute	Military maneuvers	Arizona calls out the National Guard and militia units to the border with California to protest the construction of Parker Dam and diversions from the Colorado River; dispute ultimately is settled in court.	Reisner 1986, 1993
1938	China and Japan	Military tool, Military target	Yes	Chiang Kai-shek orders the destruction of flood-control dikes of the Huayuankou section of the Huang He (Yellow) river to flood areas threatened by the Japanese army. West of Kaifeng dikes are destroyed with dynamite, spilling water across the flat plain. The flood destroyed part of the invading army and its heavy equipment was mired in thick mud, though Wuhan, the headquarters of the Nationalist government was taken in October. The waters flooded an area variously estimated as between 3,000 and 50,000 square kilometers, and killed Chinese estimated in numbers between "tens of thousands" and "one million."	Hillel 1991, Yang Lang 1989, 1994
1939–1942	Japan, China	Military target, Military tool	Yes	Japanese chemical and biological weapons activities reportedly include tests by "Unit 731" against military and civilian targets by lacing water wells and reservoirs with typhoid and other pathogens.	Harris 1994
1940–1945	Multiple parties	Military target	Yes	Hydroelectric dams routinely bombed as strategic targets during World War II.	Gleick 1993

Date	Parties Involved	Basis of Conflict	Violent Conflict or in the Context of Violence?	Description	Sources
1943	Britain, Germany	Military target	Yes	British Royal Air Force bombed dams on the Möhne, Sorpe, and Eder Rivers, Germany (May 16, 17). Möhne Dam breech killed 1,200, destroyed all downstream dams for 50 km. The flood that occurred after breaking the Eder dam reached a peak discharge of 8,500 m³/s, which is nine times higher than the highest flood observed. Many houses and bridges were destroyed. 68 were killed.	Kirschner 1949, Semann 1950
1944	Germany, Italy, Britain, United States	Military tool	Yes	German forces used waters from the Isoletta Dam (Liri River) in January and February to successfully destroy British assault forces crossing the Garigliano River (downstream of Liri River). The German Army then dammed the Rapido River, flooding a valley occupied by the American Army.	Corps of Engineers 1953
1944	Germany, Italy, Britain, United States	Military tool	Yes	German Army flooded the Pontine Marches by destroying drainage pumps to contain the Anzio beachhead established by the Allied landings in 1944. Over 40 square miles of land were flooded; a 30-mile stretch of landing beaches was rendered unusable for amphibious support forces.	Corps of Engineers 1953
1944	Germany, Allied forces	Military tool	Yes	Germans flooded the Ay River, France (July) creating a lake two meters deep and several kilometers wide, slowing an advance on Saint Lo, a German communications center in Normandy.	Corps of Engineers 1953
1944	Germany, Allied forces	Military tool	Yes	Germans flooded the Ill River Valley during the Battle of the Bulge (winter 1944-45) creating a lake 16 kilometers long, 3-6 kilometers wide, and 1-2 meters deep, greatly delaying the American Army's advance toward the Rhine.	Corps of Engineers 1953
1945	Romania, Germany	Military target	Yes	The only known German tactical use of biological warfare was the pollution of a large reservoir in northwestern Bohemia with sewage in May 1945.	SIPRI 1971

Date	Parties Involved	Basis of Conflict	Violent Conflict or in the Context of Violence?	Description	Sources
1947 onwards	Bangladesh, India	Development dispute	No	Partition divides the Ganges River between Bangladesh and India; construction of the Farakka barrage by India, beginning in 1962, increases tension; short-term agreements settle dispute in 1977–82, 1982–84, and 1985–88, and thirty-year treaty is signed in 1996.	Butts 1997, Samson & Charrier 1997
1947–1960s	India, Pakistan	Development dispute	No	Partition leaves Indus basin divided between India and Pakistan; disputes over irrigation water ensue, during which India stems flow of water into irrigation canals in Pakistan; Indus Waters Agreement reached in 1960 after 12 years of World Bank-led negotiations.	Bingham *et al.* 1994, Wolf 1997
1948	Arabs, Israelis	Military tool	Yes	Arab forces cut of West Jerusalem's water supply in first Arab-Israeli war.	Wolf 1995, 1997
1950s	Korea, United States, others	Military target	Yes	Centralized dams on the Yalu River serving North Korea and China are attacked during Korean War.	Gleick 1993
1951	Korea, United Nations	Military tool and Military target	Yes	North Korea released flood waves from the Hwachon Dam damaging floating bridges operated by UN troops in the Pukhan Valley. U.S. Navy plans were then sent to destroy spillway crest gates.	Corps of Engineers 1953
1951	Israel, Jordan, Syria	Military tool, Development disputes	Yes	Jordan makes public its plans to irrigate the Jordan Valley by tapping the Yarmouk River; Israel responds by commencing drainage of the Huleh swamps located in the demilitarized zone between Israel and Syria; border skirmishes ensue between Israel and Syria.	Wolf 1997, Samson & Charrier 1997

Date	Parties Involved	Basis of Conflict	Violent Conflict or in the Context of Violence?	Description	Sources
1953	Israel, Jordan, Syria	Development dispute, Military target	Yes	Israel begins construction of its National Water Carrier to transfer water from the north of the Sea of Galilee out of the Jordan basin to the Negev Desert for irrigation. Syrian military actions along the border and international disapproval lead Israel to move its intake to the Sea of Galilee.	Naff and Matson 1984, Samson & Charrier 1997
1958	Egypt, Sudan	Military tool, Development dispute	Yes	Egypt sends an unsuccessful military expedition into disputed territory amidst pending negotiations over the Nile waters, Sudanese general elections, and an Egyptian vote on Sudan-Egypt unification; Nile Water Treaty signed when pro-Egyptian government elected in Sudan.	Wolf 1997
1960s	North Vietnam, United States	Military target	Yes	Irrigation water supply systems in North Vietnam are bombed during Vietnam War. 661 sections of dikes damaged or destroyed.	IWTC 1967, Gleick 1993, Zemmali 1995
1962	Israel, Syria	Development dispute, Military target	Yes	Israel destroys irrigation ditches in the lower Tarfiq in the demilitarized zone. Syria complains.	Naff and Matson 1984
1962 to 1967	Brazil; Paraguay	Military tool, Development dispute	Military maneuvers	Negotiations between Brazil and Paraguay over the development of the Paraná River are interrupted by a unilateral show of military force by Brazil in 1962, which invades the area and claims control over the Guaira Falls site. Military forces were withdrawn in 1967 following an agreement for a joint commission to examine development in the region.	Murphy and Sabadell 1986

Date	Parties Involved	Basis of Conflict	Violent Conflict or in the Context of Violence?	Description	Sources
1963-1964	Ethiopia, Somalia	Development dispute, Military tool	Yes	Creation of boundaries in 1948 leaves Somali nomads under Ethiopian rule; border skirmishes occur over disputed territory in Ogaden desert where critical water and oil resources are located; cease-fire is negotiated only after several hundred are killed.	Wolf 1997
1964	Cuba, United States	Military tool	No	On February 6, 1964, the Cuban government ordered the water supply to the U.S. Naval Base at Guantanamo Bay cut off.	Guantanamo Bay Gazette. 1964.
1964	Israel, Syria	Military target	Yes	Headwaters of the Dan River on the Jordan River are bombed at Tell El-Qadi in a dispute about sovereignty over the source of the Dan.	Naff and Matson 1984
1965	Zambia, Rhodesia, Great Britain	Military target	No	President Kenneth Kaunda calls on British government to send troops to Kariba Dam to protect it from possible saboteurs from Rhodesian government.	Chenje 2001
1965	Israel, Palestinians	Terrorism	Yes	First attack ever by the Palestinian National Liberation Movement Al-Fatah is on the diversion pumps for the Israeli National Water Carrier. Attack fails.	Naff and Matson 1984, Dolatyar 1995
1965-1966	Israel, Syria	Military tool, Development dispute	Yes	Fire is exchanged over "all-Arab" plan to divert the Jordan River headwaters (Hasbani and Banias) and presumably preempt Israeli National Water Carrier; Syria halts construction of its diversion in July 1966.	Wolf 1995, 1997
1966-1972	Vietnam, US	Military tool	Yes	U.S. tries cloud-seeding in Indochina to stop flow of materiel along Ho Chi Minh trail.	Plant 1995

Date	Parties Involved	Basis of Conflict	Violent Conflict or in the Context of Violence?	Description	Sources
1967	Israel, Syria	Military target and tool	Yes	Israel destroys the Arab diversion works on the Jordan River headwaters. During Arab-Israeli War Israel occupies Golan Heights, with Banias tributary to the Jordan; Israel occupies West Bank.	Gleick 1993, Wolf 1995, 1997, Wallenstein & Swain 1997
1969	Israel, Jordan	Military target and tool	Yes	Israel, suspicious that Jordan is overdiverting the Yarmouk, leads two raids to destroy the newly-built East Ghor Canal; secret negotiations, mediated by the US, lead to an agreement in 1970.	Samson & Charrier 1997
1970	United States	Terrorism	No: Threat	The Weathermen, a group opposed to American imperialism and the Vietnam war, allegedly attempted to obtain biological agents to contaminate the water supply systems of US urban centers.	Kupperman and Trent 1979, Eitzen and Takafuji 1997, Purver 1995
1970s	Argentina, Brazil, Paraguay	Development dispute	No	Brazil and Paraguay announce plans to construct a dam at Itaipu on the Paraná River, causing Argentina concern about downstream environmental repercussions and the efficacy of their own planned dam project downstream. Argentina demands to be consulted during the planning of Itaipu but Brazil refuses. An agreement is reached in 1979 that provides for the construction of both Brazil and Paraguay's dam at Itaipu and Argentina's Yacyreta dam.	Wallenstein & Swain 1997
1972	United States	Terrorism	No: Threat	Two members of the right-wing "Order of the Rising Sun" are arrested in Chicago with 30-40 kg of typhoid cultures that are allegedly to be used to poison the water supply in Chicago, St. Louis, and other cities. It was felt that the plan would have been unlikely to cause serious health problems due to chlorination of the water supplies.	Eitzen and Takafuji 1997

Date	Parties Involved	Basis of Conflict	Violent Conflict or in the Context of Violence?	Description	Sources
1972	United States	Terrorism	No: Threat	Reported threat to contaminate water supply of New York City with nerve gas.	Purver 1995.
1972	North Vietnam	Military target	Yes	United States bombs dikes in the Red River delta, rivers, and canals during massive bombing campaign.	Columbia Electronic Encyclopedia 2000
1973	Germany	Terrorism	No: Threat	Threat by a biologist in Germany to contaminate water supplies with bacilli of anthrax and botulinum toxin unless he was paid $8.5 million	Jenkins and Rubin 1978. Kupperman and Trent 1979
1974	Iraq, Syria	Military target, Military tool, Development dispute	Military maneuvers	Iraq threatens to bomb the al-Thawra dam in Syria and massed troops along the border, alleging that the dam had reduced the flow of Euphrates River water to Iraq.	Gleick 1994
1975	Iraq, Syria	Development dispute, Military tool	Military maneuvers	As upstream dams are filled during a low-flow year on the Euphrates, Iraqis claim that flow reaching its territory is "intolerable" and asks the Arab League to intervene. Syrians claim they are receiving less than half the river's normal flow and pull out of an Arab League technical committee formed to mediate the conflict. In May Syria closes its airspace to Iraqi flights and both Syrian and Iraqi reportedly transfer troops to their mutual border. Saudi Arabia successfully mediates the conflict.	Gleick 1993, 1994, Wolf 1997
1975	Angola, South Africa	Military goal, military target	Yes	South African troops move into Angola to occupy and defend the Ruacana hydropower complex, including the Gové Dam on the Kunene River. Goal is to take possession of and defend the water resources of southwestern Africa and Namibia.	Meissner 2000
1977	United States	Terrorism	Yes	Contamination of a North Carolina reservoir with unknown	Clark 1980,

Date	Parties Involved	Basis of Conflict	Violent Conflict or in the Context of Violence?	Description	Sources
				materials. According to Clark: "Safety caps and valves were removed, and poison chemicals were sent into the reservoir....Water had to be brought in."	Purver 1995
1978-onwards	Egypt, Ethiopia	Development dispute, Political tool	No	Long standing tensions over the Nile, especially the Blue Nile, originating in Ethiopia. Ethiopia's proposed construction of dams on the headwaters of the Blue Nile leads Egypt to repeatedly declare the vital importance of water. "The only matter that could take Egypt to war again is water" (Anwar Sadat-1979). "The next war in our region will be over the waters of the Nile, not politics" (Boutrous Ghali-1988).	Gleick 1991, 1994
1978-1984	Sudan	Development dispute, Military target, Terrorism	Yes	Demonstrations in Juba, Sudan in 1978 opposing the construction of the Jonglei Canal led to the deaths of two students. Construction of the Jonglei Canal in the Sudan was forcibly suspended in 1984 following a series of attacks on the construction site.	Suliman 1998; Keluel-Jang 1997
1980s	Mozambique, Rhodesia/Zimbabwe, South Africa	Military target, Terrorism	Yes	Regular destruction of power lines from Cahora Bassa Dam during fight for independence in the region. Dam targeted by RENAMO (the Mozambican National Resistance).	Chenje 2001
1981	Iran, Iraq	Military target and tool	Yes	Iran claims to have bombed a hydroelectric facility in Kurdistan, thereby blacking out large portions of Iraq, during the Iran-Iraq War.	Gleick 1993
1980-1988	Iran, Iraq	Military tool	Yes	Iran diverts water to flood Iraqi defense positions.	Plant 1995
1982	United States	Terrorism	No: Threat	Los Angeles police and the FBI arrest a man who was preparing to poison the city's water supply with a biological agent.	Livingston 1982, Eitzen and Takafuji 1997

Date	Parties Involved	Basis of Conflict	Violent Conflict or in the Context of Violence?	Description	Sources
1982	Israel, Lebanon, Syria	Military tool	Yes	Israel cuts off the water supply of Beirut during siege.	Wolf 1997
1981–1982	Angola	Military target, Military tool	Yes	Water infrastructure, including dams and the major Cunene-Cuvelai pipeline, was targeted during the conflicts in Namibia and Angola in the 1980s.	Turton 2005
1982	Guatemala	Development dispute	Yes	177 civilians killed in Rio Negro over opposition to Chixoy hydroelectric dam.	Levy 2000
1983	Israel	Terrorism	No	The Israeli government reports that it had uncovered a plot by Israeli Arabs to poison the water in Galilee with "an unidentified powder."	Douglass and Livingstone 1987
1984	United States	Terrorism	Yes	Members of the Rajneeshee religious cult contaminate a city water supply tank in The Dalles, Oregon, using Salmonella. A community outbreak of over 750 cases occurred in a county that normally reports fewer than five cases per year.	Clark and Deininger 2000
1985	United States	Terrorism	No	Law enforcement authorities discovered that a small survivalist group in the Ozark Mountains of Arkansas known as The Covenant, the Sword, and the Arm of the Lord (CSA) had acquired a drum containing 30 gallons of potassium cyanide, with the apparent intent to poison water supplies in New York, Chicago, and Washington, D.C. CSA members devised the scheme in the belief that such attacks would make the Messiah return more quickly by punishing unrepentant sinners. The objective appeared to be mass murder in the name of a divine mission rather than to change government policy. The amount of poison possessed by the group is believed to have been insufficient to contaminate the water supply of even one city.	Tucker 2000, NTI 2005

Date	Parties Involved	Basis of Conflict	Violent Conflict or in the Context of Violence?	Description	Sources
1986	North Korea, South Korea	Military tool	No	North Korea's announcement of its plans to build the Kumgansan hydroelectric dam on a tributary of the Han River upstream of Seoul raises concerns in South Korea that the dam could be used as a tool for ecological destruction or war.	Gleick 1993
1986	Lesotho, South Africa	Military goal, Development dispute	Yes	South Africa supports coup in Lesotho over support for ANC and anti-apartheid, and water. New government in Lesotho then quickly signs Lesotho Highlands water agreement.	American University 2000b
1986	Lesotho, South Africa	Development dispute, Military goal	Yes	Bloodless coup by Lesotho's defense forces, with support from South Africa, lead to immediate agreement with South Africa for water from the Highlands of Lesotho, after 30 previous years of unsuccessful negotiations. There is disagreement over the degree to which water was a motivating factor for either party.	Mohamed 2001
1988	Angola, South Africa, Cuba	Military goal, Military target	Yes	Cuban and Angolan forces launch an attack on Calueque Dam via land and then air. Considerable damage inflicted on dam wall; power supply to dam cut. Water pipeline to Owamboland cut and destroyed.	Meissner 2000
1990	South Africa	Development dispute	No	Pro-apartheid council cuts off water to the Wesselton township of 50,000 blacks following protests over miserable sanitation and living conditions.	Gleick 1993
1990	Iraq, Syria, Turkey	Development dispute, Military tool	No	The flow of the Euphrates is interrupted for a month as Turkey finishes construction of the Ataturk Dam, part of the Grand Anatolia Project. Syria and Iraq protest that Turkey now has a weapon of war. In mid-1990 Turkish president Turgut Ozal threatens to restrict water flow to Syria to force it to withdraw support for Kurdish rebels operating in southern Turkey.	Gleick 1993 & 1995
1991-	Karnataka,	Development	Yes	Violence erupts when Karnataka rejects an Interim Order	Gleick 1993,

Date	Parties Involved	Basis of Conflict	Violent Conflict or in the Context of Violence?	Description	Sources
present	Tamil Nadu (India)	dispute		handed down by the Cauvery Waters Tribunal, set up by the Indian Supreme Court. The Tribunal was established in 1990 to settle two decades of dispute between Karnataka and Tamil Nadu over irrigation rights to the Cauvery River.	Butts 1997, American University 2000a
1991	Iraq, Kuwait, US	Military target	Yes	During the Gulf War, Iraq destroys much of Kuwait's desalination capacity during retreat.	Gleick 1993
1991	Canada	Terrorism	No: Threat	A threat is made via an anonymous letter to contaminate the water supply of the city of Kelowna, British Columbia, with "biological contaminates." The motive was apparently "associated with the Gulf War." The security of the water supply was increased in response and no group was identified as the perpetrator.	Purver 1995
1991	Iraq, Turkey, United Nations	Military tool	Yes	Discussions are held at the United Nations about using the Ataturk Dam in Turkey to cut off flows of the Euphrates to Iraq.	Gleick 1993
1991	Iraq, Kuwait, US	Military target	Yes	Baghdad's modern water supply and sanitation system are intentionally and unintentionally damaged by Allied coalition. "Four of seven major pumping stations were destroyed, as were 31 municipal water and sewerage facilities – 20 in Baghdad, resulting in sewage pouring into the Tigris. Water purification plants were incapacitated throughout Iraq" (Arbuthnot 2000). In the first eight months of 1991, after Iraq's water infrastructure was damaged by the Persian Gulf War, the New England Journal of Medicine reported that nearly 47,000 more children than normal died in Iraq and the country's infant mortality rate doubled to 92.7 per 1,000 live births.	Gleick 1993, Arbuthnot 2000, Barrett 2003

Date	Parties Involved	Basis of Conflict	Violent Conflict or in the Context of Violence?	Description	Sources
1992	Czechoslovakia, Hungary	Political tool, Development dispute	Military maneuvers	Hungary abrogates a 1977 treaty with Czechoslovakia concerning construction of the Gabcikovo/Nagymaros project based on environmental concerns. Slovakia continues construction unilaterally, completes the dam, and diverts the Danube into a canal inside the Slovakian republic. Massive public protest and movement of military to the border ensue; issue taken to the International Court of Justice.	Gleick 1993
1992	Turkey	Terrorism	Yes	Lethal concentrations of potassium cyanide are reported discovered in the water tanks of a Turkish Air Force compound in Istanbul. The Kurdish Workers' Party (PKK) claimed credit.	Chelyshev 1992
1992	Bosnia, Bosnian Serbs	Military tool	Yes	The Serbian siege of Sarajevo, Bosnia and Herzegovina, includes a cutoff of all electrical power and the water feeding the city from the surrounding mountains. The lack of power cuts the two main pumping stations inside the city despite pledges from Serbian nationalist leaders to United Nations officials that they would not use their control of Sarajevo's utilities as a weapon. Bosnian Serbs take control of water valves regulating flow from wells that provide more than 80 percent of water to Sarajevo; reduced water flow to city is used to 'smoke out' Bosnians.	Burns 1992, Husarska 1995
1993-present	Iraq	Military tool	No	To quell opposition to his government, Saddam Hussein reportedly poisons and drains the water supplies of southern Shiite Muslims, the Ma'dan. The marshes of southern Iraq are intentionally targeted. The European Parliament and UN Human Rights Commission deplore use of water as weapon in region.	Gleick 1993, American University 2000c, National Geographic News 2001

Date	Parties Involved	Basis of Conflict	Violent Conflict or in the Context of Violence?	Description	Sources
1993	Iran	Terrorism	No	A report suggests that proposals were made at a meeting of fundamentalist groups in Tehran, under the auspices of the Iranian Foreign Ministry, to poison water supplies of major cities in the West "as a possible response to Western offensives against Islamic organizations and states."	Haeri 1993
1993	Yugoslavia	Military target and tool	Yes	Peruca Dam intentionally destroyed during war.	Gleick 1993
1994	Moldavia	Terrorism	No: Threat	Reported threat by Moldavian General Nikolay Matveyev to contaminate the water supply of the Russian 14th Army in Tiraspol, Moldova, with mercury.	Purver 1995
1995	Ecuador, Peru	Military and political tool	Yes	Armed skirmishes arise in part because of disagreement over the control of the headwaters of Cenepa River. Wolf argues that this is primarily a border dispute simply coinciding with location of a water resource.	Samson & Charrier 1997, Wolf 1997
1997	Singapore, Malaysia	Political tool	No	Malaysia supplies about half of Singapore's water and in 1997 threatened to cut off that supply in retribution for criticisms by Singapore of policy in Malaysia.	Zachary 1997
1998	Tajikistan	Terrorism, Political tool	No: Threat	On November 6, a guerrilla commander threatened to blow up a dam on the Kairakkhum channel if political demands are not met. Col. Makhmud Khudoberdyev made the threat, reported by the ITAR-Tass News Agency.	WRR 1998
1998	Angola	Military and political tool	Yes	In September 1998, fierce fighting between UNITA and Angolan government forces broke out at Gove Dam on the Kunene River for control of the installation.	Meissner 2001

Date	Parties Involved	Basis of Conflict	Violent Conflict or in the Context of Violence?	Description	Sources
1998/1994	United States	Cyber-terrorism	No	The Washington Post reports a 12-year old computer hacker broke into the SCADA computer system that runs Arizona's Roosevelt Dam, giving him complete control of the dam's massive floodgates. The cities of Mesa, Tempe, and Phoenix, Arizona are downstream of this dam. No damage was done. This report turns out to be incorrect. A hacker did break into the computers of an Arizona water facility, the Salt River Project in the Phoenix area. But he was 27, not 12, and the incident occurred in 1994, not 1998. And while clearly trespassing in critical areas, the hacker never could have had control of any dams–leading investigators to conclude that no lives or property were ever threatened.	Gellman 2002, Lemos 2002
1998	Democratic Republic of Congo	Military target, Terrorism	Yes	Attacks on Inga Dam during efforts to topple President Kabila. Disruption of electricity supplies from Inga Dam and water supplies to Kinshasa	Chenje 2001, Human Rights Watch 1998
1998 to 2000	Eritrea and Ethiopia	Military target	Yes	Water pumping plants and pipelines in the border town of Adi Quala were destroyed during the civil war between Eritrea and Ethiopia.	ICRC 2003
1999	Lusaka, Zambia	Terrorism, Political tool	Yes	Bomb blast destroyed the main water pipeline, cutting off water for the city of Lusaka, population 3 million.	FTGWR 1999
1999	Yugoslavia	Military target	Yes	Belgrade reported that NATO planes had targeted a hydroelectric plant during the Kosovo campaign.	Reuters 1999a
1999	Bangladesh	Development dispute, Political tool	Yes	50 hurt during strikes called to protest power and water shortages. Protest led by former Prime Minister Begum Khaleda Zia over deterioration of public services and in law and order.	Ahmed 1999

Date	Parties Involved	Basis of Conflict	Violent Conflict or in the Context of Violence?	Description	Sources
1999	Yugoslavia	Military target	Yes	NATO targets utilities and shuts down water supplies in Belgrade. NATO bombs bridges on Danube, disrupting navigation.	Reuters 1999b
1999	Yugoslavia	Political tool	Yes	Yugoslavia refuses to clear war debris on Danube (downed bridges) unless financial aid for reconstruction is provided; European countries on Danube fear flooding due to winter ice dams will result. Diplomats decry environmental blackmail.	Simons 1999
1999	Kosovo	Political tool	Yes	Serbian engineers shut down water system in Pristina prior to occupation by NATO.	Reuters 1999c
1999	South Africa	Terrorism	Yes	A home-made bomb was discovered at a water reservoir at Wallmansthal near Pretoria. It was thought to have been meant to sabotage water supplies to farmers.	Pretoria Dispatch 1999
1999	Angola	Terrorism, Political tool	Yes	100 bodies were found in four drinking water wells in central Angola.	International Herald Tribune 1999
1999	Puerto Rico, U.S.	Political tool	No	Protesters blocked water intake to Roosevelt Roads Navy Base in opposition to U.S. military presence and Navy's use of the Blanco River, following chronic water shortages in neighboring towns.	New York Times 1999
1999	China	Development dispute; terrorism	Yes	Around Chinese New Years, farmers from Hebei and Henan Provinces fought over limited water resources. Heavy weapons, including mortars and bombs, were used and nearly 100 villagers were injured. Houses and facilities were damaged and the total loss reached one million $US. Parties involved: Huanglongkou Village, Shexian County, Hebei Province and Gucheng Village, Linzhou City, Henan Province	China Water Resources Daily 2002

Date	Parties Involved	Basis of Conflict	Violent Conflict or in the Context of Violence?	Description	Sources
1999	East Timor	Military tool, Terrorism	Yes	Militia opposing East Timor independence kill pro-independence supporters and throw bodies in water well.	BBC 1999
1999	Yemen	Development dispute	Yes	700 soldiers were sent to quell fighting that claimed six lives and injured 60 others in clashes that erupted between two villages fighting over a local spring near Ta'iz. The village of Al-Marzuh believed it was entitled to exclusive rights from a spring because it was located on their land; the neighboring village of Quradah believed their rights to the water was affirmed in a 50-year-old court verdict. The dispute erupted in violence. President Ali Abdullah Saleh intervened by summoning the sheikhs of the two villages to the capital, and sorted out the problem by dividing the water into halves.	Al-Qadhi 2006
1998-1999	Kosovo	Terrorism, Political tool	Yes	Contamination of water supplies/wells by Serbs disposing of bodies of Kosovar Albanians in local wells. Other reports of Yugoslav federal forces poisoning wells with carcasses and hazardous materials.	CNN 1999, Hickman 1999.
1999 to 2000	Namibia, Botswana, Zambia	Military goal: Development dispute	No	Sedudu/Kasikili Island, in the Zambezi/Chobe River. Dispute over border and access to water. Presented to the International Court of Justice	ICJ 1999
2000	Ethiopia	Development dispute	Yes	One man stabbed to death during fight over clean water during famine in Ethiopia	Sandrasagra 2000
2000	Central Asia: Kyrgyzstan, Kazakhstan Uzbekistan	Development dispute	No	Kyrgyzstan cuts off water to Kazakhstan until coal is delivered; Uzbekistan cuts off water to Kazakhstan for non-payment of debt.	Pannier 2000

Date	Parties Involved	Basis of Conflict	Violent Conflict or in the Context of Violence?	Description	Sources
2000	Belgium	Terrorism	Yes	In July, workers at the Cellatex chemical plant in northern France dumped 5,000 liters of sulfuric acid into a tributary of the Meuse River when they were denied workers' benefits. A French analyst pointed out that this was the first time "the environment and public health were made hostage in order to exert pressure, an unheard-of situation until now."	Christian Science Monitor. 2000
2000	Hazarajat, Afghanistan	Development dispute	Yes	Violent conflicts broke out over water resources in the villages Burna Legan and Taina Legan, and in other parts of the region, as drought depleted local resources.	Cooperation Center for Afghanistan 2000
2000	India: Gujarat	Development dispute	Yes	Water riots reported in some areas of Gujarat to protest against authority's failure to arrange adequate supply of tanker water. Police are reported to have shot into a crowd at Falla village near Jamnagar, resulting in the death of three and injuries to 20 following protests against the diversion of water from the Kankavati dam to Jamnagar town.	FTGWR 2000
2000	Kenya	Development dispute	Yes	A clash between villagers and thirsty monkeys left eight apes dead and ten villagers wounded. The duel started after water tankers brought water to a drought-stricken area and monkeys desperate for water attacked the villagers.	BBC 2000, Okoko 2000
2000	Australia	Cyber-terrorism	Yes	In Queensland, Australia, on April 23rd, 2000, police arrested a man for using a computer and radio transmitter to take control of the Maroochy Shire wastewater system and release sewage into parks, rivers, and property.	Gellman 2002

Date	Parties Involved	Basis of Conflict	Violent Conflict or in the Context of Violence?	Description	Sources
2000	China	Development dispute	Yes	Civil unrest erupted over use and allocation of water from Baiyangdian Lake – the largest natural lake in northern China. Several people died in riots by villagers in July 2000 in Shandong after officials cut off water supplies. In August 2000, six died when officials in the southern province of Guangdong blew up a water channel to prevent a neighboring county from diverting water.	Pottinger 2000
2001	Israel, Palestine	Terrorism, Military target	Yes	Palestinians destroy water supply pipelines to West Bank settlement of Yitzhar and to Kibbutz Kisufim. Agbat Jabar refugee camp near Jericho disconnected from its water supply after Palestinians looted and damaged local water pumps. Palestinians accuse Israel of destroying a water cistern, blocking water tanker deliveries, and attacking materials for a wastewater treatment project.	Israel Line 2001a,b; ENS 2001a.
2001	Pakistan	Development dispute, Terrorism	Yes	Civil unrest over severe water shortages caused by the long-term drought. Protests began in March and April and continued into summer. Riots, four bombs in Karachi (June 13), one death, 12 injuries, 30 arrests. Ethnic conflicts as some groups "accuse the government of favoring the populous Punjab province [over Sindh province] in water distribution."	Nadeem 2001, Soloman 2001
2001	Macedonia	Terrorism, Military target	Yes	Water flow to Kumanovo (population 100,000) cut off for 12 days in conflict between ethnic Albanians and Macedonian forces. Valves of Glaznja and Lipkovo Lakes damaged.	AFP 2001, Macedonia Information Agency 2001
2001	China	Development dispute	Yes	In an act to protest destruction of fisheries from uncontrolled water pollution, fishermen in northern Jiaxing City, Zhejiang Province, dammed the canal that carries 90 million tons of industrial wastewater per year for 23 days. The wastewater discharge into the neighboring Shengze Town, Jiangsu Province, killed fish, and threatened people's health.	China Ministry of Water Resources 2001.

Date	Parties Involved	Basis of Conflict	Violent Conflict or in the Context of Violence?	Description	Sources
2001	Philippines	Terrorism, Political tool	No	Philippine authorities shut off water to six remote southern villages yesterday after residents complained of a foul smell from their taps, raising fears Muslim guerrillas had contaminated the supplies. Abu Sayyaf guerrillas, accused of links with Saudi-born militant Osami bin Laden, had threatened to poison the water supply in the mainly Christian town of Isabela on Basilan island if the military did not stop an offensive against them.	World Environment News 2001
2001	Afghanistan	Military target	Yes	U.S. forces bombed the hydroelectric facility at Kajaki Dam in Helmand province of Afghanistan, cutting off electricity for the city of Kandahar. The dam itself was apparently not targeted.	BBC 2001, Parry 2001
2002	Nepal	Terrorism, Political Tool	Yes	The Khumbuwan Liberation Front (KLF) blew up a hydroelectric powerhouse of 250 kilowatts in Bhojpur District January 26. The power supply to Bhojpur and adjoining areas was cut off. Estimated repair time was 6 months; repair costs were estimated at 10 million Rs. By June 2002, Maoist rebels had destroyed more than seven micro-hydro projects as well as an intake of a drinking water project and pipelines supplying water to Khalanga in western Nepal.	Kathmandu Post 2002; FTGWR 2002a
2002	Rome, Italy	Terrorism	No: Threat	Italian police arrest four Moroccans allegedly planning to contaminate the water supply system in Rome with a cyanide-based chemical, targeting buildings that included the United States embassy. Ties to Al-Queda were suggested.	BBC 2002
2002	Kashmir, India	Development dispute	Yes	Two people were killed and 25 others injured in Kashmir when police fired at a group of villagers clashing over water sharing. The incident took place in Garend village in a dispute over sharing water from an irrigation stream.	The Japan Times 2002

Date	Parties Involved	Basis of Conflict	Violent Conflict or in the Context of Violence?	Description	Sources
2002	United States	Terrorism	No: Threat	Papers seized during the arrest of a Lebanese national who moved to the US and became an Imam at a Islamist mosque in Seattle included "instructions on poisoning water sources" from a London-based al-Qaida recruiter. The FBI issued a bulletin to computer security experts around the country indicating that al-Qaida terrorists may have been studying American dams and water-supply systems in preparation for new attacks. "U.S. law enforcement and intelligence agencies have received indications that al-Qaida members have sought information on Supervisory Control And Data Acquisition (SCADA) systems available on multiple SCADA-related Web sites," reads the bulletin, according to SecurityFocus. "They specifically sought information on water supply and wastewater management practices in the U.S. and abroad."	McDonnell and Meyer 2002, MSNBC 2002
2002	Colombia	Terrorism	Yes	Colombian rebels in January damaged a gate valve in the dam that supplies most of Bogota's drinking water. Revolutionary Armed Forces of Colombia (FARC), detonated an explosive device planted on a German-made gate valve located inside a tunnel in the Chingaza Dam.	Waterweek 2002
2002	Karnataka, Tamil Nadu, India	Development dispute	Yes	Continuing violence over the allocation of the Cauvery River between Karnataka and Tamil Nadu. Riots, property destruction, more than 30 injuries, arrests through September and October.	The Hindu 2002a,b, The Times of India 2002a.
2002	United States	Terrorism	No: Threat	Earth Liberation Front threatens the water supply for the town of Winter Park. Previously, this group claimed responsibility for the destruction of a ski lodge in Vail, Colorado that threatened lynx habitat.	Crecente 2002, Associated Press 2002

Date	Parties Involved	Basis of Conflict	Violent Conflict or in the Context of Violence?	Description	Sources
2003	United States	Terrorism	No: Threat	Al-Qaida threatens US water systems via call to Saudi Arabian magazine. Al-Qaida does not "rule out...the poisoning of drinking water in American and Western cities."	Associated Press 2003a, Waterman 2003, NewsMax 2003, US Water News 2003
2003	United States	Terrorism	Yes	Four incendiary devices were found in the pumping station of a Michigan water-bottling plant. The Earth Liberation Front (ELF) claimed responsibility, accusing Ice Mountain Water Company of "stealing" water for profit. Ice Mountain is a subsidiary of Nestle Waters.	Associated Press 2003b
2003	Colombia	Terrorism, development dispute	Yes	A bomb blast at the Cali Drinking Water Treatment Plant killed 3 workers May 8th. The workers were members of a trade union involved in intense negotiations over privatization of the water system.	PSI 2003
2003	Jordan	Terrorism	No: Threat	Jordanian authorities arrested Iraqi agents in connection with a botched plot to poison the water supply that serves American troops in the eastern Jordanian desert near the border with Iraq. The scheme involved poisoning a water tank that supplies American soldiers at a military base in Khao, which lies in an arid region of the eastern frontier near the industrial town of Zarqa.	MJS 2003

Date	Parties Involved	Basis of Conflict	Violent Conflict or in the Context of Violence?	Description	Sources
2003	Iraq, United States, Others	Military Target	Yes	During the U.S.-led invasion of Iraq, water systems were reportedly damaged or destroyed by different parties, and major dams were military objectives of the U.S. forces. Damage directly attributable to the war includes vast segments of the water distribution system and the Baghdad water system, damaged by a missile.	UNICEF 2003, ARC 2003
2003	Iraq	Terrorism	Yes	Sabotage/bombing of main water pipeline in Baghdad. The sabotage of the water pipeline was the first such strike against Baghdad's water system, city water engineers said. It happened around 7 in the morning, when a blue Volkswagen Passat stopped on an overpass near the Nidaa mosque and an explosive was fired at the six-foot-wide water main in the northern part of Baghdad, said Hayder Muhammad, the chief engineer for the city's water treatment plants.	Tierney and Worth 2003
2003–2004	Sudan	Military tool, Military target, Terrorism	Yes	The ongoing civil war in the Sudan has included violence against water resources. In 2003, villagers from around Tina said that bombings had destroyed water wells. In Khasan Basao they alleged that water wells were poisoned. In 2004, wells in Darfur were intentionally contaminated as part of a strategy of harassment against displaced populations.	Toronto Daily 2004, Reuters Foundation 2004
2004	Mexico	Development dispute	Yes	Two Mexican farmers argued for years over water rights to a small spring used to irrigate a small corn plot near the town of Pihuamo. In March, these farmers shot each other dead.	The Guardian 2004
2004	Pakistan	Terrorism	Yes	In military action aimed at Islamic terrorists, including Al Qaeda and the Islamic Movement of Uzbekistan, homes, schools, and water wells were damaged and destroyed.	Reuters 2004a

Date	Parties Involved	Basis of Conflict	Violent Conflict or in the Context of Violence?	Description	Sources
2004	India, Kashmir	Terrorism	Yes	Twelve Indian security forces were killed by an IED planted in an underground water pipe during "counter-insurgency operation in Khanabal area in Anantnag district."	TNN 2004
2004	Gaza Strip	Terrorism, Development dispute	Yes	The United States halts two water development projects as punishment to the Palestinian Authority for their failure to find those responsible for a deadly attack on a U.S. diplomatic convoy in October 2003.	Associated Press 2004a
2004	India	Development dispute	Yes	Four people were killed in October and more than 30 injured in November in ongoing protests by farmers over allocations of water from the Indira Ghandi Irrigation Canal in Sriganganagar district, which borders Pakistan. A curfew was imposed in the towns of Gharsana, Raola and Anoopgarh.	Indo-Asian News Service 2004
2004–2006	Somalia	Development dispute	Yes	At least 250 people killed and many more injured in clashes over water wells and pastoral lands. Villagers call it the "War of the Well," and describe "well warlords, well widows, and well warriors." A three-year drought has led to extensive violence over limited water resources, worsened by the lack of effective government and central planning.	BBC 2004, Wax 2006
2005	Kenya	Development dispute	Yes	Police were sent to the northwestern part of Kenya to control a major violent dispute between Kikuyu and Maasai groups over water. More than 20 people were killed in fighting in January. The tensions arose when Maasai herdsmen accused a local Kikuyu politician of diverting a river to irrigate his farm, depriving downstream livestock. Fighting displaced more than 2000 villagers and reflects tensions between nomadic and settled communities.	BBC 2005a, Ryu 2005

Date	Parties Involved	Basis of Conflict	Violent Conflict or in the Context of Violence?	Description	Sources
2006	Yemen	Development dispute	Yes	Local media reported a struggle between Hajja and Amran tribes over a well located between the two governorates in Yemen. According to news reports, armed clashes between the two sides forced many families to leave their homes and migrate. News reports confirmed that authorities arrested 20 people in an attempt to stop the fighting.	Al-Ariqi 2006
2006	Ethiopia	Development dispute, water scarcity	Yes	At least 12 people died and over 20 were wounded in clashes over competition for water and pasture in the Somali border region.	BBC 2006a
2006	Sri Lanka	Military tool, military target, terrorism	Yes	Tamil Tiger rebels cut the water supply to government-held villages in northeastern Sri Lanka. Sri Lankan government forces then launched attacks on the reservoir, declaring the Tamil actions to be terrorism.	BBC 2006b, 2006c
2006	Israel, Lebanon	Military target, terrorism	Yes	Hezbollah rockets damaged a wastewater treatment plant in Israel. The Lebanese government estimates that Israeli attacks damaged water systems throughout southern Lebanon, including tanks, pipes, pumping stations, and facilities along the Litani River.	Science 2006, Amnesty International 2006, Murphy 2006

Notes:

1. Conflicts may stem from the drive to possess or control another nation's water resources, thus making water systems and resources a *political or military goal*. Inequitable distribution and use of water resources, sometimes arising from a water development, may lead to *development disputes*, heighten the importance of water as a strategic goal or may lead to a degradation of another's source of water. Conflicts may also arise when water systems are used as instruments of war, either as *targets* or *tools*. These distinctions are described in detail in Gleick (1993, 1998). In 2001, the Institute began including incidents involving water and *terrorism*. We note, however, the difficulty in defining "terrorism" (as opposed to military target, tool, or goal or other category) and caution users to use care with apply these categories. We use this term when individuals or groups act against governments or official agencies.

2. Thanks to the many people who have contributed to this over time, including William Meyer who sent 9 fascinating items from the 1800s, Patrick Marsh, Hans-Juergen.Liebscher, Robert Halliday, Ma Jun, Marcus Moench, and others I've no doubt forgotten.

Sources:

Absolute Astronomy webpage. Reviewed 2006. "Incapacitating agent." http://www.absoluteastronomy.com/reference/incapacitating_agent

Agence France Press (AFP). 2001. "Macedonian troops fight for water supply as president moots amnesty." AFP, June 8, 2001. http://www.balkanpeace.org/hed/archive/june01/hed3454.shtml.

Ahmed, A. 1999. "Fifty hurt in Bangladesh strike violence." Reuters News Service, Dhaka, April 18, 1999.

Al-Ariqi, A. 2006. "Water war in Yemen." Yemen Times, Vol. 14, Issue 932, April 24, 2006. http://yementimes.com/article.shtml?i=932&p=health&a=1.

Al-Qadhi, M. 2003. "Thirst for water and development leads to conflict in Yemen." Choices. United Nations Development Programme, Vol. 12, No. 1, pp. 13-14. See also: http://yementimes.com/article.shtml?i=642&p=health&a=1.

American Red Cross (ARC). 2003. "Baghdad Hospitals Reopen But Health Care System Strained." Mason Booth, Staff Writer, RedCross.org . April 24, http://www.redcross.org/news/in/iraq/030424baghdad.html.

American University (Inventory of Conflict and the Environment ICE). 2000a. Cauvery River Dispute. http://www.american.edu/projects/mandala/TED/ice/CAUVERY.HTM.

American University (Inventory of Conflict and the Environment ICE). 2000b. Lesotho "Water Coup." http://www.american.edu/projects/mandala/TED/ice/LESWATER.HTM

American University (Inventory of Conflict and the Environment ICE). 2000c. Marsh Arabs and Iraq. http://www.american.edu/projects/mandala/TED/ice/MARSH.HTM.

Amnesty International. 2006. "Israel/Lebanon. Deliberate destructi on or "collateral damage"? Israeli attacks on civilian infrastructure." http://web.amnesty.org/library/Index/ENGMDE180072006.

Arbuthnot, F. 2000. "Allies deliberately poisoned Iraq public water supply in Gulf War." Sunday Herald (Scotland) September 17, 2000.

Associated Press. 2002. "Earth Liberation Front members threaten Colorado town's water." AP, October 15, 2002.

Associated Press. 2003a. "Water targeted, magazine reports." AP, May 29, 2003.

Associated Press. 2003b. "Incendiary devices placed at water plant." AP, September 25, 2003.

Associated Press. 2004a. "US dumps water projects in Gaza over convoy bomb." AP, May 6, 2004.

Barrett, G. 2003. "Iraq's bad water brings disease, alarms relief workers. The Olympian, Olympia Washington, Gannett News Service, June 29, http://www.theolympian.com/home/news/20030629/frontpage/39442.shtml.

Barry J.M. 1997. Rising Tide: The Great Mississippi Flood of 1927 and How it Changed America. Simon and Schuster, New York. p. 67.

Bingham, G., A. Wolf, and T. Wohlgenant. 1994. "Resolving water disputes: Conflict and cooperation in the United States, the Near East, and Asia." US Agency for International Development (USAID). Bureau for Asia and the Near East. Washington DC.

BBC 1999. "World: Asia-Pacific Timor atrocities unearthed." September 22, 1999. http://news.bbc.co.uk/hi/english/world/asia-pacific/newsid_455000/455030.stm

BBC 2000. "Kenyan monkeys fight humans for water." BBC News March 21, 2000. http://news.bbc.co.uk/1/hi/world/africa/685381.stm

BBC 2001. US 'bombed Afghan power plant.' http://news.bbc.co.uk/1/hi/world/south_asia/1632304.stm

BBC 2002. "Cyanide attack' foiled in Italy." February 20, 2002. http://news.bbc.co.uk/hi/english/world/europe/newsid_1831000/1831511.stm

BBC 2004. "Dozens dead' in Somalia clashes." http://news.bbc.co.uk/2/hi/africa/4073063.stm. BBC News World Edition online.

BBC 2005a. "Thousands flee Kenyan water clash." BBC News. January 24, 2005. http://news.bbc.co.uk/1/hi/world/africa/4201483.stm

BBC 2006a. "Somalis clash over scarce water." BBC News. February 17, 2006. http://news.bbc.co.uk/go/pr/fr/-/1/hi/world/africa/4723008.stm

BBC 2006b. "Sri Lanka forces attack reservoir." BBC News. August 7, 2006. http://news.bbc.co.uk/2/hi/south_asia/5249884.stm?ls

BBC 2006c. "Water and war in Sri Lanka." BBC News. August 3, 2006. http://news.bbc.co.uk/2/hi/south_asia/5239570.stm.

Burns, J.F. 1992. "Tactics of the Sarajevo Siege: Cut Off the Power and Water," New York Times, September 25, 1992. p. A1.

Butts, K., ed. 1997. Environmental Change and Regional Security. Carlisle, PA: Asia-Pacific Center for Security Studies, Center for Strategic Leadership, US Army War College.

Cable News Network (CNN). 1999. "U.S.: Serbs destroying bodies of Kosovo victims." May 5. www.cnn.com/WORLD/europe/9905/05/kosovo.bodies.

Chenje, M. 2001. Hydro-politics and the quest of the Zambezi River Basin Organization." In M. Nakayama (ed.) International Waters in Southern Africa. United Nations University, Tokyo, Japan.

Chelyshev, A. 1992. "Terrorists Poison Water in Turkish Army Cantonment." Telegraph Agency of the Soviet Union (TASS), 29 March.

China Ministry of Water Resources. 2001. http://shuizheng.chinawater.com.cn/ssjf/20021021/200210160087.htm (the website of the Policy and Regulatory Department).

China Water Resources Daily 2002. Villagers fight over water resources. 24 October 2002. Citation provided by Ma Jun, personal communication.

Christian Science Monitor. 2000. "Ecoterrorism as negotiating tactic." July 21, 2000, p. 8.

Clark, R.C. 1980. Technological Terrorism. Devin-Adair, Old Greenwich, Connecticut.

Clark, R.M. and R.A. Deininger. 2000. Protecting the Nation's Critical Infrastructure: The Vulnerability of U.S. Water Supply Systems. Journal of Contingencies and Crisis Management. Vol. 8, No. 2, pp. 73-80.

Columbia Electronic Encyclopedia 2000. "Vietnam: History." Available at http://www.infoplease.com/ce6/world/A0861793.html.

Columbia Encyclopedia 2000. "Netherlands." 6th Edition. Columbia Encyclopedia available at http://www.bartleby.com/65/ne/Nethrlds.html

Cooperation Center for Afghanistan. 2000. The Social Impact of Drought in Hazarajat. http://www.ccanata.com/impact.html

Corps of Engineers. 1953. "Applications of Hydrology in Military Planning and Operations and Subject Classification Index for Military Hydrology Data." Military Hydrology R&D Branch, Engineering Division, Corps of Engineers, Department of the Army, Washington.

Crecente, B.D. 2002. "ELF targets water: Group threatens eco-terror attack on Winter Park tanks." Rocky Mountain News, October 15, 2002. http://www.rockymountainnews.com/drmn/state/article/0,1299,DRMN_21_1479883,00.html

Daniel, C. (ed.). 1995. Chronicle of the 20th Century. Dorling Kindersley Publishing, Inc., New York.

Dolatyar, M. 1995. "Water diplomacy in the Middle East." In E. Watson (editor) The Middle Eastern Environment. John Adamson Publishing, London. 256 pp.

Douglass, J.D. and N.C. Livingstone. 1987. America the Vulnerable: The Threat of Chemical and Biological Warfare. Lexington Books, Lexington, Massachusetts.

Drower, M.S. 1954. "Water-supply, irrigation, and agriculture." In C. Singer, E.J. Holmyard, and A.R. Hall (ed.) A History of Technology. Oxford University Press, New York.

Dutch Water Line. 2002. Information on the historical use of water in defense of Holland. http://www.xs4all.nl/~pho/Dutchwaterline/dutchwaterl.htm.

Eitzen, E.M. and E.T. Takafuji. 1997. "Historical Overview of Biological Warfare." In Textbook of Military Medicine, Medical Aspects of Chemical and Biological Warfare. Published by the Office of The Surgeon General, Department of the Army, USA. Pages 415-424.

ENS: Environment News Service. 2001a. "Environment a weapon in the Israeli-Palestinian conflict." February 5, 2001, http://www.ens-newswire.com/ens/feb2001/2001-02-05-01.asp.

Fatout, P. 1972. Indiana Canals. Purdue University Studies, West Lafayette, Indiana, pp. 158-162.

Ferguson, R. Brian. 2001. "The Birth of War." Natural History, Vol. 122, No.6 pp. 28-35 (July-August 2003).

Fickle, J.E. 1983. "The 'people' versus 'progress': Local opposition to the construction of the Wabash and Erie Canal." Old Northwest, Vol. 8, No. 4, pp. 309-328.

Financial Times Global Water Report. 1999. "Zambia: Water Cutoff." FTGWR Issue 68, p. 15 (March 19, 1999).
Financial Times Global Water Report. 2000. "Drought in India comes as no surprise." FTGWR Issue 94, p. 14 (April 28, 2000).

Financial Times Global Water Report. 2002a. "Maoists destroy Nepal's infrastructure." FTGWR, Issue 146, pp. 4-5 (May 17, 2002).

Fonner, D.K. 1996. Scipio Africanus. Military History Magazine March 1996. Cited in http://historynet.com/mh/blscipioafricanus/index1.html

Forkey, N.S. 1998. "Damming the dam: Ecology and community in Ops Township, Upper Canada." Canadian Historical Review, Vol. 79, No. 1, pp. 68-99.

Gellman, B. 2002. "Cyber-attacks by Al Qaeda feared." Washington Post, June 27, 2002, page A1.

Gleick, P.H. 1991. "Environment and security: The clear connections." Bulletin of the Atomic Scientists. April:17-21.

Gleick, P.H. 1993. "Water and conflict: Fresh water resources and international security." International Security 18, Vol. 1, pp. 79-112.

Gleick, P.H. 1994. "Water, war, and peace in the Middle East." Environment, Vol. 36, No. 3, pp.6-on. Heldref Publishers, Washington.

Gleick, P.H. 1995. "Water and Conflict: Critical Issues." Presented to the 45th Pugwash Conference on Science and World Affairs. Hiroshima, Japan: 23-29 July.

Gleick, P.H. 1998. "Water and conflict." In The World's Water 1998-1999. Island Press, Washington.

Gowan, H. 2004. Hannibal Barca and the Punic Wars." Website. http://www.barca.fsnet.co.uk/. Reviewed March 2005.

Grant, U.S. 1885. Personal Memoirs of U.S. Grant. C.L. Webster, New York. ["On the second of February, [1863] this dam, or levee, was cut,...The river being high the rush of water through the cut was so great that in a very short time the entire obstruction was washed away.... As a consequence the country was covered with water."]

Green Cross International. The Conflict Prevention Atlas: http://www.greencrossinternational.net/GreenCrossPrograms/waterres/gcwater/report.html

Guantanamo Bay Gazette. 1964. The History of Guantanamo Bay: An Online Edition. http://www.gtmo.net/gazz/hisidx.htm. Chapter XXI: The 1964 Water Crisis. http://www.gtmo.net/gazz/HISCHP21.HTM.

Guardian. 2004. "Water duel kills elderly cousins." The Guardian Newspapers Limited. March 11, 2004.

Haeri, S. 1993. "Iran: Vehement Reaction." Middle East International (19 March), p. 8.

Harris, S.H. 1994. Factories of Death: Japanese Biological Warfare 1932-1945 and the American Cover-up. Routledge, New York, N.Y.

Hatami, H. and Gleick, P. 1994. Chronology of Conflict over Water in the Legends, Myths, and History of the Ancient Middle East. In "Water, war, and peace in the Middle East." Environment, Vol. 36, No. 3, pp.6-on. Heldref Publishers, Washington.

Hickman, D.C. 1999. A Chemical and Biological Warfare Threat: USAF Water Systems at Risk." Couterproliferation Paper No. 3. uSAF Counterproliferation Center, Air War College, Maxwell Air Force Base, Alabama.

Hillel, D. 1991. "Lash of the Dragon." Natural History (August), pp. 28-37.

Hindu, The. 2002a. "Ryots on the rampage in Mandya." The Hindu, India's National Newspaper. October 31, 2002. http://www.hinduonnet.com/thehindu/2002/10/31/stories/2002103106680100.htm

Hindu, The. 2002b. "Farmers go berserk; MLA's house attacked." The Hindu, India's National Newspaper, October 30, 2002. http://www.hinduonnet.com/thehindu/2002/10/30/stories/2002103004870400.htm

Honan, W.H. 1996. "Scholar sees Leonardo's influence on Machiavelli." The New York Times (December 8), p. 18.

Human Rights Watch. 1998. Human Rights Watch Condemns Civilian Killings by Congo Rebels. http://www.hrw.org/press98/aug/congo827.htm

Husarska, A. 1995. "Running dry in Sarajevo: Water fight." The New Republic. July 17 & 24.

InfoRoma. 2004. "Roman Aqueducts." http://www.inforoma.it/feature.php?lookup=aqueduct. Viewed March 2005.

Indo-Asian News Service. 2004. "Curfew imposed in three Rajasthan towns." http://www.hindustantimes.com/news/181_1136315,000900010008.htm. Hindustan Times. December 4, 2004. Also, see http://news.newkerala.com/india-news/?action=fullnews&id=46359. India News at newkerala.com.

International Committee of the Red Cross. 2003. Eritrea: ICRC repairs war-damaged health centre and water system. 15 Dec 2003. ICRC News No. 03/158. http://www.alertnet.org/thenews/fromthefield/107148342038.htm.

International Court of Justice. 1999. International Court of Justice Press Communiqué 99/53, Kasikili Island/Sedudu Island (Botswana/Namibia). The Hague, Holland 13 December 1999, p. 2 (http://www.icj-cij.org/icjwww/ipresscom/ipress1999/ipresscom9953_ibona_19991213.htm.)

International Herald Tribune. 1999. "100 bodies found in well." International Herald Tribune, August 14-15. p.4.

Israel Line. 2001a. "Palestinians loot water pumping center, cutting of supply to refugee camp." Israel Line (http://www.israel.org/mfa/go.asp?MFAH0dmp0), downloaded January 5, 2001, http://www.mfa.gov.il/mfa/go.asp?MFAH0iy50.

Israel Line. 2001a. "Palestinians vandalize Yitzhar water pipe." Israel Line, January 9, 2001, http://www.mfa.gov.il/mfa/go.asp?MFAH0izu0.

IWCT. 1967. International War Crimes Tribunal "Some Facts on Bombing of Dikes." http://www.infotrad.clara.co.uk/antiwar/warcrimes/index.html.

Japan Times. 2002. "Kashmir water clash." The Japan Times, May 27, 2002, p. 3.

Jenkins, B.M. and A.P. Rubin, 1978. "New Vulnerabilities and the Acquisition of New Weapons by Nongovernment Groups." In Evans, A.E. and J.F. Murphy (eds.) Legal Aspects of International Terrorism. Lexington Books, Lexington, Massachusetts. pp. 221-276.

Kathmandu Post. 2002. "KLF destroys micro hydro plant." The Kathmandu Post, January 28, 2002. http://www.nepalnews.com.np/contents/englishdaily/ktmpost/2002/jan28/index.htm

Kirschner, O. 1949. "Destruction and Protection of Dams and Levees." Military Hydrology, Research and Development Branch, U.S. Corps of Engineers, Department of the Army, Washington District. From Schweizerische Bauzeitung 14 March 1949, Translated by H.E. Schwarz, Washington.

Keluel-Jang, S.A. 1997. "Alier and the Jonglei Canal." Southern Sudan Bulletin, Vol. 2, No. 3 (January) (from www.sufo.demon.co.uk/poli007.htm).

Kupperman, R.H. and D.M. Trent, 1979. Terrorism: Threat, Reality, Response. Hoover Institution Press, Stanford, California.

Lemos, R. 2002. Safety: Assessing the infrastructure risk. CNET/new.com. http://news.com.com/2009-1001_3-954780.html . August 26th.

Levy, K. 2000. "Guatemalan dam massacre survivors seek reparations from financiers." World Rivers Review, International Rivers Network, Berkeley, California. December 2000, pp. 12-13.

Livingston, N.C. 1982. The War Against Terrorism. Lexington Books, Lexington and Toronto, Canada.

Lockwood, R.P. Reviewed April 2006. "The battle over the Crusades." http://www.catholicleague.org/research/battle_over_the_crusades.htm.

Macedonia Information Agency. 2001. "Humanitarian catastrophe averted in Kumanovo and Lipkovo." Republic of Macedonia Agency of Information Archive. June 18, 2001. http://wwww.reliefweb.int/w/rwb.nsf/0/dbd4ef105d93da4ac1256a6f005bc328?OpenDocument .

McDonnell, P.J. and J. Meyer. 2002. "Links to Terrorism Probed in Northwest." Los Angeles Times, 13 July 2002.

Meissner, R. 2000. "Hydropolitical hotspots in Southern Africa: Will there be a water war? The case of the Kunene river." In H. Solomon and A. Turton (editors). *Water Wars: Enduring Myth or Impending Reality?* Africa Dialogue Monograph Series No. 2. Accord, Creda Communications, KwaZulu-Natal, South Africa, pp. 103-131.

Meissner, R. 2001. "Interaction and existing constraints in international river basins: The case of the Kunene River Basin." In M. Nakayama (ed.) *International Waters in Southern Africa.* United Nations University, Tokyo, Japan.

Milwaukee Journal Sentinel. 2003. "Jordan foils Iraqi plot to poison U.S. troops' water, officials say." April 1, 2003. **http://www.jsonline.com/news/gen/apr03/130338.asp.**

Mohamed, A.E. 2001. "Joint development and cooperation in international water resources: The case of the Limpopo and Orange River Basins in Southern Africa." In M. Nakayama (ed.) *International Waters in Southern Africa.* United Nations University, Tokyo, Japan.

Moorehead, A. 1960. *The White Nile.* Penguin Books, England.

MSNBC 2002. "FBI says al-Qaida after water supply." Numerous wire reports, see, for example, http://www.ionizers.org/water-terrorism.html.

Murphy, K. 2006. "Old feud over Lebanese river takes new turn. Israel's airstrikes on canals renew enduring suspicions that it covets water from the Litani." August 10, 2006. http://www.latimes.com/news/nationworld/world/la-fg-litani10aug10,1,3447228.story?coll=la-headlines-world

Murphy, I.L. and J. E. Sabadell. 1986. "International river basins: A policy model for conflict resolution." *Resources Policy.* Vol. 12, No. 1, pp. 133-144. Butterworth and Co. Ltd., United Kingdom.

Museum of the City of New York (MCNY). No date. The Greater New York Consolidation Timeline. http://www.mcny.org/Exhibitions/GNY/timeline.htm

Nadeem, A. 2001. "Bombs in Karachi Kill One. *Associated Press.* downloaded June 13, 2001. At http://dailynews.yahoo.com/h/ap/20010613/wl/pakistan_strike_3.html

Naff, T. and R.C. Matson (editors). 1984. *Water in the Middle East: conflict or cooperation?* Westview Press, Boulder, Colorado.

National Geographic News. 2001. "Ancient Fertile Crescent almost gone, satellite images show." May 18, 2001. http://news.nationalgeographic.com/news/2001/05/0518_crescent.html

New Scofield Reference Bible. 1967. C.I. Schofield, Editor, Oxford University press, Oxford, United Kingdom.

New York Times. 1999. "Puerto Ricans protest Navy's use of water." The New York Times, October 31, p. 30.

NewsMax. 2003. "Al-Qaida Threat to U.S. Water Supply. NewsMax Wires, May 29, 2003. http://www.newsmax.com/archives/articles/2003/5/28/202658.shtml.

NTI Nuclear Threat Initiative. 2005. A Brief History of Chemical Warfare. http://www.nti.org/h_learnmore/cwtutorial/chapter02_02.html

Okoko, T.O. 2000. "Monkeys, Humans Fight over Drinking Water." Panafrican News Agency March 21, 2000.

Pannier, B. 2000. "Central Asia: Water becomes a political issue." Radio Free Europe, www.rferl.org/nca/features/2000/08/F.RU.000803122739.html.
Parry, R.L. 2001. "UN fears 'disaster' over strikes near huge dam." The Independent, London, November 8.

Plant, G. 1995. "Water as a weapon in war." Water and War, Symposium on Water in Armed Conflicts, Montreux 21-23 November 1994, Geneva, ICRC.

Pottinger, M. 2000. "Major Chinese lake disappearing in water crisis." Reuters Science News, http://us.cnn.com/2000/NATURE/12/20/china.lake.reut/

Pretoria Dispatch Online. 1999. "Dam bomb may be 'aimed at farmers'." http://www.dispatch.co.za/1999/07/21/southafrica/RESEVOIR.HTM. (July 21).

Priscoli, J.D. "Water and Civilization: Conflict, Cooperation and the Roots of a New Eco Realism." A Keynote Address for the 8th Stockholm World Water Symposium, 10-13 August 1998. http://www.genevahumanitarianforum.org/docs/Priscoli.pdf.

PSI. 2003. "Urgent action: Bomb blast kills 3 workers at the Cali Water Treatment Plant." Public Services International. www.world-psi.org. Also at http://209.238.219.111/Water.htm

Purver, R. 1995. Chemical and Biological Terrorism: The Threat According to the Open Literature. Canadian Security Intelligence Service, Ottawa, Canada. http://www.csis.gc.ca/en/publications/other/c_b_terrorism01.asp.

Rasch, P.J. 1968. "The Tularosa Ditch War." New Mexico Historical Review, Vol. 43, No. 3, pp. 229-235.

Reisner, M. 1986, 1993. Cadillac Desert: The American West and its Disappearing Water. Penguin Books, New York.

Reuters. 1999a. "Serbs Say NATO Hit Refugee Convoys." April 14, 1999. http://dailynews.yahoo.com/headlines/ts/story.html?s=v/nm/19990414/ts/yugoslavia_192.html. http://www.uia.ac.be/u/carpent/kosovo/messages/397.html

Reuters 1999b. "NATO Keeps Up Strikes But Belgrade Quiet." June 5, 1999. Downloaded June 1999. http://dailynews.yahoo.com/headlines/wl/story.html?s=v/nm/19990605/wl/yugoslavia_strikes_129.html.

Reuters 1999c. "NATO Builds Evidence Of Kosovo Atrocities." June 17, 1999. Downloaded June 1999. http://dailynews.yahoo.com/headlines/ts/story.html?s=v/nm/19990617/ts/yugoslavia_leadall_171.html.

Reuters. 2004a. "Al Qaeda spy chief killed in Pakistani raid." Reuters Yahoo.

Reuters Foundation. 2004. Darfur: "2.5 million people will require food aid in 2005." http://www.medair.org/en_portal/medair_news?news=258. R. Schofield. November 22, 2004.

Rome Guide. 2004. "Fontana di Trevi: History." http://web.tiscali.it/romaonlineguide/Pages/eng/rbarocca/sBMy5.htm. Viewed March 2005.

Rowe, W.T. 1988. "Water control and the Qing political process: The Fankou Dam controversy, 1876-1883." Modern China, Vol. 14, No. 4, pp. 353-387.

Ryu, A. 2005. "Water rights dispute sparks ethnic clashes in Kenya's Rift Valley." Voice of America, http://www.voanews.com/english/archive/2005-03/2005-03-21-voa28.cfm.

Samson, P. and B. Charrier. 1997. "International freshwater conflict: Issues and prevention strategies." Green Cross International. http://www.greencrossinternational.net/GreenCrossPrograms/waterres/gcwater/report.html

Sandrasagra, M.J. 2000. "Development Ethiopia: Relief agencies warn of major food crisis." Inter Press Service. April 11, 2000.

Scheiber, H.N. 1969. Ohio Canal Era. Ohio University Press, Athens, Ohio, pp. 174-175.

Science. 2006. "Tallying Mideast damage." Science, Vol. 313, Issue 5793, p. 1549.

Semann, D. 1950. Die Kriegsbeschädigungen der Edertalspermauer, die Wiederherstellungsarbeiten und die angestellten Untersuchungen über die Standfestigkeit der Mauer. - Die Wasserwirtschaft 41. Jg., Nr. 1 u. 2.

Shapiro, C. 2004. "A search for flaws deep in the heart of the Surry reactor." The Virginia-Pilot. December 6, 2004. http://home.hamptonroads.com/stories/print.cfm?story=78992&ran=226100.

Simons, M. 1999. "Serbs refuse to clear bomb-littered river." New York Times, October 24, 1999.

Strategy Page. 2006. Reviewed April 2006. http ://www.strategypage.com/articles/biotoxin_files/BIOTOXINSINWARFARE.asp

Stockholm International Peace Research Institute (SPIRI). 1971. The Rise of CB Weapons: The Problem of Chemical and Biological Warfare. Humanities Press, New York, NY

.Soloman, A. 2001. "Policeman dies as blasts rock strike-hit Karachi." Reuters, June 13, 2001. at http://dailynews.yahoo.com/h/nm/20010613/ts/pakistan_strike_dc_1.html. http://www.labline.de/indernet/partikel/karachi/bombse.htm

Steinberg, T.S. 1990. "Dam-breaking in the nineteenth-century Merrimack Valley." Journal of Social History, Vol. 24, No. 1, pp. 25-45.

Styran, R.M. and R.R. Taylor. 2001. 'The Great Swivel Link': Canada's Welland Canal. The Champlain Society, Toronto, Canada.

Suliman, M. 1998. "Resource access: A major cause of armed conflict in the Sudan. The case of the Nuba Mountains." Institute for African Alternatives, London, UK (from http://srdis.ciesin.org/cases/Sudan-Paper.html.)

Thatcher, J. 1827. A Military Journal During the American Revolutionary War, From 1775 to 1783. Second Edition, Revised and Corrected. Cottons and Barnard. Boston, Massachusetts. (from http://www.fortklock.com/journal1777.htm.)

Tierney, J. and R.F. Worth. 2003. "Attacks in Iraq May Be Signals of New Tactics." The New York Times, August 18, 2003. Page 1. Also at http://www.nytimes.com/2003/08/18/international/worldspecial/18IRAQ.html?hp

Times of India. 2002a. "Cauvery row: Farmers renew stir." October 20, 2002. http://timesofindia.indiatimes.com/cms.dll/html/uncomp/articleshow?art_id=26586125.

Times News Network (TNN). 2004. "IED was planted in underground pipe." http://timesofindia.indiatimes.com/articleshow/947432.cms December 5, 2004.

Toronto Daily. 2004. "Darfur: 'Too many people killed for no reason.'" Amnesty International Index: AFR 54/008/2004, 3 February 2004.

Tucker, J.B. ed. 2000. Toxic Terror: Assessing Terrorist Use of Chemical and Biological Weapons. MIT Press, Cambridge, Massachusetts.

Turton, A. 2005. "A Critical Assessment of the Basins at Risk in the Southern African Hydropolitical Complex." Workshop on the Management of International Rivers and Lakes Hosted by the Third World Centre for Water Management & Helsinki University of Technology, 17-19 August 2005 Helsinki, Finland. Council for Scientific and Industrial Research (CSIR). African Water Issues Research Unit (AWIRU) CSIR Report Number: ENV-P-CONF 2005-001, Pretoria, South Africa.

UNICEF 2003. "Iraq: Cleaning up neglected, damaged water system, clearing away garbage." News Note Press Release, May 27. http://www.unicef.org/media/media_6998.html.

US Water News. 2003. "Report suggests al-Qaida could poison U.S. water." US Water News Online. June. http://www.uswaternews.com/archives/arcquality/3repsug6.html.

Vanderwood, P.J. 1969. Night riders of Reelfoot Lake. Memphis State University Press, Memphis, Tennessee.

Wallenstein, P. and A. Swain. 1997. "International freshwater resources - Conflict or cooperation?" Comprehensive Assessment of the Freshwater Resources of the World: Stockholm: Stockholm Environment Institute.

Walters, E. 1948. Joseph Benson Foraker: Uncompromising Republican. Ohio History Press, Columbus, Ohio, pp. 44-45.

Waterman, S. 2003. "Al-Qaida threat to U.S. water supply." United Press International (UPI), May 28, 2003.

Waterweek. 2002. "Water facility attacked in Colombia." Waterweek, American Water Works Association. January 2002. http://www.awwa.org/advocacy/news/020602.cfm.

Wax, E. 2006. "Dying for water in Somalia's drought: Amid anarchy, warlords hold precious resource." Washington Post. April 14, 2006. p. A1. http://www.washingtonpost.com/wp-dyn/content/article/2006/04/13/AR2006041302116.html.

Wolf, A.T. 1995. Hydropolitics along the Jordan River: Scarce Water and its Impact on the Arab-Israeli Conflict. United Nations University Press, Tokyo, Japan.

Wolf, A. T. 1997. "'Water wars' and water reality: Conflict and cooperation along international waterways." NATO Advanced Research Workshop on Environmental Change, Adaptation, and Human Security. Budapest, Hungary. 9-12 October.

World Environment News. 2001. "Philippine rebels suspected of water 'poisoning.'" http://www.planetark.org/avantgo/dailynewsstory.cfm?newsid=12807.

World Rivers Review (WRR). 1998. "Dangerous Dams: Tajikistan" Volume 13, No. 6, p. 13 (December).

Yang Lang. 1989/1994. "High Dam: The Sword of Damocles." In Dai Qing (ed.), Yangtze! Yangtze! Probe International, Earthscan Publications, London, United Kingdom. pp. 229-240.

Zachary G.P. 1997. "Water pressure: Nations scramble to defuse fights over supplies." Wall Street Journal, December 4, p. A17.

Zemmali, H. 1995. "International humanitarian law and protection of water." Water and War, Symposium on Water in Armed Conflicts, Montreux 21-23 November 1994, Geneva, ICRC.

T - #0093 - 101024 - C0 - 234/156/20 [22] - CB - 9781578085118 - Gloss Lamination